Cambridge IGCSE ™

Co-ordinated Sciences Biology

STUDENT'S BOOK

Sarah Jinks, Mike Smith, Sue Kearsey, Jackie Clegg and Gareth Price

William Collins' dream of knowledge for all began with the publication of his first book in 1819. A self-educated mill worker, he not only enriched millions of lives, but also founded a flourishing publishing house. Today, staying true to this spirit, Collins books are packed with inspiration, innovation and practical expertise. They place you at the centre of a world of possibility and give you exactly what you need to explore it.

Collins. Freedom to teach.
Published by Collins
An imprint of HarperCollins*Publishers*
The News Building,
1 London Bridge Street,
London, SE1 9GF, UK

HarperCollins*Publishers*
Macken House, 39/40 Mayor Street Upper,
Dublin 1, D01 C9W8, Ireland

Browse the complete Collins catalogue at
collins.co.uk

10 9 8 7 6 5 4 3 2 1

ISBN 978-0-00-854592-5

British Library Cataloguing-in-Publication Data
A catalogue record for this publication is available from the British Library.

Authors: Sarah Jinks, Mike Smith, Sue Kearsey, Jackie Clegg and Gareth Price
Publisher: Elaine Higgleton
Product manager: Joanna Ramsay
Content editor: Renée Lewis
Development editor: Gillian Lindsey
Proofreader and answer checker: Aidan Gill
Cover illustration: Ann Paganuzzi
Cover designer: Gordon MacGilp
Typesetter: Jouve India Private Limited
Production controller: Lyndsey Rogers
Printed and bound in the UK using 100% Renewable Electricity at CPI Group (UK) Ltd

This book is produced from independently certified FSC™ paper to ensure responsible forest management.

For more information visit: www.harpercollins.co.uk/green

Endorsement indicates that a resource has passed Cambridge International's rigorous quality-assurance process and is suitable to support the delivery of a Cambridge International syllabus. However, endorsed resources are not the only suitable materials available to support teaching and learning, and are not essential to be used to achieve the qualification. Resource lists found on the Cambridge International website will include this resource and other endorsed resources.

Any example answers to questions taken from past question papers, practice questions, accompanying marks and mark schemes included in this resource have been written by the authors and are for guidance only. They do not replicate examination papers. In examinations the way marks are awarded may be different. Any references to assessment and/or assessment preparation are the publisher's interpretation of the syllabus requirements. Examiners will not use endorsed resources as a source of material for any assessment set by Cambridge International.

While the publishers have made every attempt to ensure that advice on the qualification and its assessment is accurate, the official syllabus, specimen assessment materials and any associated assessment guidance materials produced by the awarding body are the only authoritative source of information and should always be referred to for definitive guidance. Cambridge International recommends that teachers consider using a range of teaching and learning resources based on their own professional judgement of their students' needs.

Cambridge International has not paid for the production of this resource, nor does Cambridge International receive any royalties from its sale. For more information about the endorsement process, please visit www.cambridgeinternational.org/endorsed-resources

Cambridge International copyright material in this publication is reproduced under licence and remains the intellectual property of Cambridge Assessment International Education.

The publishers gratefully acknowledge the permission granted to reproduce the copyright material in this book. Every effort has been made to trace copyright holders and to obtain their permission for the use of copyright material. The publishers will gladly receive any information enabling them to rectify any error or omission at the first opportunity.

Contents

Getting the best from the book

Welcome to Collins Cambridge IGCSE™ Co-ordinated Biology.

This textbook has been designed to help you to succeed in this strand of your Cambridge IGCSE Co-ordinated Sciences syllabus study.

Just as there are 19 sections in this strand of the syllabus, there are 19 sections in the textbook. Each section in the textbook covers the essential knowledge and skills you need. The textbook also has some useful features which have been designed to help you understand all the aspects of Biology you will need to know for this syllabus.

SAFETY IN THE SCIENCE LESSON

This book is a textbook, not a laboratory or practical manual. As such, you should not interpret any information in this book that relates to practical work as including comprehensive safety instructions. Your teachers will provide full guidance for practical work and cover rules that are specific to your school.

A brief introduction gives context to the science covered in the section.

Starting points will help you to revise previous learning and see what you already know about the ideas in the section.

Section contents shows the syllabus topics covered in the section.

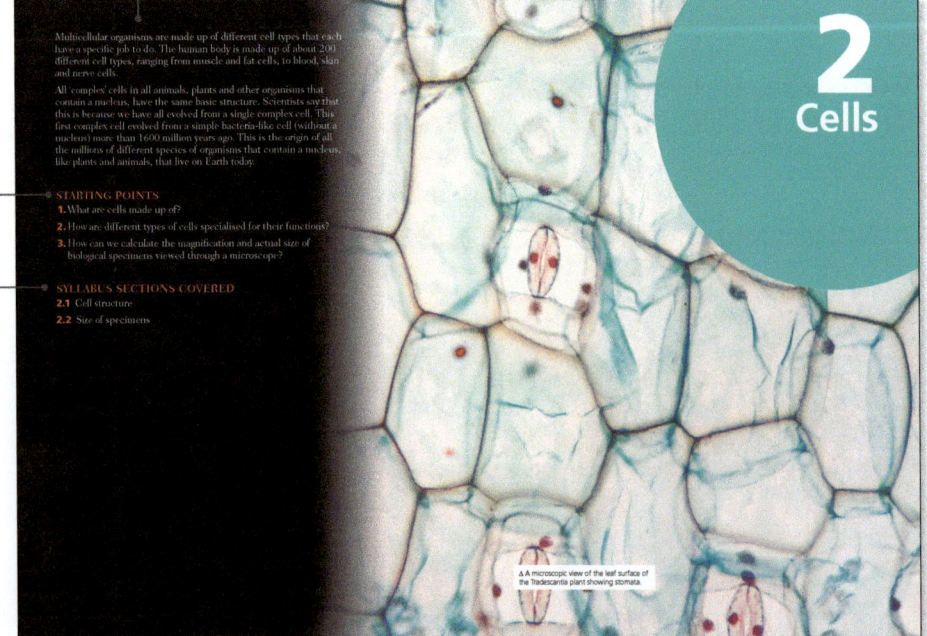

Knowledge check reminds you of the ideas you should have already encountered in previous work before starting the section.

Learning objectives show you what the syllabus requires in the section.

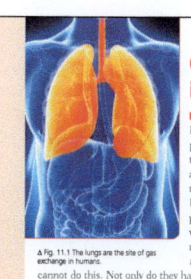

Gas exchange in humans

INTRODUCTION

Respiration uses oxygen from the air and produces carbon dioxide that is returned to the environment. These gases must get into and out of the body fast enough to support the rate at which respiration needs to work. For single-celled organisms this isn't a problem. They have a large surface area to volume ratio, and diffusion across the cell membrane can supply and remove the gases at a fast enough rate. Larger organisms cannot do this. Not only do they have a much smaller external surface area to volume ratio, which slows the rate of diffusion, but many of them also live on land, where the delicate surface required for gas exchange would dry out if it was directly exposed to the external environment. Different groups of organisms have different solutions to these problems but all involve structures with a large surface area. Plants exchange gases inside the leaf; insects have internal tubes (a tracheal system) inside the body where they exchange gases; fish have gills; and many vertebrates, including humans, have lungs.

△ Fig. 11.1 The lungs are the site of gas exchange in humans.

KNOWLEDGE CHECK
✓ Animals use oxygen from the air inspired and give out the carbon dioxide they produce in the air they expire.
✓ Humans use lungs for breathing.

LEARNING OBJECTIVES
✓ Identify in diagrams and images the following parts of the breathing system: lungs, diaphragm, ribs, intercostal muscles, larynx, trachea, bronchi, bronchioles, alveoli and associated capillaries.
✓ SUPPLEMENT Describe the features of gas exchange surfaces in humans, limited to: large surface area, thin surface, good blood supply, and good ventilation with air.
✓ Investigate the differences in composition between inspired and expired air using limewater as a test for carbon dioxide.
✓ Describe the differences in composition between inspired and expired air, limited to: oxygen, carbon dioxide, and water vapour.
✓ SUPPLEMENT Explain the differences in composition between inspired and expired air.
✓ Investigate and describe the effects of physical activity on the rate and depth of breathing.
✓ SUPPLEMENT Explain the link between physical activity and the rate and depth of breathing in terms of: an increased carbon dioxide concentration in the blood, which is detected by the brain, leading to an increased rate and greater depth of breathing.

The human breathing system

Breathing is the way that oxygen is taken into our bodies and carbon dioxide is removed. When we breathe, air is moved into and out of our lungs. This involves different parts of the breathing system within the chest.

When we breathe in, air enters through the nose and mouth. In the nose the air is moistened and warmed. The air passes over the **larynx**, where it may be used to make sounds, for example when we talk. The air travels down the **trachea** to the lungs. The air enters the lungs through the **bronchi** (singular: **bronchus**), which branch and divide to form a network of **bronchioles**.

At the end of the bronchioles are air sacs. The bulges on an air sac are called **alveoli** (singular: **alveolus**). The alveoli are covered in tiny blood capillaries. This is where oxygen and carbon dioxide are exchanged between the blood and the air in the lungs.

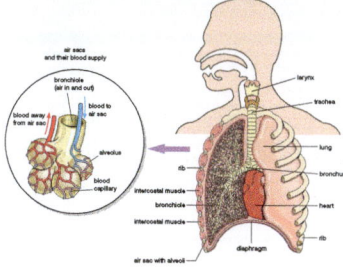

△ Fig. 11.2 The human breathing system.

SUPPLEMENT

Gas exchange

Animals need to exchange gases with the environment, to supply oxygen for respiration in cells, and to remove the waste product of respiration – carbon dioxide. These gases are exchanged at surfaces by diffusion. So **gas exchange surfaces**, such as in the human lungs, need adaptations to maximise the rate at which diffusion occurs.

An effective gas exchange surface has:
• a large surface area

Clearly differentiated material helps you focus your learning. Supplement content is labelled for students studying the extended syllabus (consisting of all core and supplement content). Content that is not labelled as supplement is core material and for all students.

Examples of investigations are included with questions matched to the practical skills you will need to learn.
It is not expected you will need to perform or learn all the methods of the example investigations

Oxygen

Active living cells respire and the most useful form of respiration, aerobic respiration, requires oxygen. Seeds can use anaerobic respiration for a short while, but the rate at which energy is released is very slow (not useful in an actively growing organism) and the by-products are toxic. That is why most seeds will only germinate successfully if there is plenty of oxygen in the soil.

△ Fig. 15.19 Waterlogged soil excludes oxygen, making it difficult for seeds to germinate and grow.

Developing practical skills
We can investigate the particular conditions for germination.

Devise and plan investigations
1. Using the apparatus shown, write a plan to investigate the effect of:
   a) light
   b) water
   c) temperature
   on the germination of seeds. Think carefully about what controls to use in each case.

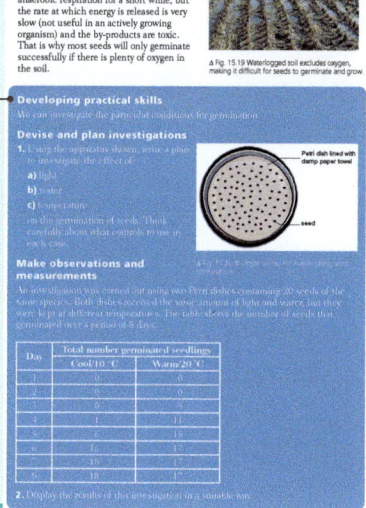

Petri dish lined with damp paper towel

seed

Make observations and measurements
An investigation was carried out using two Petri dishes containing 20 seeds of the same species. Both dishes received the same amount of light and water, but they were kept at different temperatures. The table shows the number of seeds that germinated over a period of 8 days.

Day	Total number germinated seedlings	
	Cool 10 °C	Warm 20 °C
1	0	0
2	0	0
3	0	5
4	0	8
5	2	11
6	5	15
7	15	17
8	18	17

2. Display the results of this investigation in a suitable way.

Analyse and interpret data
3. Describe the patterns shown by these results.
4. Give a conclusion from these results.
5. Explain the results using your scientific knowledge.

QUESTIONS
1. What is *germination*?
2. Explain the effects that the following conditions have on the germination of seeds:
   a) oxygen
   b) water
   c) warmth.

SEXUAL REPRODUCTION IN HUMANS
Male reproductive system

A human male has two **testes** (singular: testis) in which sperm are produced. The testes are supported outside the body in the **scrotum** to keep them cooler, because at higher temperatures fewer sperm are produced.

Sperm ducts carry the sperm from the testes to the **penis**. The **prostate gland** and seminal vesicles together produce the liquid in which the sperm are able to swim. Semen is the mixture of sperm cells and fluids.

Semen passes along the sperm ducts to the **urethra** to outside the body. The urethra also carries urine from the bladder to outside the body. When the man is sexually excited, large spaces in the penis fill with blood. This causes the penis to become larger and stiffer causing an erection. At the same time a muscle ring (sphincter) at the top of the urethra contracts, preventing urine entering the urethra from the bladder.

The erection makes it possible for the man to insert his penis into the vagina of the woman for sexual intercourse. Rapid contractions of muscles in the penis during ejaculation send the sperm shooting out into the vagina.

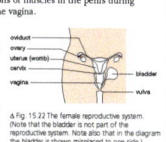

△ Fig. 15.21 The male reproductive system. (Note that the bladder is not part of the reproductive system.)

Female reproductive system

The two **ovaries** are the organs in humans that produce the **egg cells**. They are positioned within the abdominal cavity, either side of the **uterus** and joined to it by the **oviducts**.

Every month from puberty until menopause, when a woman is around 50 years old, one ovary usually releases one egg, which travels down the oviduct to the uterus (womb). If it is not

△ Fig. 15.22 The female reproductive system. (Note that the bladder is not part of the reproductive system. Note also that in the diagram the bladder is shown misplaced to one side.)

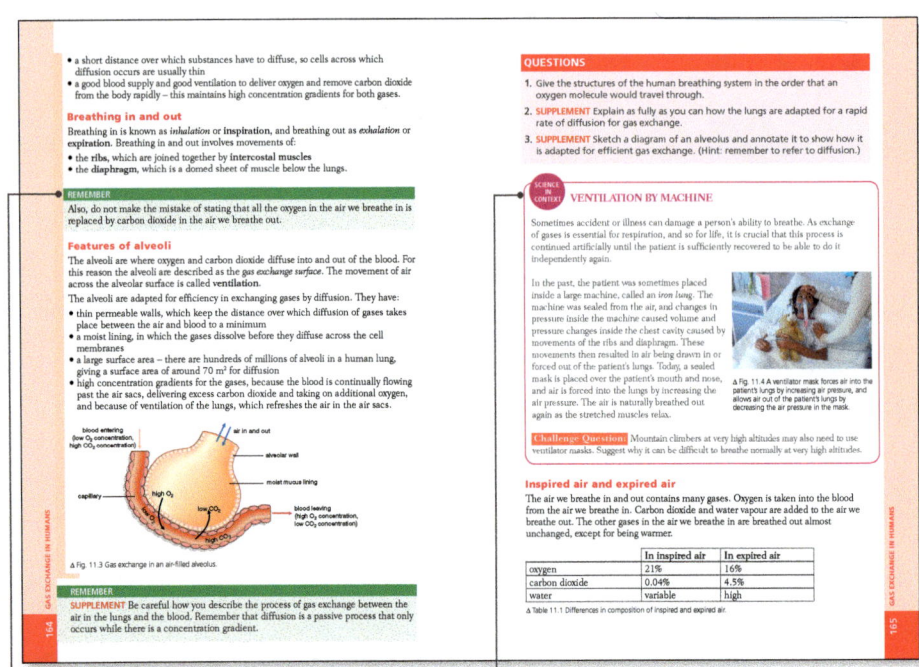

- a short distance over which substances have to diffuse, so cells across which diffusion occurs are usually thin
- a good blood supply and good ventilation to deliver oxygen and remove carbon dioxide from the body rapidly – this maintains high concentration gradients for both gases.

Breathing in and out

Breathing in is known as *inhalation* or *inspiration*, and breathing out is *exhalation* or *expiration*. Breathing in and out involves movements of:
- the **ribs**, which are joined together by **intercostal muscles**
- the **diaphragm**, which is a domed sheet of muscle below the lungs.

REMEMBER

Also, do not make the mistake of stating that all the oxygen in the air we breathe in is replaced by carbon dioxide in the air we breathe out.

Features of alveoli

The alveoli are where oxygen and carbon dioxide diffuse into and out of the blood. For this reason the alveoli are described as the *gas exchange surface*. The movement of air across the alveolar surface is called **ventilation**.

The alveoli are adapted for efficiency in exchanging gases by diffusion. They have:
- thin permeable walls, which keep the distance over which diffusion of gases takes place between the air and blood to a minimum
- a moist lining, in which the gases dissolve before they diffuse across the cell membranes
- a large surface area – there are hundreds of millions of alveoli in a human lung, giving a surface area of around 70 m² for diffusion
- high concentration gradients for the gases, because the blood is continually flowing past the air sacs, delivering excess carbon dioxide and taking on additional oxygen, and because of ventilation of the lungs, which refreshes the air in the air sacs.

△ Fig. 11.3 Gas exchange in an air-filled alveolus.

REMEMBER

SUPPLEMENT Be careful how you describe the process of gas exchange between the air in the lungs and the blood. Remember that diffusion is a passive process that only occurs while there is a concentration gradient.

QUESTIONS

1. Give the structures of the human breathing system in the order that an oxygen molecule would travel through.
2. **SUPPLEMENT** Explain as fully as you can how the lungs are adapted for a rapid rate of diffusion for gas exchange.
3. **SUPPLEMENT** Sketch a diagram of an alveolus and annotate it to show how it is adapted for efficient gas exchange. (Hint: remember to refer to diffusion.)

SCIENCE IN CONTEXT **VENTILATION BY MACHINE**

Sometimes accident or illness can damage a person's ability to breathe. As exchange of gases is essential for respiration, and so for life, it is crucial that this process is continued artificially until the patient is sufficiently recovered to be able to do it independently again.

In the past, the patient was sometimes placed inside a large machine, called an *iron lung*. The machine was sealed from the air, and changes in pressure inside the machine caused volume and pressure changes inside the chest cavity caused by movements of the ribs and diaphragm. These movements then resulted in air being drawn in or forced out of the patient's lungs. Today, a sealed mask is placed over the patient's mouth and nose, and air is forced into the lungs by increasing the air pressure. The air is naturally breathed out again as the stretched muscles relax.

△ Fig. 11.4 A ventilator mask forces air into the patient's lungs by increasing air pressure, and allows air out of the patient's lungs by decreasing the air pressure in the mask.

Challenge Question: Mountain climbers at very high altitudes may also need to use ventilator masks. Suggest why it can be difficult to breathe normally at very high altitudes.

Inspired air and expired air

The air we breathe in and out contains many gases. Oxygen is taken into the blood from the air we breathe in. Carbon dioxide and water vapour are added to the air we breathe out. The other gases in the air we breathe in are breathed out almost unchanged, except for being warmer.

	In inspired air	In expired air
oxygen	21%	16%
carbon dioxide	0.04%	4.5%
water	variable	high

△ Table 11.1 Differences in composition of inspired and expired air.

Remember boxes provide tips and guidance to help you during your course.

Science in context boxes put the ideas you are learning into real-life context. The content in these boxes is beyond the requirements of the syllabus. However, they do provide interesting examples of scientific application that are designed to enhance your understanding and a challenge question to encourage you to think more deeply beyond the syllabus content.

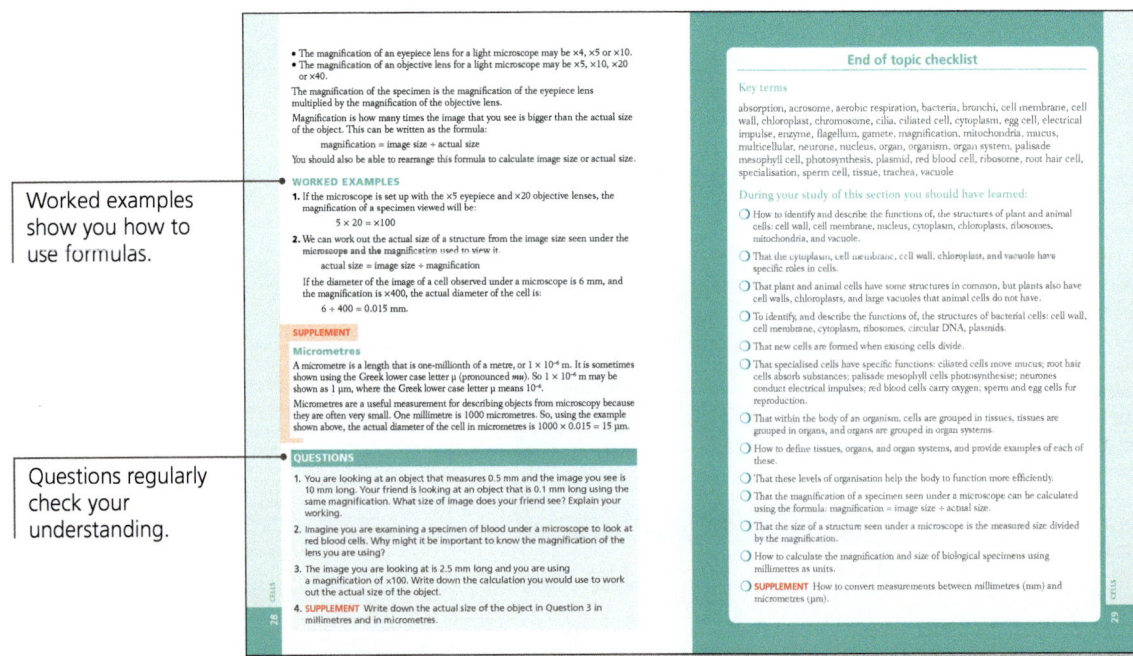

- The magnification of an eyepiece lens for a light microscope may be ×4, ×5 or ×10.
- The magnification of an objective lens for a light microscope may be ×5, ×10, ×20 or ×40.

The magnification of the specimen is the magnification of the eyepiece lens multiplied by the magnification of the objective lens.

Magnification is how many times the image you see is bigger than the actual size of the object. This can be written as the formula:

magnification = image size ÷ actual size

You should also be able to rearrange this formula to calculate image size or actual size.

WORKED EXAMPLES

1. If the microscope is set up with the ×5 eyepiece and ×20 objective lenses, the magnification of a specimen viewed will be:

5 × 20 = ×100

2. We can work out the actual size of a structure from the image size seen under the microscope and the magnification used to view it.

actual size = image size ÷ magnification

If the diameter of the image of a cell observed under a microscope is 6 mm, and the magnification is ×400, the actual diameter of the cell is:

6 ÷ 400 = 0.015 mm.

SUPPLEMENT

Micrometres

A micrometre is a length that is one-millionth of a metre, or 1 × 10⁻⁶ m. It is sometimes shown using the Greek lower case letter μ (pronounced *mu*). So 1 × 10⁻⁶ m may be shown as 1 μm, where the Greek lower case letter μ means 10⁻⁶.

Micrometres are a useful measurement for describing objects from microscopy because they are often very small. One millimetre is 1000 micrometres. So, using the example shown above, the actual diameter of the cell in micrometres is 1000 × 0.015 = 15 μm.

QUESTIONS

1. You are looking at an object that measures 0.5 mm and the image you see is 10 mm long. Your friend is looking at an object that is 0.1 mm long using the same magnification. What size of image does your friend see? Explain your working.
2. Imagine you are examining a specimen of blood under a microscope to look at red blood cells. Why might it be important to know the magnification of the lens you are using?
3. The image you are looking at is 2.5 mm long and you are using a magnification of ×100. Write down the calculation you would use to work out the actual size of the object.
4. **SUPPLEMENT** Write down the actual size of the object in Question 3 in millimetres and in micrometres.

End of topic checklist

Key terms

absorption, acrosome, aerobic respiration, bacteria, bronchi, cell membrane, cell wall, chloroplast, chromosome, cilia, ciliated cell, cytoplasm, egg cell, electrical impulse, enzyme, flagellum, gamete, magnification, mitochondria, mucus, multicellular, neurone, nucleus, organ, organism, organ system, palisade mesophyll cell, photosynthesis, plasmid, red blood cell, ribosome, root hair cell, specialisation, sperm cell, tissue, trachea, vacuole

During your study of this section you should have learned:

○ How to identify and describe the functions of, the structures of plant and animal cells: cell wall, cell membrane, nucleus, cytoplasm, chloroplasts, ribosomes, mitochondria, and vacuole.

○ That the cytoplasm, cell membrane, cell wall, chloroplast, and vacuole have specific roles in cells.

○ That plant and animal cells have some structures in common, but plants also have cell walls, chloroplasts, and large vacuoles that animal cells do not have.

○ To identify, and describe the functions of, the structures of bacterial cells: cell wall, cell membrane, cytoplasm, ribosomes, circular DNA, plasmids.

○ That new cells are formed when existing cells divide.

○ That specialised cells have specific functions: ciliated cells move mucus; root hair cells absorb substances; palisade mesophyll cells photosynthesise; neurones conduct electrical impulses; red blood cells carry oxygen; sperm and egg cells for reproduction.

○ That within the body of an organism, cells are grouped in tissues, tissues are grouped in organs, and organs are grouped in organ systems.

○ How to define tissues, organs, and organ systems, and provide examples of each of these.

○ That these levels of organisation help the body to function more efficiently.

○ That the magnification of a specimen seen under a microscope can be calculated using the formula: magnification = image size ÷ actual size.

○ That the size of a structure seen under a microscope is the measured size divided by the magnification.

○ How to calculate the magnification and size of biological specimens using millimetres as units.

○ **SUPPLEMENT** How to convert measurements between millimetres (mm) and micrometres (μm).

Worked examples show you how to use formulas.

Questions regularly check your understanding.

End of topic questions allow you to apply the knowledge and understanding you have learned in the topic to answer the questions.

Key terms for the section are defined in the glossary.

A full checklist of all the information you need to cover the biology strand of the syllabus content covered in each topic.

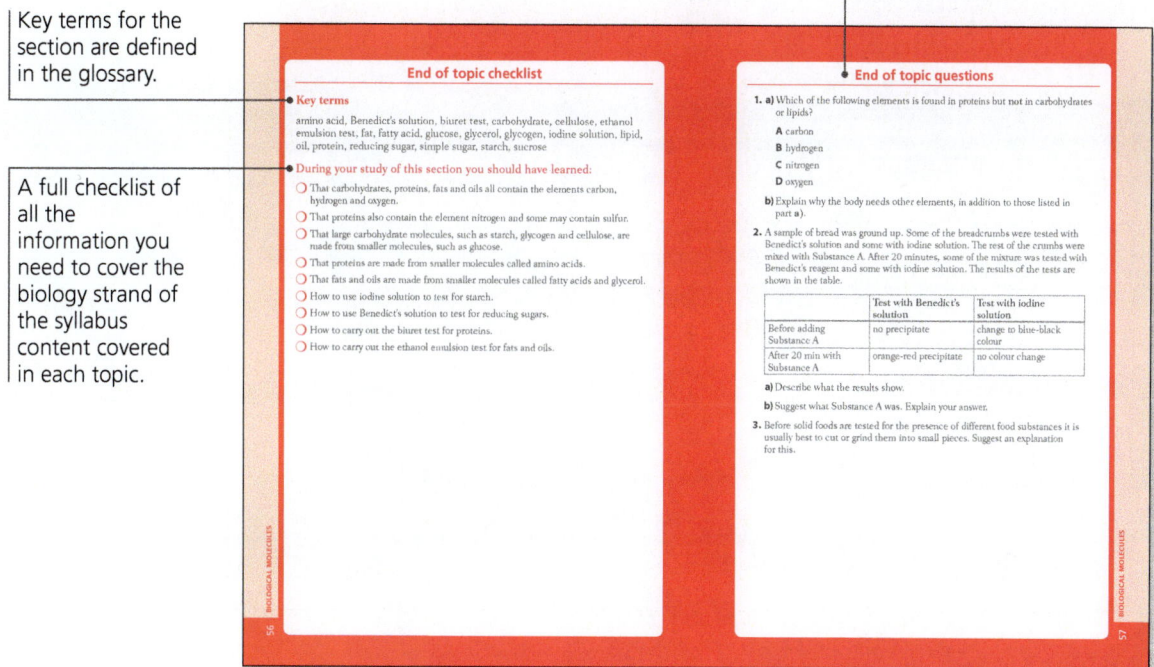

Practice questions provide a checkpoint for your learning and help you prepare for your exams.

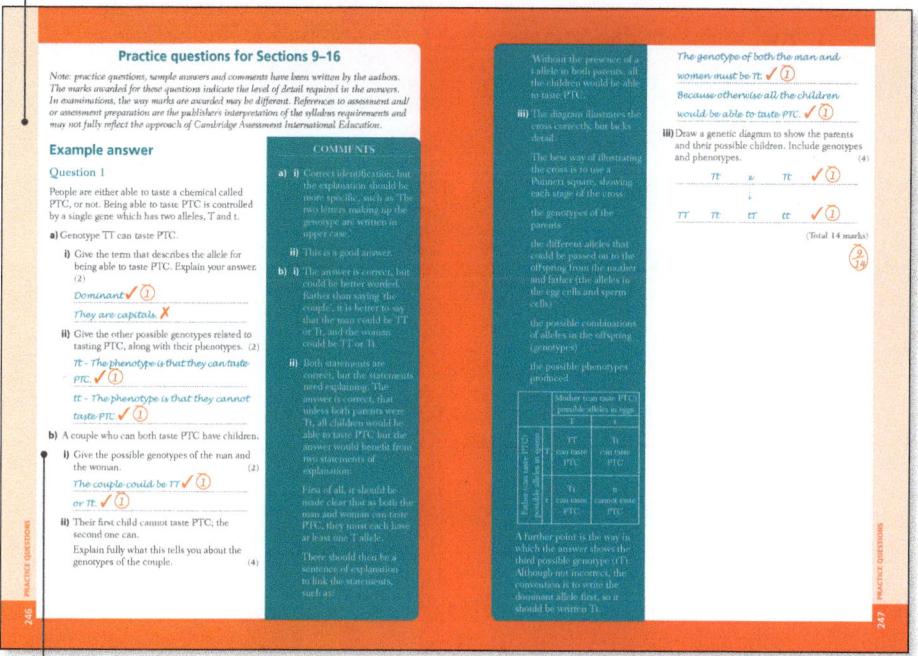

Sample answers and comments have been written by the authors to illustrate a possible response and show where it answers the question with the level of detail required or how it could be improved. Practice questions and sample answers have been written by the authors. In examinations, the way marks are awarded may be different.

Around 1.9 million living species have been identified on Earth. Around 50 000 of these species are classified as plants and around 1.37 million species are classified as animals. Over 66 000 of the animal species are vertebrates (animals with bony skeletons) and the rest are invertebrates (animals without backbones), of which the majority (around 1 million species) are insects.

It is difficult to know how many species are still to be discovered, although it is thought that about 15 000 new species are discovered around the world every year. The smaller the organism, the greater the chance that there are species we don't yet know about. So, although around 4000 species of bacteria have been identified, there could be many more species of bacteria than of all the other kinds of organism put together.

STARTING POINTS

1. What do we mean when we say that something is alive?

2. How do the different characteristics of life look different in different organisms?

SYLLABUS SECTIONS COVERED

1.1 Characteristics of living organisms

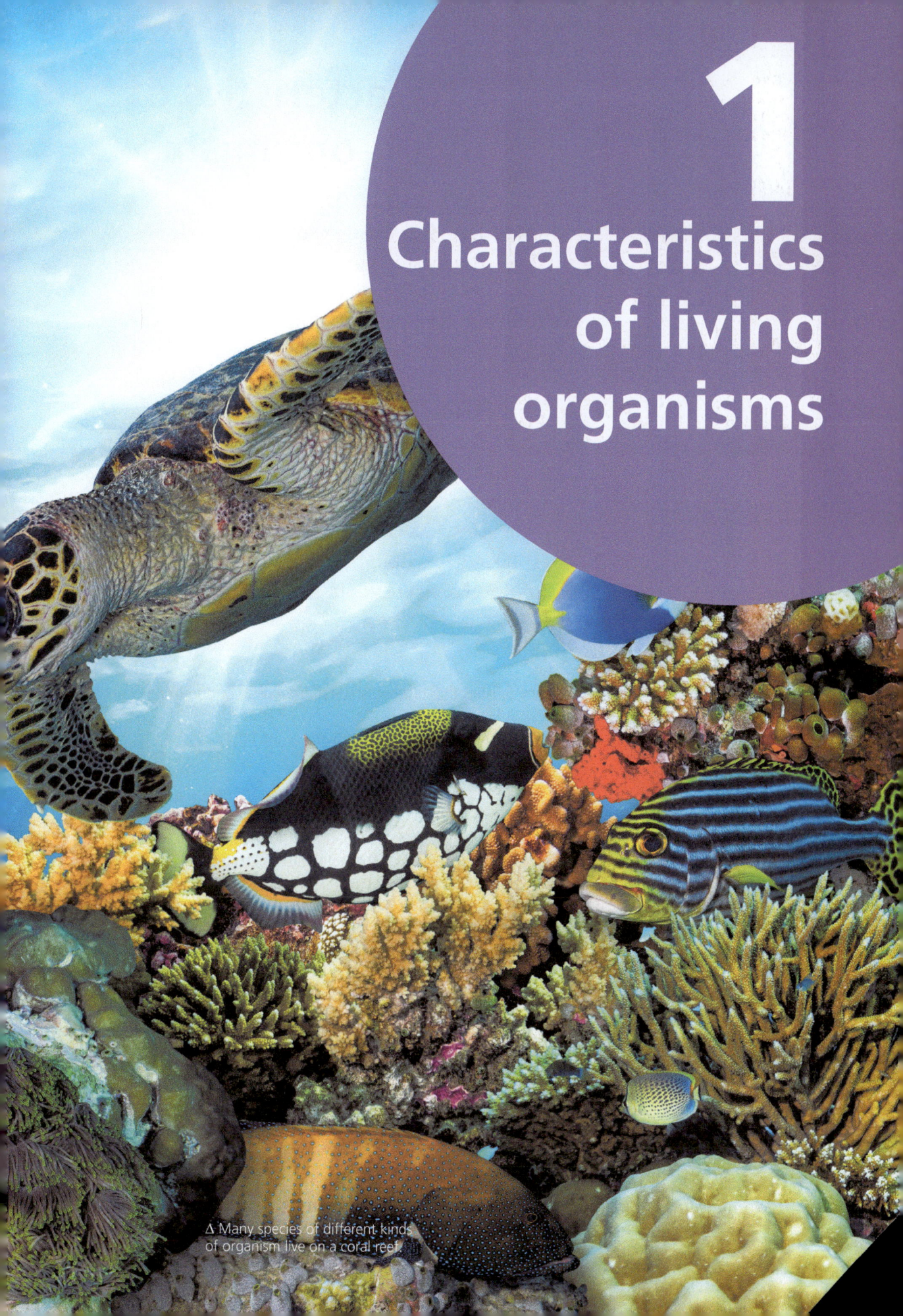

1

Characteristics of living organisms

△ Many species of different kinds of organism live on a coral reef.

Δ Fig. 1.1 Tiny tardigrades (about 1 mm long) are one of the toughest organisms known. They can survive temperatures below −200 °C, 10 days in the vacuum of space and over 10 years without water!

Characteristics of living organisms

INTRODUCTION

Deciding whether something is alive is one of the most important starting points of biology. Scientists have decided on around seven characteristics that help decide whether something is alive. However, this is not always as straightforward as we might think, and trying to decide what the characteristics of life are can be difficult. Viruses are problematic to categorise as they share some of the characteristics of life, but not others. While scientists say that they are not alive, viruses are still studied within biology and they carry out some of the characteristics of life.

KNOWLEDGE CHECK

✓ Living organisms show a range of characteristics that distinguish them from dead or non-living material.

✓ The life processes are supported by the cells, tissues, organs and systems of the body.

LEARNING OBJECTIVES

✓ Describe the characteristics of living organisms by defining: movement as an action by an organism or part of an organism causing a change of position or place; respiration as the chemical reactions in cells that break down nutrient molecules and release energy for metabolism; sensitivity as the ability to detect and respond to changes in the internal or external environment; growth as a permanent increase in size and dry mass; reproduction as the processes that make more of the same kind of organism; excretion as the removal of the waste products of metabolism and substances in excess of requirements; nutrition as the taking in of materials for energy, growth and development.

CHARACTERISTICS OF LIVING ORGANISMS

There are seven life processes that most living organisms will show at some time during their life.

- **Movement**: Organisms may move their entire body so that it changes position or place. Organisms may also move parts of their body. For example, **plants** may move their leaves in response to external stimuli such as light, while structures in the **cytoplasm** of all living cells move.
- **Respiration**: This is a series of chemical reactions inside living cells that break down nutrient molecules and release energy. The energy released from respiration is

used for all the chemical reactions that help to keep the body alive. Together, these reactions are known as **metabolism**.

- **Sensitivity**: Organisms are able to detect (or sense) and respond to changes in the environment around them. For example, we see, hear and respond to touch. Organisms can also detect and respond appropriately to changes inside their bodies (the internal environment).

△ Fig. 1.2 Sunflowers respond to light by tracking the Sun across the sky during the day.

- **Growth**: This is the permanent increase in size of an organism. Growth is also often defined as an increase in dry mass (mass without water content) of cells or the whole body of an organism. This is because total mass can vary, depending on how much the organism eats and drinks. Dry mass only measures the amount by which the body increases in size when nutrients are taken into the cells and used to increase their number and size.

- **Reproduction**: This includes the processes that result in making more individuals of that kind of organism, such as making gametes and the fertilisation of those gametes.

- **Excretion**: This is the removal from the body of substances that are toxic (poisonous) and may damage cells if they stay in the body. Living cells produce many products from the metabolic reactions that take place inside them. Some of these are waste products – materials that the body does not use; for example, **animals** cannot use the carbon dioxide produced during respiration. As these waste products may also be toxic they must be removed from the body by excretion. Organisms also excrete substances that are in **excess**, where there is more in the body than is needed.

- **Nutrition**: This is the taking of nutrients into the body. Nutrients are the raw materials needed by the cells to release energy and to make more cells for growth, development and repair. Plant

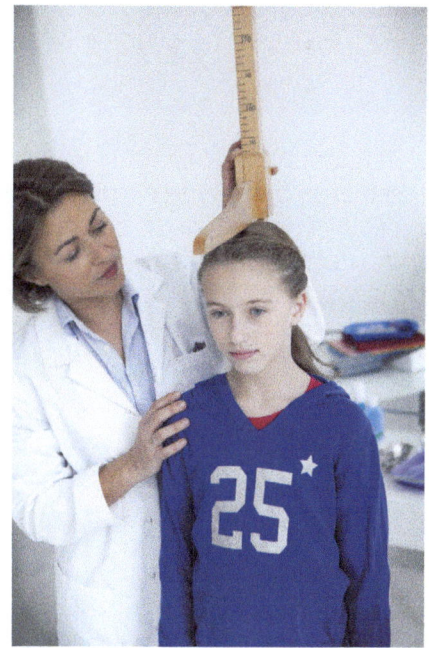

△ Fig. 1.3 Growth of a child can be measured by recording their change in height over time.

nutrition requires light, carbon dioxide, water and mineral ions, such as nitrate and magnesium. Animal nutrition requires organic compounds such as carbohydrates and proteins, mineral ions such as iron and calcium, and usually water.

All these characteristics will be described in greater detail in later sections in this book.

QUESTIONS

1. For each of the seven characteristics, give one example for:

   a) a human

   b) an animal of your choice

   c) a plant.

2. For each of the seven characteristics, explain why they are essential to a living organism.

An easy way to remember all seven processes is to take the first letter from each process. This spells Mrs Gren. Alternatively you may wish to make up a sentence in which each word begins with same letter as one of the processes, for example: My Revision System Gets Really Entertaining Now.

End of topic checklist

Key terms

animal, cytoplasm, excess, excretion, growth, metabolism, movement, nutrition, plant, reproduction, respiration, sensitivity

During your study of this section you should have learned:

○ About the seven characteristics of life: movement, respiration, sensitivity, growth, reproduction, excretion and nutrition.

End of topic questions

1. State and describe the seven processes of life.

2. Give two life processes necessary for an organism to release energy.

3. Explain why dry mass is often used to measure growth.

4. When you place a crystal of copper(II) sulfate in a saturated solution of the same compound, the crystal will increase in size. Does this mean that the crystal is alive? Explain your answer.

5. Plants cannot move about, as animals can. Does that mean animals are more alive than plants? Explain your answer.

6. During winter, an oak tree in the UK will lose its leaves and not grow. Is the tree still living during this time? Explain your answer using all the characteristics of life.

7. A student investigated the growth of sunflower seedlings. The student put three seeds into a pot filled with soil. Every day for the next 30 days they watered the pot with 5ml of water and measured the height of each seedling. The results are shown in the table.

Height of seedlings (cm)							
Plant pot	Day 0	Day 5	Day 10	Day 15	Day 20	Day 25	Day 30
1	0	0.3	1.6	2.2	2.7	3.2	3.3
2	0	0.2	1.8	2.3	2.5	3.0	3.2
3	0	0.5	1.5	2.0	2.8	3.3	3.2
mean	0.0	0.3					

a) Calculate the mean height of the seedlings for days 10, 15, 20, 25 and 30.

b) Plot a line graph of the mean height of the seedlings (vertical axis) against time (horizontal axis). Join the points with ruled, straight lines.

c) Describe in as much detail as possible how the mean height of the seedlings changed between days 0 and 30.

d) Between which days was the growth fastest?

e) Between which days was the growth the slowest? Show how you worked this out.

Multicellular organisms are made up of different cell types that each have a specific job to do. The human body is made up of about 200 different cell types, ranging from muscle and fat cells, to blood, skin and nerve cells.

All 'complex' cells in all animals, plants and other organisms that contain a nucleus, have the same basic structure. Scientists say that this is because we have all evolved from a single complex cell. This first complex cell evolved from a simple bacteria-like cell (without a nucleus) more than 1600 million years ago. This is the origin of all the millions of different species of organisms that contain a nucleus, like plants and animals, that live on Earth today.

STARTING POINTS

1. What are cells made up of?

2. How are different types of cells specialised for their functions?

3. How can we calculate the magnification and actual size of biological specimens viewed through a microscope?

SYLLABUS SECTIONS COVERED

2.1 Cell structure

2.2 Size of specimens

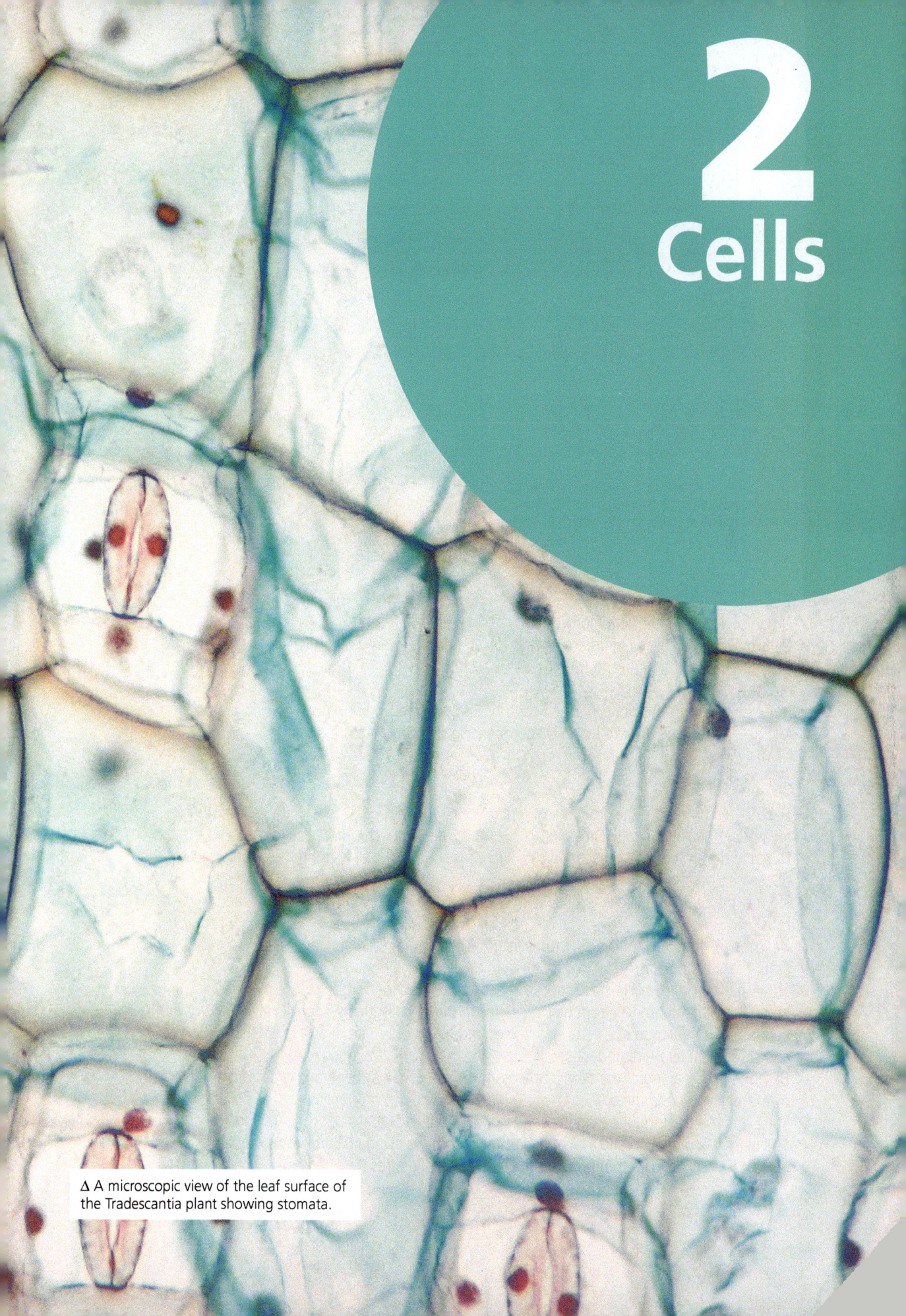

2
Cells

△ A microscopic view of the leaf surface of the Tradescantia plant showing stomata.

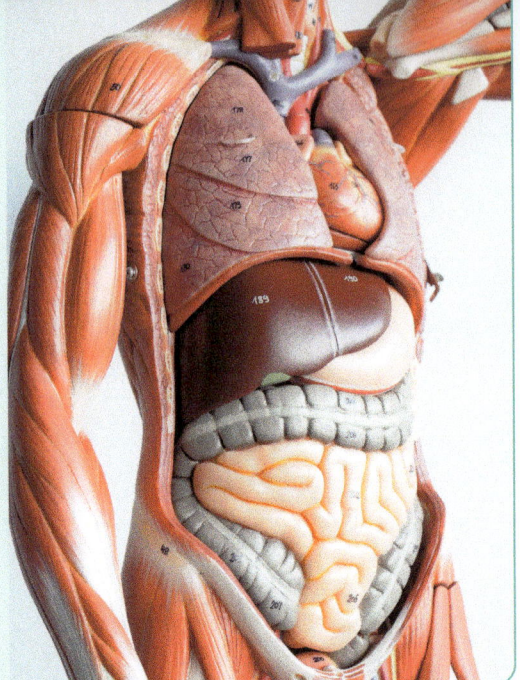

△ Fig. 2.1 The human body is made up of several systems of grouped organs. Each one of these organs is made of specialised cells, which enable the body systems to complete different tasks.

Cells

INTRODUCTION

Bringing together similar activities that have the same purpose can make things much more efficient. For example, bringing teachers and students together in a school helps more students to learn more quickly than if each teacher travelled to each student's home for lessons. The same is true in the body. Having groups of similar cells in the same place as a tissue, and grouping tissues into organs, helps the body carry out all the life processes much more efficiently and so stay alive.

KNOWLEDGE CHECK

✓ Organisms are formed from many cells.
✓ Cells may be specialised in different ways to carry out different functions.
✓ Cells are grouped together to form tissues, which together form organs, which together form organ systems.
✓ How the body systems are organised contributes to the seven life processes.
✓ Microscopes can be used to magnify specimens so we can see more detail.

LEARNING OBJECTIVES

✓ Describe and compare the structure of a plant cell with an animal cell, limited to: cell wall, cell membrane, nucleus, cytoplasm, chloroplasts, ribosomes, mitochondria, vacuoles.
✓ Describe the structure of a bacterial cell, limited to: cell wall, cell membrane, cytoplasm, ribosomes, circular DNA, plasmids.
✓ Identify the cell structures listed above in diagrams and images of plant, animal, and bacterial cells.
✓ Describe the functions of the structures listed above in plant, animal, and bacterial cells.
✓ State that new cells are produced by division of existing cells.
✓ State that specialised cells have specific functions, limited to: ciliated cells – movement of mucus in the trachea and bronchi; root hair cells – absorption; palisade mesophyll cells – photosynthesis; neurones – conduction of electrical impulses; red blood cells – transport of oxygen; sperm and egg cells (gametes) – reproduction.
✓ Describe the meaning of the terms: cell, tissue, organ, organ system, and organism as illustrated by examples given in the syllabus.
✓ State and use the formula: magnification = image size ÷ actual size.
✓ Calculate magnification and size of biological specimens using millimetres as units.
✓ **SUPPLEMENT** Convert measurements between millimetres (mm) and micrometres (μm).

CELL STRUCTURE

Animal and plant cells

The diagrams in Fig. 2.2 show a typical animal cell and typical plant cells. Although there are differences, the cells have many features in common.

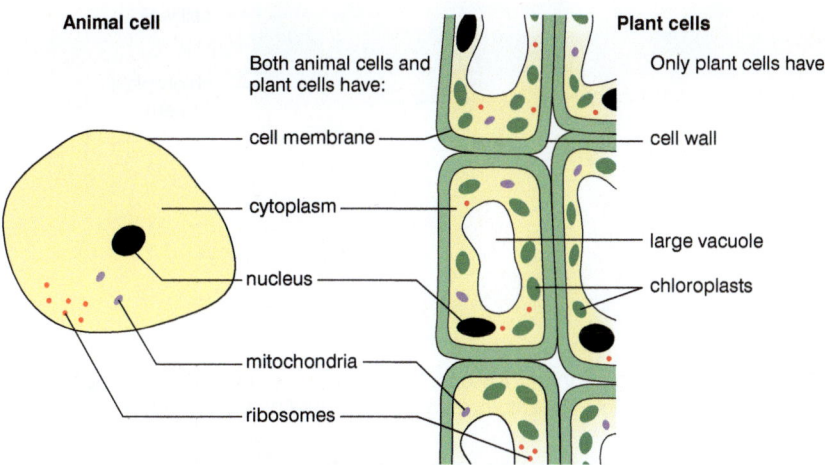

△ Fig. 2.2 The basic structures of an animal cell (e.g. liver cell) and plant cells (e.g. palisade mesophyll cells).

All living organisms are made of cells. Some, such as bacteria, most protoctists and some fungi, are formed from a single cell; others, such as the majority of plants and animals, are **multicellular**, with a body made of many cells. All animal and plant cells have certain features in common:

- a **cell membrane** surrounding the cell
- **cytoplasm** inside the cell, in which all the other structures are found
- a large **nucleus**
- small structures called **mitochondria** (singular: mitochondrion)
- even smaller structures called **ribosomes**.

A typical animal cell is a human liver cell.

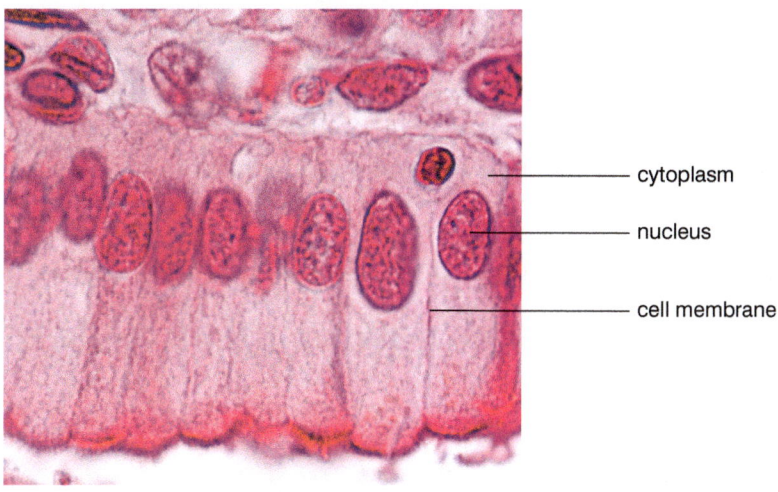

△ Fig. 2.3 Structures in animal cells seen using a light microscope. Note these cells have been stained to make some structures easier to see (mitochondria and ribosomes are too small to see clearly with a light microscope).

nucleus

cell wall

chloroplast (green)

△ Fig. 2.4 Structures in plant cells seen using a light microscope. Note that the cell membrane and vacuole are difficult to distinguish in this image. The chloroplasts are supported by the cytoplasm.

Plant cells also have features that are not found in animal cells, such as:

- a **cell wall** surrounding the cell membrane
- a large central **vacuole**
- green **chloroplasts** found in some, but not all, plant cells.

A typical plant cell is a palisade cell in the upper part of a leaf.

Functions of cell structures

Each structure in a cell has a particular role.

- The cell membrane holds the cell together and controls substances entering and leaving the cell.
- The cytoplasm supports many small cell structures and is where many different chemical processes happen. It contains water, and many solutes are dissolved in it.
- The nucleus contains genetic material in the **chromosomes** (see Section 16). These control how a cell grows and works. The nucleus also controls cell division.
- The mitochondria are the site of **aerobic respiration** which releases energy for use in other cell processes (see Section 12), such as for contraction of muscle cells, or for the production of enzymes or hormones in gland cells. Cells that are adapted for one of these purposes have much larger numbers of mitochondria than other cells.
- Ribosomes are tiny structures found in all cells. They are the sites where cells make proteins.
- The plant cell wall is made of **cellulose**, which gives the cell extra support and defines its shape.
- The plant vacuole contains cell sap. The vacuole is used for storage of some materials, and to support the shape of the cell. If there is not enough cell sap in the vacuole, the whole plant may wilt.
- Chloroplasts contain the green pigment **chlorophyll**, which absorbs the light energy that plants need to make food in the process known as **photosynthesis**.

Mitochondria and ribosomes are too small to be clearly seen with a light microscope. However, they can be seen with an electron microscope which has a much greater magnification.

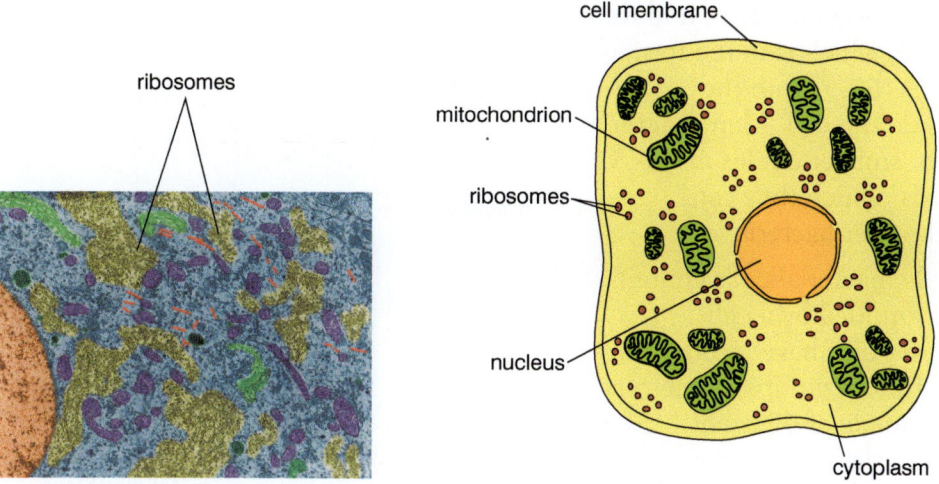

Δ Fig. 2.5 Left: A false colour electron micrograph of a neuron cell body showing nucleus (peach), mitochondria (purple) and Ribosomes (yellow). Right: diagram of an animal cell showing nucleus, mitochondria and ribosomes, not drawn to scale.

Δ Fig. 2.6 Electron micrograph (image taken with an electron microscope) of muscle tissue, showing mitochondria (dark red) that supply energy for cell contraction, and a diagram of a section through one mitochondrion.

SCIENCE IN CONTEXT   **ARTIFICIAL CELLS**

Scientists have discovered so much about how the structures of cells are formed and work together that they are starting to create artificial cells. This has great potential for medicine, because these cells could be used, for example, to deliver drugs inside the body directly to the cells that need them. They could also be used in biotechnology, for example to make fuels that could replace fossil fuels.

Challenge Question: Suggest an advantage of using artificial cells rather than naturally occurring ones.

Bacterial cells

Bacteria are single-celled organisms (see Section 1) and have cells that are much smaller and simpler than those of the other kingdoms, for example animals and plants.

Bacterial cells have no nucleus, and their genetic material, which is in the form of a circle of DNA, lies free in the cytoplasm inside the cell. Many bacteria also have some smaller circles of DNA, called **plasmids**, which they can exchange with other bacteria.

Like other cells, bacterial cells contain cytoplasm surrounded by a cell membrane. They also have a cell wall, although in different groups of bacteria the cell wall is made of different chemicals. Bacteria have ribosomes although they are slightly smaller than those of other cells.

Bacteria do not have mitochondria, but carry out respiration within the cytoplasm.

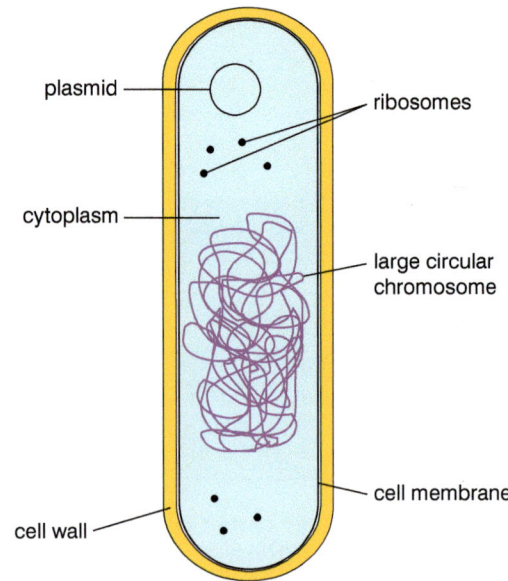

△ Fig. 2.7 Generalised structure of a bacterial cell.

BACTERIAL PLASMIDS

Bacterial plasmids have become very useful to us in genetic modification. Not all bacteria have plasmids, but those that do transfer these small circles of genetic material to other bacteria quite easily. Plasmids may even be transferred between bacteria of different species. This is not reproduction, as the transfer is not of the main genetic material and does not lead directly to the production of new individuals. However, this kind of transfer may be important in the spread of antibiotic resistance between bacterial species, because some of the genes for antibiotic resistance are found in the plasmids.

Challenge Question: Many bacteria can transfer genes from one bacterial species to another. However, organisms in the other kingdoms cannot transfer genes between different species. Explain why not.

Developing practical skills

Fig 2.8 shows a view of some blood cells seen through a light microscope.

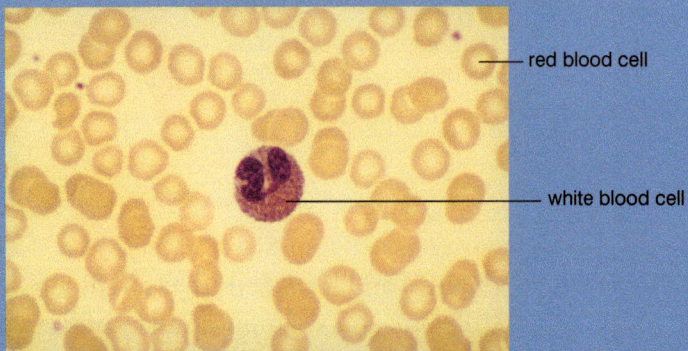

△ Fig. 2.8 A light micrograph (image taken with a light microscope) of blood cells.

QUESTIONS

1. a) Using the light micrograph in Fig. 2.3, make a careful drawing of one of the cells using a sharpened pencil to make clear lines.

b) Label your drawing to show the three key structures of animal cells.

2. a) State three cell structures that are found in plant cells but not in animal cells.

b) State three cell structures that are found in bacterial cells but not in animal cells.

3. State the part of a plant cell that does the following:

a) carries out photosynthesis

b) contains cell sap

c) stops the cell swelling if it takes in a lot of water.

4. Describe the roles of mitochondria and ribosomes in cells.

5. Suggest why the structure of bacterial cells was not well understood before the development of the electron microscope.

Cell growth

New cells are produced by the division of existing cells. This is how single-celled organisms reproduce. This is also how multicellular organisms grow. As multicellular organisms produce new cells these become specialised into different types. There is more about cell division in Section 16.

Cell specialisation for function

Different types of cells carry out different jobs. Cells have specific features that allow them to carry out their specific functions. This is called **specialisation**. Good examples of specialised cells are:

- ciliated cells, neurones (nerve cells), red blood cells, and sperm and egg cells in humans
- root hair cells, palisade mesophyll cells, and xylem vessels (see Section 8) in flowering plants.

SCIENCE IN CONTEXT

STEM CELLS

Every tissue in the human body contains a small number of unspecialised cells. These are called stem cells, and their role is to divide and produce new specialised cells within the tissue, for growth and repair. Scientists are investigating how stem cells could be given to people to mend tissue that the body cannot mend, such as the spinal cord after an accident in which it is cut. This could make it possible for a person who is paralysed following an accident to move their whole body again.

Challenge Question: Explain why stem cells could be used to help someone with spinal cord damage.

Ciliated cells

Cilia are tiny hair-like projections that cover the surfaces of certain types of cells. Cilia can move and the cells can coordinate this movement to produce waves that pass over the cells. These waves of moving cilia can move liquid in particular directions.

Ciliated cells in the lining of the respiratory tract move a liquid called **mucus**. Tiny particles of dust or bacteria that are trapped in the mucus are carried along in this flow and pass up the **bronchi** and **trachea**. They are then emptied, along with the mucus, into the throat, where they are swallowed and pass into the stomach. In this way the ciliated cells keep the lungs clean. Smoking reduces the effectiveness of these cilia, which explains why smokers often have a cough – because they cannot easily clear away the dirty mucus that collects in their lungs.

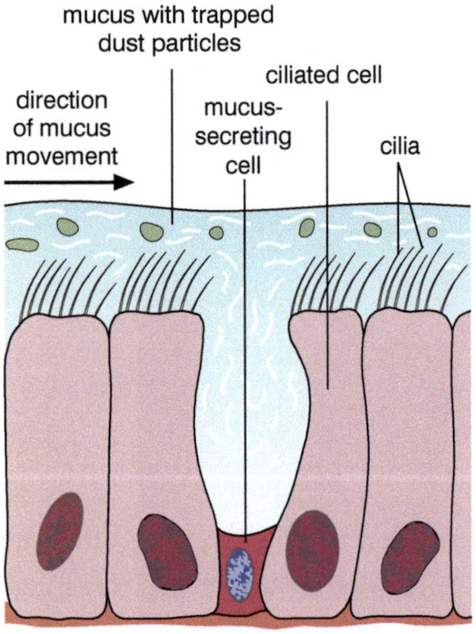

mucus with trapped dust particles

ciliated cell

direction of mucus movement

mucus-secreting cell

cilia

△ Fig. 2.9 Secreting cells produce mucus that traps particles in the lungs. Cilia sweep the mucus out of the lungs and into the throat, where it is swallowed.

Root hair cells

In many plants, water and mineral ions are absorbed from the soil by root hairs, which penetrate the spaces between soil particles. These hairs are very fine extensions of the **root hair cells** on the root surface, just behind the growing tip of a root. The elongated shape of the cells increases the surface area available for **absorption** of water and dissolved mineral ions. As they age, root hairs develop a waterproof layer and become non-functional. New root hairs are constantly growing as the root pushes through the soil. There is more about root hair cells in Section 8.

△ Fig. 2.10 The root hair cells greatly increase the surface area for absorption near the tips of roots.

Palisade mesophyll cells

Palisade mesophyll cells are plant cells found in the upper part of a leaf. They have all the features of a plant cell (see Fig 2.4) but contain a large number of chloroplasts. This is because most photosynthesis carried out by a plant happens in these cells. There is more about palisade mesophyll cells in Section 6.

Neurones (nerve cells)

Neurones (nerve cells) are specialised for conducting **electrical impulses** around the body. Many have a long fibre that can carry the impulses a long distance, such as from the toe to the base of the spine. The ends of the cell have many endings. Some of these connect to other neurones, so that the impulses can be carried to other parts of the body, such as the brain. Other endings connect to sense organs, muscles or glands, so that we can sense changes and respond to them. There is more about neurones in Section 13.

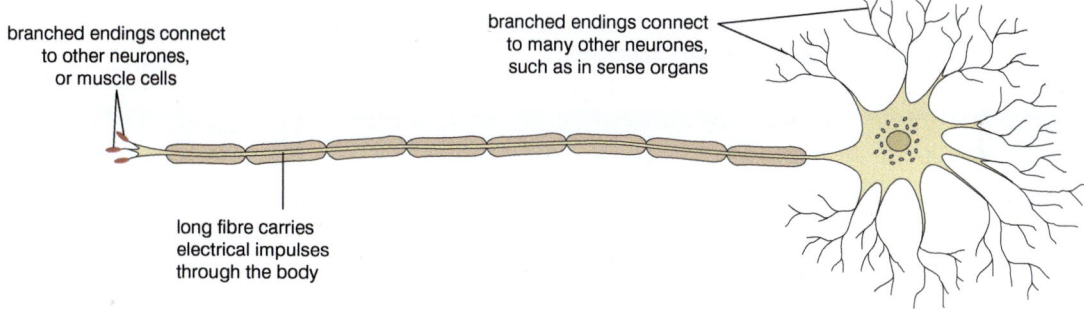

branched endings connect to other neurones, or muscle cells

branched endings connect to many other neurones, such as in sense organs

long fibre carries electrical impulses through the body

△ Fig. 2.11 This is one kind of neurone. All neurones carry electrical impulses.

Red blood cells

Red blood cells in mammals are unusual in that they do not have a nucleus. The whole of the cell is filled with a chemical called haemoglobin, which can pick up oxygen in the lungs and release it near the cells that need it deep inside the body. The shape of the cell means that the innermost part of the red blood cell is never far away from the outside, so diffusion of oxygen in and out happens very rapidly. Red blood cells are made in bone marrow and last only 120 days before they are destroyed in the spleen and liver. As they have no nucleus they cannot divide. There is more about red blood cells in Section 9.

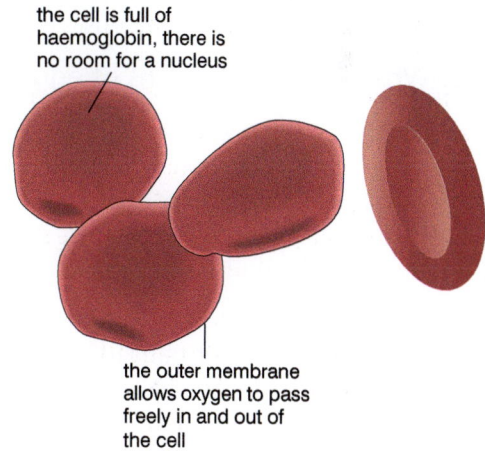

the cell is full of haemoglobin, there is no room for a nucleus

the outer membrane allows oxygen to pass freely in and out of the cell

△ Fig. 2.12 Red blood cells are specialised for carrying oxygen.

Human sex cells

The human sex cells (**gametes**) are the **sperm cell** and the **egg cell**. Sperm cells and egg cells have particular forms that are adapted to their roles in reproduction (see Section 15).

Sperm cells are relatively small compared with an egg cell. The human egg cell is a large cell and is almost visible without a microscope.

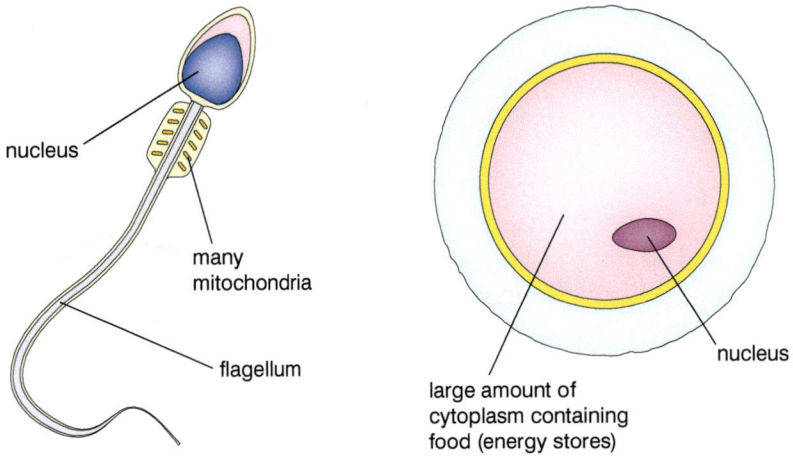

△ Fig. 2.13 Diagrams of a human sperm cell (left) and human egg cell (right). Note these are not drawn to scale. An egg cell is around 20 times larger than a single sperm cell.

QUESTIONS

1. Where would you find the following cells and what do they do?

    a) ciliated cells

    b) neurones

    c) red blood cells

    d) root hair cells

    e) palisade mesophyll cells

2. What are the functions of egg and sperm cells?

Levels of organisation

Cells in multicellular organisms, such as most plants and animals, rarely work independently of all other cells.

- Similar cells are grouped in **tissues**, to perform a shared function, such as muscle cells in a human, or palisade cells in a plant leaf.
- Different tissues are grouped together to form **organs**, which carry out specific functions, for example the heart in a human, or the leaf in a plant.
- Organs are arranged in **organ systems**, which carry out major body functions, such as the circulatory system, reproductive system, nervous system, and respiratory system in humans.
- All the different organ systems form the whole **organism**.

For example, the heart is an organ in the human circulatory system that pumps blood around the body. The heart can contract to pump because it is formed from muscle tissue, which contains muscle cells that are specially adapted to contract.

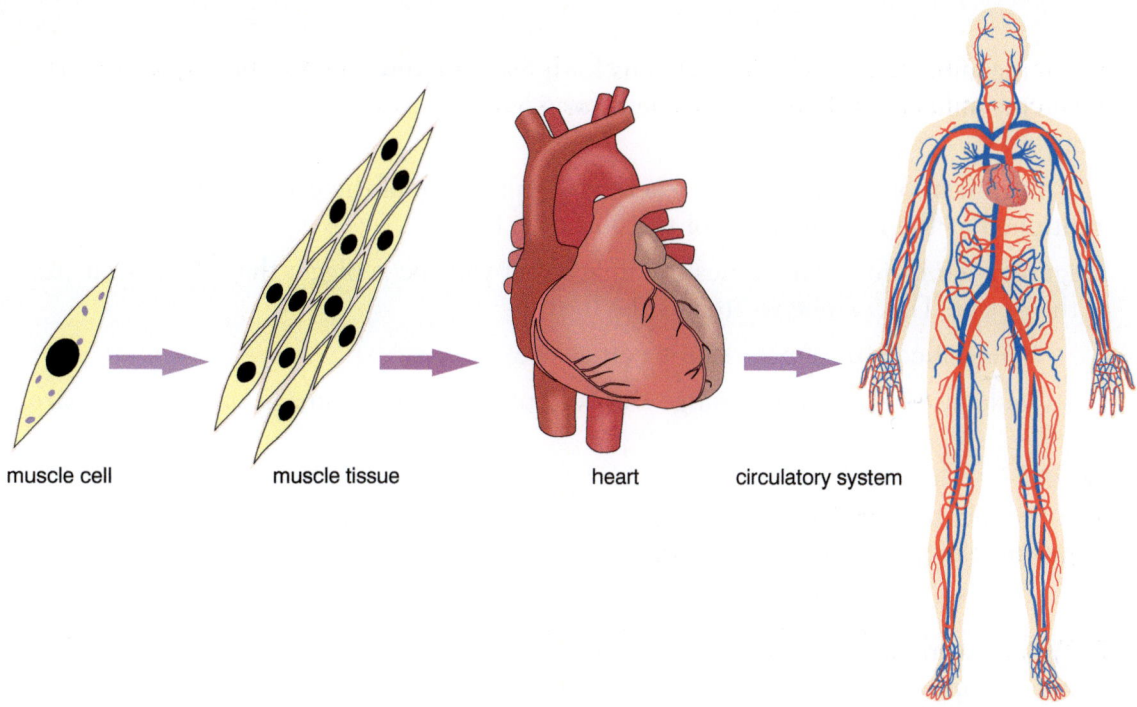

muscle cell   muscle tissue   heart   circulatory system

△ Fig. 2.14 The human body is organised at cell level. An individual muscle cell forms part of muscle tissue, which may be found in the heart, which is part of the circulatory system, which is one of the systems forming the whole organism.

REMEMBER

As you study body systems in more detail through your course, try to identify the organs, tissues, and cell types involved in each system.

QUESTIONS

1. Give two examples of each of the following in the human body:

   a) tissue   **b)** organ   **c)** system.

2. Give two examples of each of the following in a plant:

   a) tissue   **b)** organ.

SIZE OF SPECIMENS
Magnification

Many of the structures that we study in biology are too small to be seen just using our eyes. We can use magnifying glasses and microscopes to examine details of plant and animal cells, and to take pictures and draw the diagrams you see in this book. But often we want to know the actual size of the specimen we are looking at. If we know the magnification we are using to look at a specimen then we can work out the size of a structure.

When using a microscope, the **magnification** of a specimen is calculated from the eyepiece and the objective lenses used to view it.

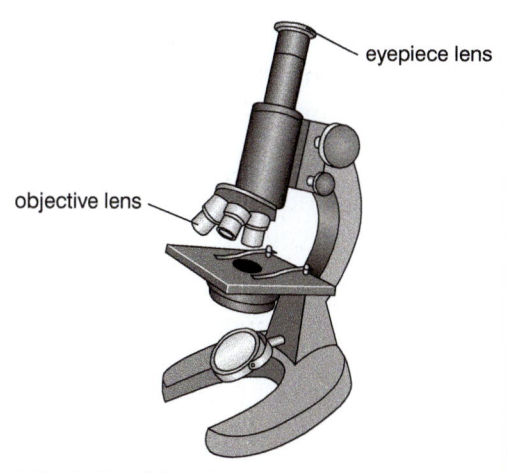

eyepiece lens

objective lens

△ Fig. 2.15 A light microscope.

- The magnification of an eyepiece lens for a light microscope may be ×4, ×5 or ×10.
- The magnification of an objective lens for a light microscope may be ×5, ×10, ×20 or ×40.

The magnification of the specimen is the magnification of the eyepiece lens multiplied by the magnification of the objective lens.

Magnification is how many times the image that you see is bigger than the actual size of the object. This can be written as the formula:

magnification = image size ÷ actual size

You should also be able to rearrange this formula to calculate image size or actual size.

WORKED EXAMPLES

1. If the microscope is set up with the ×5 eyepiece and ×20 objective lenses, the magnification of a specimen viewed will be:

5 × 20 = ×100

2. We can work out the actual size of a structure from the image size seen under the microscope and the magnification used to view it.

actual size = image size ÷ magnification

If the diameter of the image of a cell observed under a microscope is 6 mm, and the magnification is ×400, the actual diameter of the cell is:

6 ÷ 400 = 0.015 mm.

SUPPLEMENT

Micrometres

A micrometre is a length that is one-millionth of a metre, or 1×10^{-6} m. It is sometimes shown using the Greek lower case letter µ (pronounced *mu*). So 1×10^{-6} m may be shown as 1 µm, where the Greek lower case letter µ means 10^{-6}.

Micrometres are a useful measurement for describing objects from microscopy because they are often very small. One millimetre is 1000 micrometres. So, using the example shown above, the actual diameter of the cell in micrometres is $1000 \times 0.015 = 15$ µm.

QUESTIONS

1. You are looking at an object that measures 0.5 mm and the image you see is 10 mm long. Your friend is looking at an object that is 0.1 mm long using the same magnification. What size of image does your friend see? Explain your working.

2. Imagine you are examining a specimen of blood under a microscope to look at red blood cells. Why might it be important to know the magnification of the lens you are using?

3. The image you are looking at is 2.5 mm long and you are using a magnification of ×100. Write down the calculation you would use to work out the actual size of the object.

4. SUPPLEMENT Write down the actual size of the object in Question 3 in millimetres and in micrometres.

End of topic checklist

Key terms

absorption, acrosome, aerobic respiration, bacteria, bronchi, cell membrane, cell wall, chloroplast, chromosome, cilia, ciliated cell, cytoplasm, egg cell, electrical impulse, enzyme, flagellum, gamete, magnification, mitochondria, mucus, multicellular, neurone, nucleus, organ, organism, organ system, palisade mesophyll cell, photosynthesis, plasmid, red blood cell, ribosome, root hair cell, specialisation, sperm cell, tissue, trachea, vacuole

During your study of this section you should have learned:

○ How to identify and describe the functions of, the structures of plant and animal cells: cell wall, cell membrane, nucleus, cytoplasm, chloroplasts, ribosomes, mitochondria, and vacuole.

○ That the cytoplasm, cell membrane, cell wall, chloroplast, and vacuole have specific roles in cells.

○ That plant and animal cells have some structures in common, but plants also have cell walls, chloroplasts, and large vacuoles that animal cells do not have.

○ To identify, and describe the functions of, the structures of bacterial cells: cell wall, cell membrane, cytoplasm, ribosomes, circular DNA, plasmids.

○ That new cells are formed when existing cells divide.

○ That specialised cells have specific functions: ciliated cells move mucus; root hair cells absorb substances; palisade mesophyll cells photosynthesise; neurones conduct electrical impulses; red blood cells carry oxygen; sperm and egg cells for reproduction.

○ That within the body of an organism, cells are grouped in tissues, tissues are grouped in organs, and organs are grouped in organ systems.

○ How to define tissues, organs, and organ systems, and provide examples of each of these.

○ That these levels of organisation help the body to function more efficiently.

○ That the magnification of a specimen seen under a microscope can be calculated using the formula: magnification = image size ÷ actual size.

○ That the size of a structure seen under a microscope is the measured size divided by the magnification.

○ How to calculate the magnification and size of biological specimens using millimetres as units.

○ **SUPPLEMENT** How to convert measurements between millimetres (mm) and micrometres (μm).

~ /Mehma,

End of topic questions

1. Draw up a table to compare the structures found in plant and animal cells.

2. Which of the following structures is only found in bacterial cells?

A cell wall

B nucleus

C plasmid

D ribosome

3. Describe the functions of the following cell structures:

a) nucleus *DNA Inst.*

b) cell membrane *Controls what comes in and out,*

c) cytoplasm. *Where everything is c*

4. Here are some examples of statements written by students. Each statement contains an error. Identify the error and rewrite the statement so that it is correct.

a) Animal cells are surrounded by a cell ~~wall~~ *membrane* that controls what enters and leaves the cell.

b) All ~~plant~~ cells contain ~~chloroplasts~~. *nucleus*

c) ~~Both animal cells and~~ plant cells contain a large central vacuole in the middle of the cell. *only*

5. Red blood cells are unusual because they contain no nucleus. When they are damaged, they have to be replaced with new cells from the bone marrow. Explain how this is different from other cells.

6. Draw up a table with the following headings.

Organ system	Function	Organs in this system	Tissues in these organs	Cells in these tissues
Lungs.	*Breath,*			
Heart	*Keeping alive,*			
Brain	*Everything, control body*			

Complete the table, using three examples of systems in the human body.

7. Explain the advantage of cell specialisation and organisation for multicellular organisms.

8. Put the following in order of size, starting with the largest:

cell   organ   organism   organ system   tissue

9. Define each of these words:

a) *tissue*

b) *organ*

c) *organ system.*

10. SUPPLEMENT An egg cell is observed using a microscope with magnification ×100. The actual diameter of the egg is 0.1 mm. Calculate the diameter of the image in micrometres.

11. A student investigated how temperature affects diffusion using beetroot. Beetroot cells contain a purple-red pigment that can diffuse out of the cell when the cell membrane is damaged. The student placed cubes of beetroot of the same size into separate test tubes of water, which were placed in water baths at four different temperatures.

After leaving the cubes for 20 minutes the student shone light through each test tube and used a light meter to measure the percentage of light that passed through the coloured liquid in each tube. The more light that passes through the liquid, the lower the pigment concentration.

They carried out three trials at each temperature and calculated the average at each temperature. The graph shows their results.

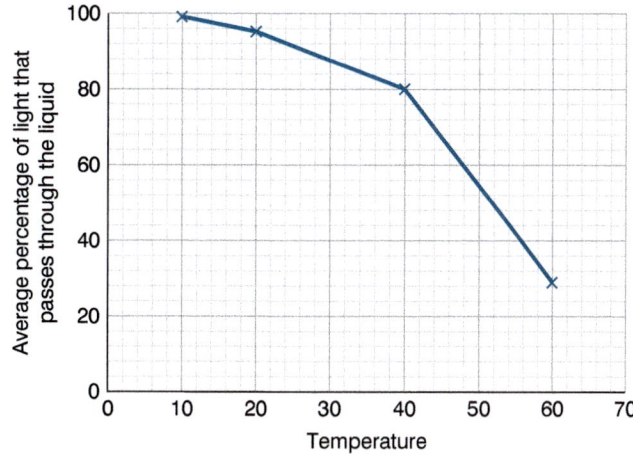

a) What should the unit be on the x axis label?

b) Use the graph to estimate the percentage of light that would pass through the liquid if the beetroot cubes had been placed in water at 50°C.

c) Use the graph to draw a conclusion

d) Suggest, with a reason, one other factor that could affect the amount of light that could pass through the liquid.

12. A student cut five potato cylinders of the same diameter from a large potato. All cylinders were cut to the same length, 40.0 mm, and placed into beakers containing different concentrations of sucrose solution. The student left the potato cylinders for three hours and then measured their lengths again using digital callipers. Their results are shown below.

Sucrose concentration (mol/dm³)	Final potato length (mm)	Change in potato length (mm)	Percentage change in length
0	41.6	1.6	4.0
0.2	40.4		
0.4	36.7	−3.3	−8.2
0.6	36.5		
0.8	33.1	−6.9	−17.3

a) Calculate the change in length for the potato cylinders in sucrose concentrations of 0.2 and 0.6 mol/dm³.

b) Calculate the percentage change in length of the potato cylinders in sucrose concentrations of 0.2 and 0.6 mol/dm³. Give your answers to 1 decimal place. Use the following equation:

$$\text{percentage change} = \frac{\text{change in length}}{\text{starting length}} \times 100$$

c) Plot a scatter graph with sucrose concentration on the x (horizontal) axis and percentage change in length on the y (vertical) axis.

d) Draw a straight line of best fit.

e) Estimate the concentration of sucrose solution at which your graph shows there would be no change in the length of the potato cylinder.

All living organisms need to be able to transport water, oxygen, carbon dioxide and other molecules around their bodies. For simple organisms such as bacteria the distances travelled are very small, but more complex animals and plants have evolved highly specialised transport mechanisms to get vital substances from one part of the organism to another. If all the blood vessels in the human body – all the arteries, veins and capillaries – were laid end to end, they would stretch for about 60 000 miles. That's nearly 100 000 km!

STARTING POINTS

1. How do substances enter and leave cells?

2. Why is water so important for living things?

3. What are the differences between diffusion, osmosis and active transport?

SYLLABUS SECTIONS COVERED

3.1 Diffusion

3.2 Osmosis

3.3 Active transport

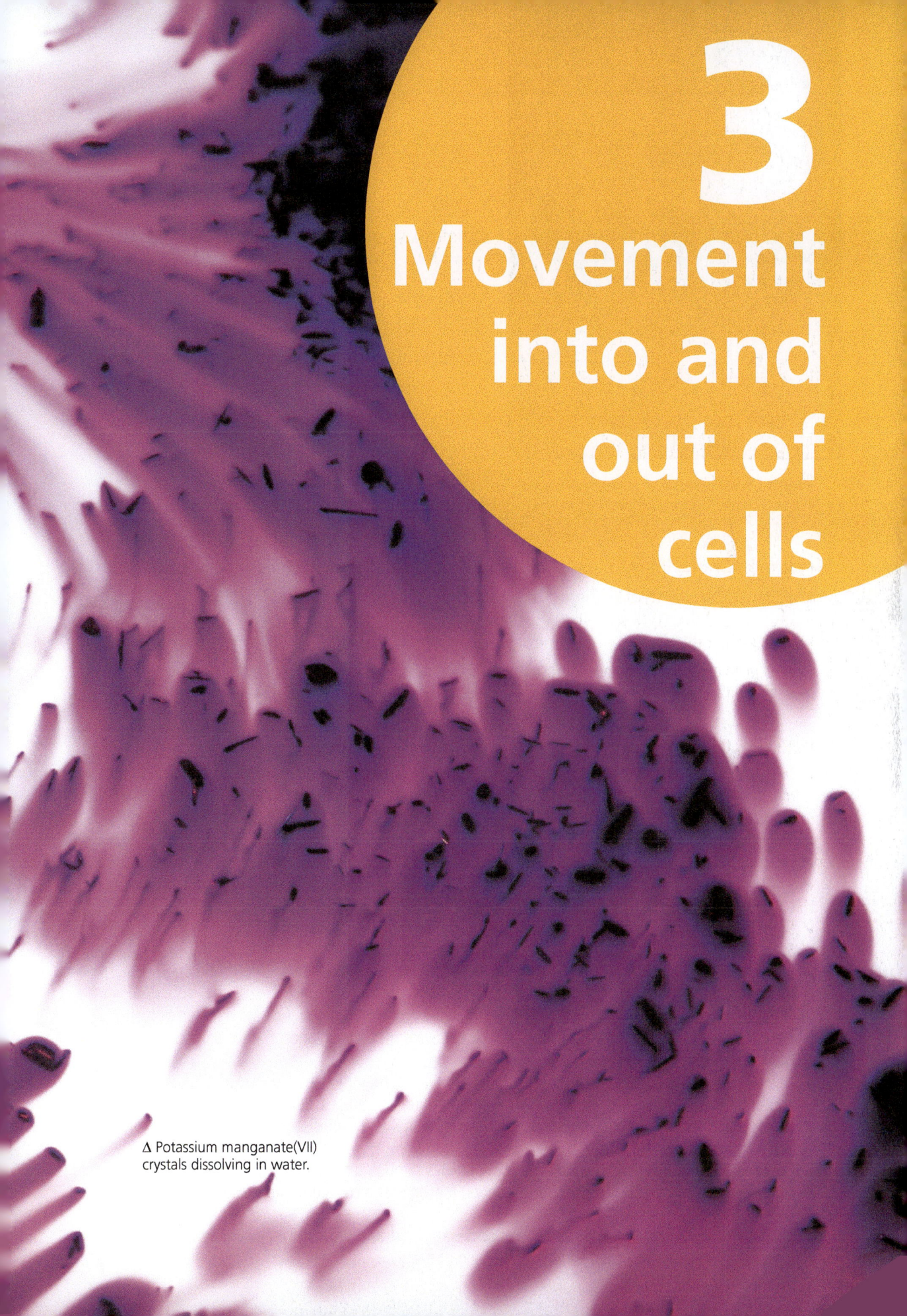

3
Movement into and out of cells

△ Potassium manganate(VII) crystals dissolving in water.

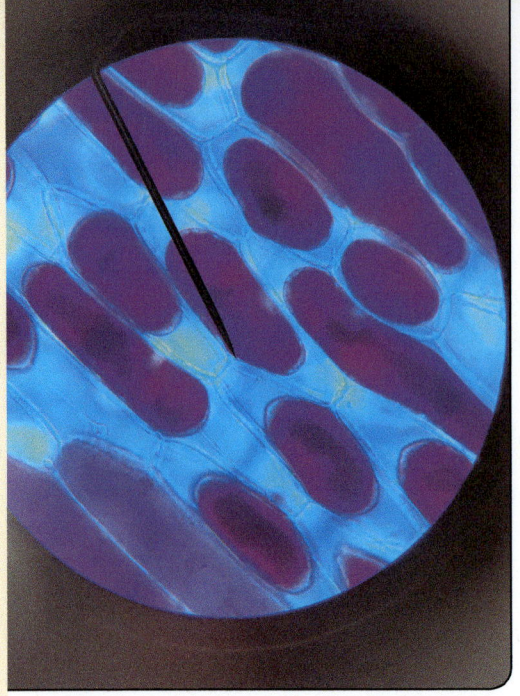

Δ Fig. 3.1 A red blood cell that has been placed in a salty solution loses water and shrinks.

Movement into and out of cells

INTRODUCTION

If you put a red blood cell into pure water, it will eventually burst open. If you place the red blood cell into a salty solution instead, it will shrink. Surrounding every cell is the cell membrane. Imagine the cell membrane as a leaky layer that is strong enough to hold all the contents in the cell together, but that allows small particles to move through it. The cell membrane also has special 'gates' that allow certain, important particles through. Different cells have different kinds of 'gate' in them. So cell membranes play an essential role in controlling what goes into and out of cells, and therefore control the way that the cell functions.

KNOWLEDGE CHECK

✓ Cells need oxygen and glucose for respiration.
✓ Cells need to get rid of waste substances, such as carbon dioxide from respiration.

LEARNING OBJECTIVES

✓ Describe diffusion as the net movement of particles from a region of their higher concentration to a region of their lower concentration (i.e. down a concentration gradient), as a result of their random movement.
✓ State that some substances move into and out of cells by diffusion through the cell membrane.
✓ Describe the importance of diffusion of gases and solutes in living organisms.
✓ **SUPPLEMENT** Investigate the factors that influence diffusion, limited to: surface area, temperature, concentration gradient, and distance.
✓ State that water diffuses through partially permeable membranes by osmosis.
✓ State that water moves into and out of cells by osmosis through the cell membrane.
✓ Investigate and describe the effects on plant tissues of immersing them in solutions of different concentrations.
✓ **SUPPLEMENT** Describe osmosis as the net movement of water molecules from a region of higher water potential (dilute solution) to a region of lower water potential (concentrated solution), through a partially permeable membrane.
✓ **SUPPLEMENT** Explain the effects on plant cells of immersing them in solutions of different concentrations by using the terms: turgid, turgor pressure, plasmolysis, flaccid.
✓ **SUPPLEMENT** Explain the importance of water potential and osmosis in the uptake and loss of water by organisms.
✓ Describe active transport as the movement of particles through a cell membrane from a region of lower concentration to a region of higher concentration (i.e. against a concentration gradient), using energy from respiration.
✓ **SUPPLEMENT** Explain the importance of active transport as a process for movement of molecules or ions across membranes, including ion uptake by root hairs.

DIFFUSION

Substances such as water, oxygen, carbon dioxide and food are made of particles (atoms, ions, and molecules).

When the particles dissolve in a liquid or a gas we call the particles dissolving **solutes**, and the substances they are dissolved in the **solvent**. Both together are called a **solution**.

In liquids and gases the particles are constantly moving around. This means that they eventually spread out evenly. For example, if you dissolve sugar in a cup of water, even if you do not stir it, the sugar molecules eventually spread throughout the liquid. This is because all the molecules are moving around, colliding with and bouncing off other particles.

- water molecule
- sugar molecule

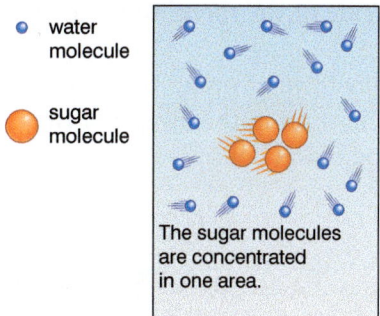

The sugar molecules are concentrated in one area.

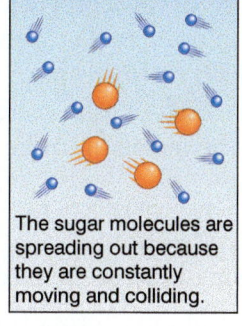

The sugar molecules are spreading out because they are constantly moving and colliding.

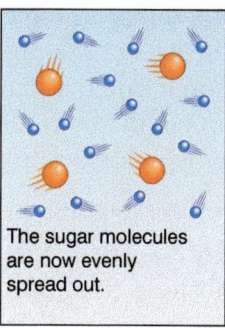

The sugar molecules are now evenly spread out.

Δ Fig. 3.2 Diffusion of sugar molecules in a solution.

The sugar molecules have spread out from an area of high concentration, when they were added to the water, to an area of low concentration. Eventually, although all the particles are still moving, the sugar molecules are evenly spread out and there is no longer a difference in concentration – a **concentration gradient** (as a result of their random movement).

Only while there is **net movement** (where there are more particles moving in one direction than another) from an area of higher concentration to an area of lower concentration is there **diffusion**.

Diffusion can be defined as the net movement of particles from a region of their higher concentration to a region of their lower concentration (i.e. down a concentration gradient).

Diffusion can only occur when there is a difference in concentration between two areas. Particles are said to move *down* their concentration gradient. This happens because of the random movement of particles.

Δ Fig. 3.3 Potassium manganate(VII) diffuses through a beaker of water as the solid crystal dissolves.

Δ Fig. 3.4 Left: as the crystal dissolves it forms a region of high concentration of the solute in the solution. Right: after a few hours, the solute is evenly spread in the water as a result of diffusion.

Diffusion is important for living organisms, as this is the process by which gases such as oxygen and carbon dioxide are exchanged between the organism and the environment. It is also the process by which **solutes** dissolved in water, such as mineral ions and food molecules, enter a living organism and how waste substances, such as urea, are excreted from the organism into the environment.

Diffusion in cells

Cells are surrounded by membranes. These membranes are leaky – they let tiny particles pass through them. Large particles can't get through, so cell membranes are said to be **partially permeable**.

Movement of particles across a cell membrane may happen more in one direction than the other if there is a difference in concentration on either side of the membrane (a concentration gradient). For example, in the blood vessels in the lungs there is a *low* oxygen concentration inside the red blood cells (because they have given up their oxygen to cells in other parts of the body) and a *high* oxygen concentration in the alveoli of the lungs. Therefore oxygen diffuses from the alveoli into the red blood cells.

Other examples of diffusion include:

- carbon dioxide entering leaf cells (see Section 6)
- digested food substances from the small intestine entering the blood (see Section 7).

Diffusion of some types of molecules and ions across the cell membrane is a **passive** process. It needs no input of energy from the cell.

Note that although diffusion requires the energy of particles in order to happen, this does not mean that diffusion is an *active* process in terms of cells (see active transport later in this section). No energy from the cell is required for particles to diffuse into or out of a cell across the cell membrane.

Diffusion can be observed simply by dropping a coloured crystal of a soluble substance such as potassium manganate(VII) into still water, and watching what happens. As the crystal dissolves, the particles of solute diffuse throughout the water until the concentration of solute is the same throughout the water.

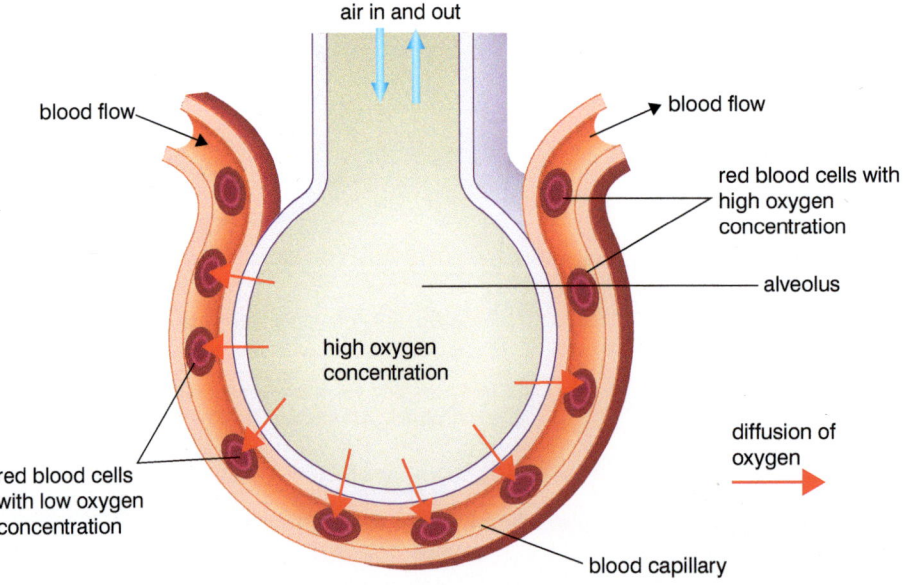

△ Fig. 3.5 In blood vessels in the lungs, oxygen diffuses down its concentration gradient from the air in the lungs into red blood cells.

KIDNEY FAILURE AND HAEMODIALYSIS

The kidneys are organs that depend on filtration and diffusion to produce urine and keep the concentration of many substances in the blood at a fairly constant level. People who suffer from kidney failure are unable to do this, and are very quickly at risk from the build-up of waste products, such as urea, in the body as a result of cell processes. In high concentrations these waste products can damage body cells and lead to death.

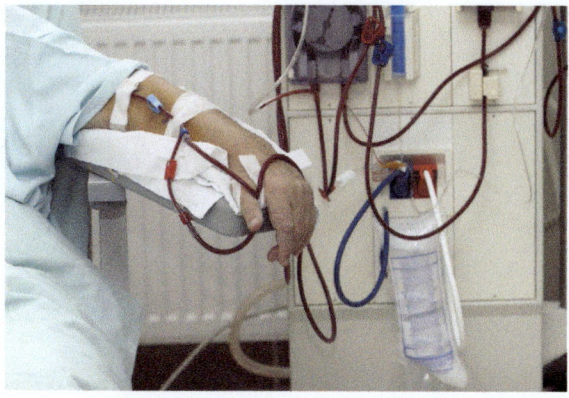

Δ Fig. 3.6 A patient undergoing kidney dialysis.

Haemodialysis is an artificial way of cleaning the blood by which substances diffuse out of the blood into dialysis fluid in a machine called a dialyser. The concentration of substances in the dialysis fluid has to be correct, so that all the waste products are removed and other substances are returned to the body at the right concentration.

Challenge Question: Explain why dialysis fluid needs to contain some substances at a similar concentration to that usually found in blood, but also some substances at a much lower concentration than usually found in blood.

SUPPLEMENT

Investigating diffusion

Diffusion in cells may be investigated by looking at coloured cells such as red onion or beetroot cells. If slices of tissue from these are left in water, after a while the coloured pigment will start to diffuse out of the cells into the water.

The rate of diffusion is affected by:

- surface area – increasing surface area increases the rate of diffusion
- temperature – increasing temperature increases the rate of diffusion
- concentration gradient – the greater the concentration gradient, the greater the rate of diffusion
- distance – increasing distance decreases the rate of diffusion.

Developing practical skills

Coloured plant tissue, such as beetroot, can be used to demonstrate diffusion.

The beetroot is cut into small cubes (approx. 1 cm³) which are first thoroughly washed in water.

A cube is then placed in clean water in a beaker as shown in Fig. 3.7.

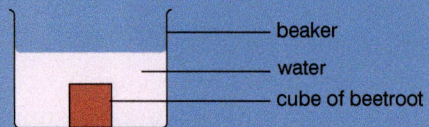

The water turns pink as the red pigment diffuses out of the cube. The time taken for the water to turn a certain shade of pink is measured. This indicates the rate of diffusion.

△ Fig. 3.7 Investigating diffusion.

This apparatus can be modified to investigate the effect of different factors on the rate of diffusion.

Devise and plan investigations

1. Explain why it is important to thoroughly wash the freshly cut cubes of beetroot before starting the investigation.

2. Outline how you could use and modify this apparatus to investigate the effects of the following factors on the rate of diffusion:

    a) surface area

    b) temperature

    c) concentration gradient

    d) distance.

Evaluate data and methods

3. As far as possible, all the cubes in an investigation should be taken from the same beetroot. Explain why.

4. Describe one way of ensuring that the same shade of pink is reached each time.

QUESTIONS

1. Explain the terms *net movement* and *diffusion.*

2. Is diffusion across a cell membrane a passive or active process? Explain your answer.

3. Explain why some particles can diffuse through cell membranes but not others.

OSMOSIS

Water is essential for living organisms and is involved in some way in almost all processes that occur in an organism, for example, in digestion, excretion and transport (see Sections 8 and 9).

Water molecules are small enough to diffuse through partially permeable membranes, such as cell membranes. However, because water molecules are so important to cells, and may be diffusing in a different direction to other molecules, this kind of diffusion has a special name – **osmosis**. Like diffusion, osmosis is a passive process and is a result of the random movement of particles.

Water molecules diffuse from a place where there is a higher concentration of water molecules (such as in a dilute sucrose sugar solution) to where there is a lower concentration of water molecules (such as in a concentrated sucrose sugar solution).

Osmosis is the diffusion (net movement) of water molecules from a region of their higher concentration to a region of their lower concentration through a partially permeable membrane.

Concentrations in solutions

Many people confuse the concentration of the solution with the concentration of the water. Remember, in osmosis it is the *water molecules* that we are considering, so you must think of the concentration of water molecules in the solution instead of the concentration of solutes dissolved in it.

- A low concentration of dissolved solutes means a high concentration of water molecules.
- A high concentration of dissolved solutes means a low concentration of water molecules.

So the water molecules are moving from a *higher concentration* (*of water molecules*) to a *lower concentration* (*of water molecules*).

SUPPLEMENT

Water potential

One way to avoid the confusion about concentration is to refer instead to **water potential**. A region of high water concentration is said to have a high water potential, and a region of low water concentration is said to have a low water potential.

Osmosis can be defined as the net movement of water molecules from a region of higher water potential (dilute solution) to a region of lower water potential (concentrated solution), through a partially permeable membrane.

Water potential can be thought of as the tendency of a solution (in a cell for example) to absorb water. Pure water has a water potential of zero. As solutes are added to the water its water potential falls – it becomes more negative. So, a concentrated sugar solution has a lower water potential (more negative) than pure water. When two regions of different water potential are separated by a partially permeable membrane, water moves from the region of higher water potential to the region of lower water potential, i.e. water molecules move down the **water potential gradient**.

Osmosis in plant cells

If a cell is placed in a solution that has a higher concentration of solute (and so a lower concentration of water molecules) than the cytoplasm inside the cell, water will leave the cell by osmosis and the cytoplasm will shrink.

If a cell is placed in a solution that has a lower concentration of solute (and so a higher concentration of water molecules) than the cytoplasm inside the cell, osmosis will result in water entering the cell.

In addition to their cell membranes, plant cells are also surrounded by cell walls that are completely permeable. This means water and solute molecules pass easily through them.

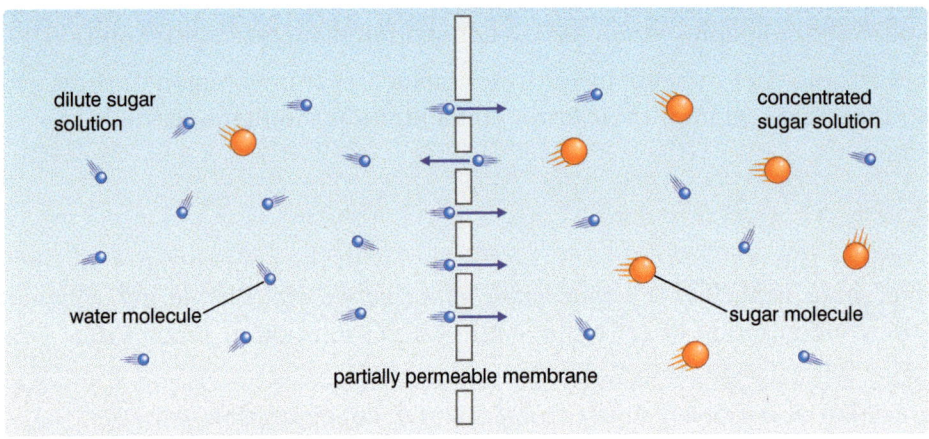

Δ Fig. 3.8 Water molecules diffuse from an area of higher concentration (of water molecules) into an area of lower concentration (of water molecules). This kind of diffusion is known as osmosis.

In a solution of high concentration of solute (low concentration of water), water will leave a plant cell by osmosis. This can be seen in plant cells as the cytoplasm shrinks inside the cell. However, the whole cell doesn't shrink, because the plant cell wall controls the structure of the cell. However, the plant as a whole will show wilting.

In a solution of low concentration of solute (high concentration of water), water will enter the plant cell by osmosis. However, the plant cell does not eventually burst.

healthy plant with cells full of water

wilting plant

Δ Fig. 3.9 When the cells of the plant are not full of water, the cell walls are not strong enough to support the plant, and the plant collapses (wilts). When the cells are full of water, the plant stands upright.

Developing practical skills

Strips of dandelion stem about 5 cm long and 3 mm wide were placed in sodium chloride solutions of different concentrations. After 10 minutes, the strips looked as shown in Fig 3.10. (Note that the outer layer of a dandelion stalk is 'waterproofed' with a waxy layer to protect it from water loss to the environment.)

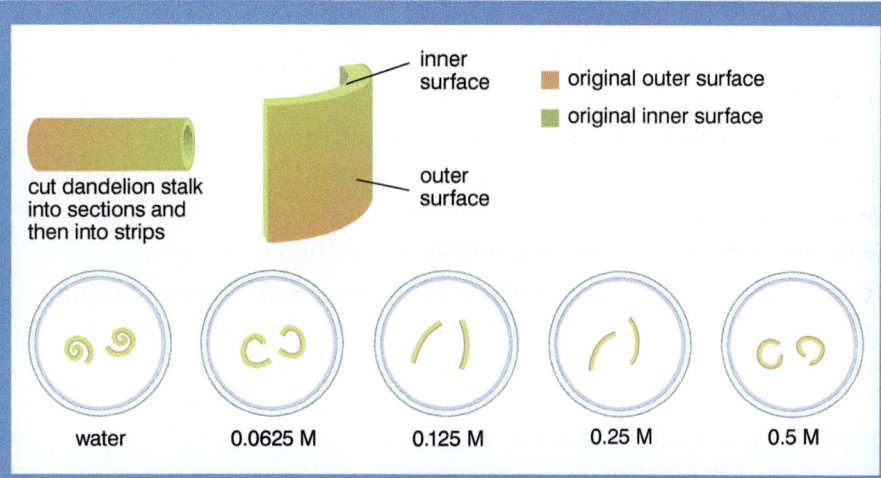

cut dandelion stalk into sections and then into strips

inner surface

original outer surface
original inner surface

outer surface

| water | 0.0625 M | 0.125 M | 0.25 M | 0.5 M |

△ Fig. 3.10 Investigating osmosis using plant tissue.

Devise and plan investigations

1. Write a plan for an experiment to carry out this investigation. Your plan should include:

   a) instructions on how to prepare the stem samples

   b) instructions on how to keep the stem samples until the experiment starts.

Make observations and measurements

2. Using the diagram, describe the results of this investigation.

Analyse and interpret data

3. Explain as fully as you can the results of this investigation.

4. Use the results to suggest the normal concentration of the cell cytoplasm. Explain your answer.

REMEMBER

Think about diffusion and osmosis in terms of particles and their concentration gradients. Be clear that, even when diffusion and osmosis stop because there is no concentration gradient, the particles in the solution continue to move – there is just no longer any net movement.

QUESTIONS

1. Explain why the properties of water are essential for living organisms.

2. Describe one example of the role of water in transport inside living organisms.

3. a) State what happens during osmosis.

   b) SUPPLEMENT Give a more detailed description of osmosis, using ideas about concentration and water potential.

4. Explain how osmosis is

   a) similar to diffusion   **b)** different from diffusion.

5. Describe the role of the plant cell wall in supporting a plant that has been well watered.

Turgid and flaccid cells

A **turgid** plant cell is one that is full of water. The pressure of the water in the cytoplasm in a turgid cell against the cell wall is called **turgor pressure.** This pressure prevents any more water entering the cell by osmosis, even if it is in a solution that has a higher water concentration than the water concentration of the cytoplasm.

A plant cell that has lost water so that the cytoplasm no longer exerts pressure against the cell wall is said to be **flaccid**. When the cytoplasm has shrunk enough to start pulling away from the inside of the wall the cell is said to be **plasmolysed**. The shrinking of the cell contents away from the cell wall is called **plasmolysis**. A plant with many flaccid or plasmolysed cells will show wilting.

cell membrane pressed
up against cell wall

cell membrane
has shrunk away
from cell wall

cell wall

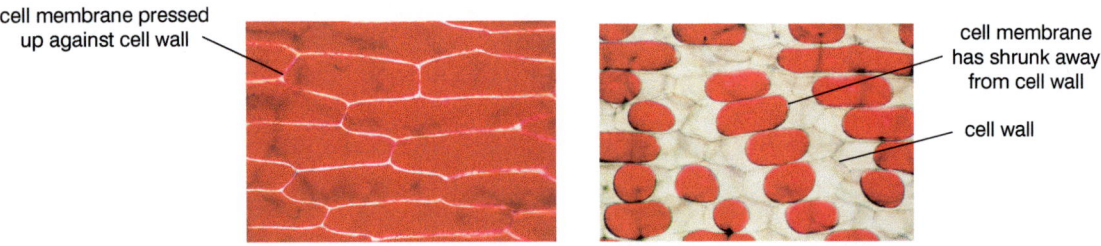

Δ Fig. 3.11 Cells from a red onion. Left: fully turgid cells. Right: plasmolysed cells.

SCIENCE IN CONTEXT

STOMATA

Stomata (single: stoma) are the holes in the surface of a leaf (usually the undersurface) that allow air to move into and out of the leaf. This provides the oxygen for respiring cells and carbon dioxide for photosynthesising cells, and allows water vapour that has evaporated from cell surfaces inside the leaf to diffuse out into the atmosphere. (There is more about stomata and photosynthesis in Section 6.)

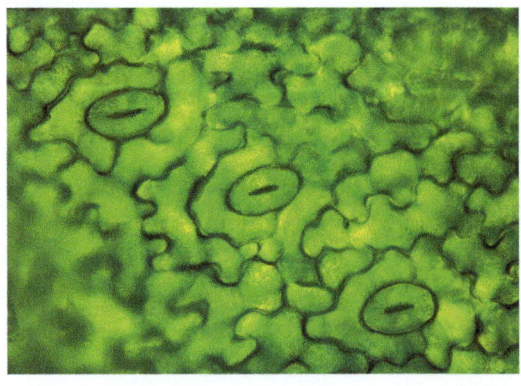

Δ Fig. 3.12 Each stoma is surrounded by two guard cells.

Each stoma is surrounded by two guard cells. These control the opening and closing of the stoma. Usually stomata are open during the day, and close at night. The stoma opens and closes as the guard cells change shape. During the day the guard cells gain water from surrounding cells as a result of osmosis. This makes the cells turgid and, because the inner edge of the guard cell does not stretch, the cells curve and create a space between them – that is the stoma. During the night, the guard cells lose water by osmosis. The cells lose their turgidity and collapse a little, closing the stoma between them.

Challenge Question: Unlike the other cells on the surface of a leaf, guard cells contain chloroplasts and so can photosynthesise making sugar. Suggest how this could lead to the guard cells becoming turgid during the day.

The importance of water potential and osmosis for animal cells

Animal cells also gain and lose water as a result of osmosis. However, the effects of this can be more dramatic than in plant cells because animal cells do not have a supporting cell wall.

If you place an animal cell (such as a red blood cell) into a solution of lower water potential, it will lose water by osmosis and become misshapen, like the cell shown in the photograph at the start of this section (Fig. 3.1).

If you place an animal cell into a solution of higher water potential, it will gain water by osmosis. Even when it is full of water, it will continue to take in water if the water potential is higher outside the cell than inside. The cell membrane is not strong, so if it is stretched too far it will burst.

SCIENCE IN CONTEXT — OVERHYDRATION

It is possible (although not easy) to die as a result of drinking too much water. If you drink a large volume of water very quickly, the water moves into the blood – and from there into cells – by osmosis.

The cells in the brain are particularly at risk because the brain has a limited space within the skull. If the brain cells swell quickly as a result of osmosis, the increased brain volume can cut off the blood supply, and brain cells start to die. Fortunately, we have an effective mechanism for getting rid of extra water from the body in urine. So overhydration is rare.

Challenge Question: If you did have to rehydrate quickly by drinking a lot of water, you could do so more safely by drinking water containing dissolved sugar. Explain why this would be safer than drinking a lot of pure water.

QUESTIONS

1. Explain the meaning of the following terms:

   a) flaccid

   b) turgid

   c) plasmolysis

   d) turgor pressure.

2. Describe the uptake of water from the soil by plant roots in terms of water potential.

3. Draw a labelled diagram to explain what happens to a red blood cell when placed in a solution that has a lower water potential than the cytoplasm of the cell.

ACTIVE TRANSPORT

Sometimes cells need to absorb molecules or ions from a region of their low concentration into a region of their higher concentration. For example, **root hair cells** take in nitrate ions from the soil even though the concentration of these ions is higher in the plant cells than in the soil.

The cells use energy to absorb these substances from a region of lower concentration to a region of higher concentration, so this is an **active** process and is called **active transport**. The energy comes from respiration.

Active transport can be defined as the movement of particles through a cell membrane from a region of lower concentration to a region of higher concentration (i.e. against a concentration gradient), using energy from respiration.

SUPPLEMENT

Active transport is used when substances need to be absorbed against their concentration gradient, for example, the uptake of mineral ions from the soil by plant root hairs (see Sections 6 and 8). It is also important when all of a substance present needs to be absorbed, for example the absorption of all the glucose from digested food in the small intestine.

SCIENCE IN CONTEXT **METABOLIC POISONS**

The energy for active transport comes from cell respiration. A simple test to show whether a substance is being absorbed by an active process or a passive process is to treat the cells with a metabolic poison that stops the cells respiring. For example, treating root hair cells with cyanide stops the uptake of nitrate ions but doesn't affect osmosis.

Challenge Question: Explain why a metabolic poison does not affect osmosis.

QUESTIONS

1. Explain what is meant by *active transport*.

2. **SUPPLEMENT** Give one example of active transport in plant cells and one in animal cells, and explain why this is important for the organism.

End of topic checklist

Key terms

active, active transport, concentration gradient, diffusion, net movement, osmosis, partially permeable, passive, root hair cells, solute, solution, solvent

SUPPLEMENT flaccid, plasmolysed, plasmolysis, turgid, turgor pressure, water potential, water potential gradient

During your study of this section you should have learned:

○ That diffusion is the net movement of particles from a region of their higher concentration to a region of their lower concentration, and is a passive process.

○ That many substances move into and out of cells by diffusion through the cell membrane and that this is essential for the life processes of living organisms.

○ **SUPPLEMENT** How to investigate the factors affecting diffusion: surface area, temperature, concentration gradient and distance.

○ That osmosis is the diffusion of water through partially permeable membranes, for example cell membranes.

○ **SUPPLEMENT** That osmosis is the net movement of water molecules from a region of higher water potential (dilute solution) to a region of lower water potential (concentrated solution), through a partially permeable membrane.

○ How to investigate the effects of osmosis on plant tissues that have been left in solutions of different concentrations.

○ **SUPPLEMENT** How to explain the terms: turgid, turgor pressure, plasmolysis, flaccid.

○ **SUPPLEMENT** How to explain the importance of water potential and osmosis in the uptake and loss of water by organisms.

○ That active transport uses energy from respiration to move particles across a cell membrane from a region of lower concentration to a region of higher concentration (i.e. against a concentration gradient).

○ **SUPPLEMENT** To explain the importance of active transport in living organisms, including the specific example of ion uptake by plant roots.

End of topic questions

1. Which of the following does **not** describe diffusion?

    A caused by the random movement of particles

    B movement down a concentration gradient

    C needs energy from respiration

    D passive process

2. For each of the following, state whether it is an example of diffusion, osmosis, or neither.

    a) Carbon dioxide entering a leaf when it is photosynthesising.

    b) Food entering your stomach when you swallow.

    c) A dried-out piece of celery swelling up when placed in a bowl of water.

3. An old-fashioned way of killing slugs in the garden is to sprinkle salt on them. This kills the slugs by drying them out. Explain why this works.

4. Explain the importance of water for living organisms.

5. **SUPPLEMENT** Explain the differences between turgid cells and flaccid cells, and the effects they have on the whole plant.

6. **SUPPLEMENT a)** If you measured the rate of respiration of plant root hair cells, would you expect it to be

       • the same as

       • more than or

       • less than

    the rate of respiration of other plant cells?

    b) Explain your answer to part **a)**.

7. **SUPPLEMENT** Nitrate ions are essential for plant growth. Nitrate concentrations are usually at a lower concentration in water in the soil than in the roots.

    a) State the type of transport plants must use to absorb nitrates.

    b) Explain why this type of transport must be used, and what the disadvantage is.

The human body is made of a complex combination of carbohydrates, fats and proteins. The body can make around 90 000 different proteins. Blood alone contains over 600 different types of fat and there are also many different types of carbohydrate.

These molecules come together to form even more complex structures in cells, tissues and organs which in turn leads to a huge variety of organisms.

These different categories of molecules have different structures which means we can use simple scientific tests to test for their presence in different substances, such as different food types.

STARTING POINTS

1. What substances are our bodies made up of?

2. How do we test for the presence of different food substances?

SYLLABUS SECTIONS COVERED

4.1 Biological molecules

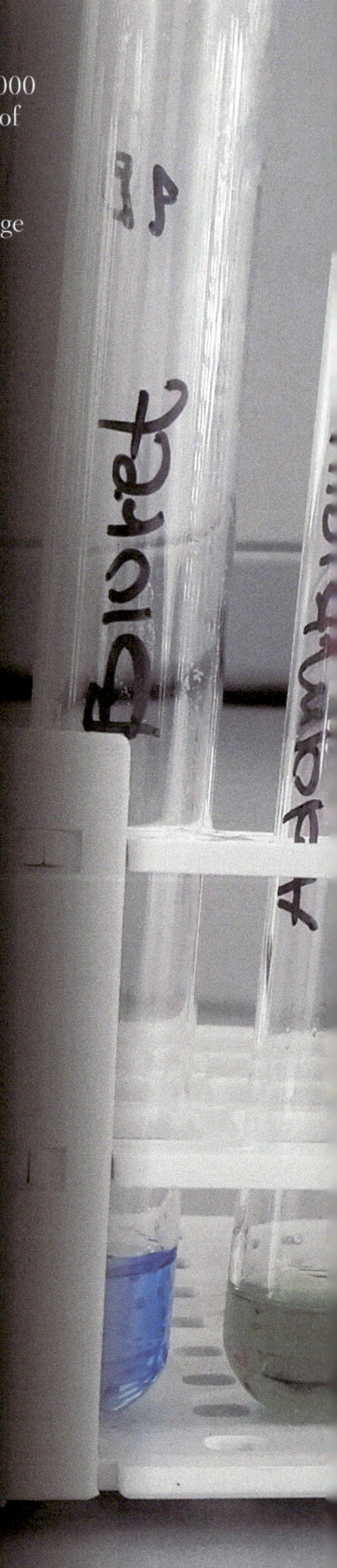

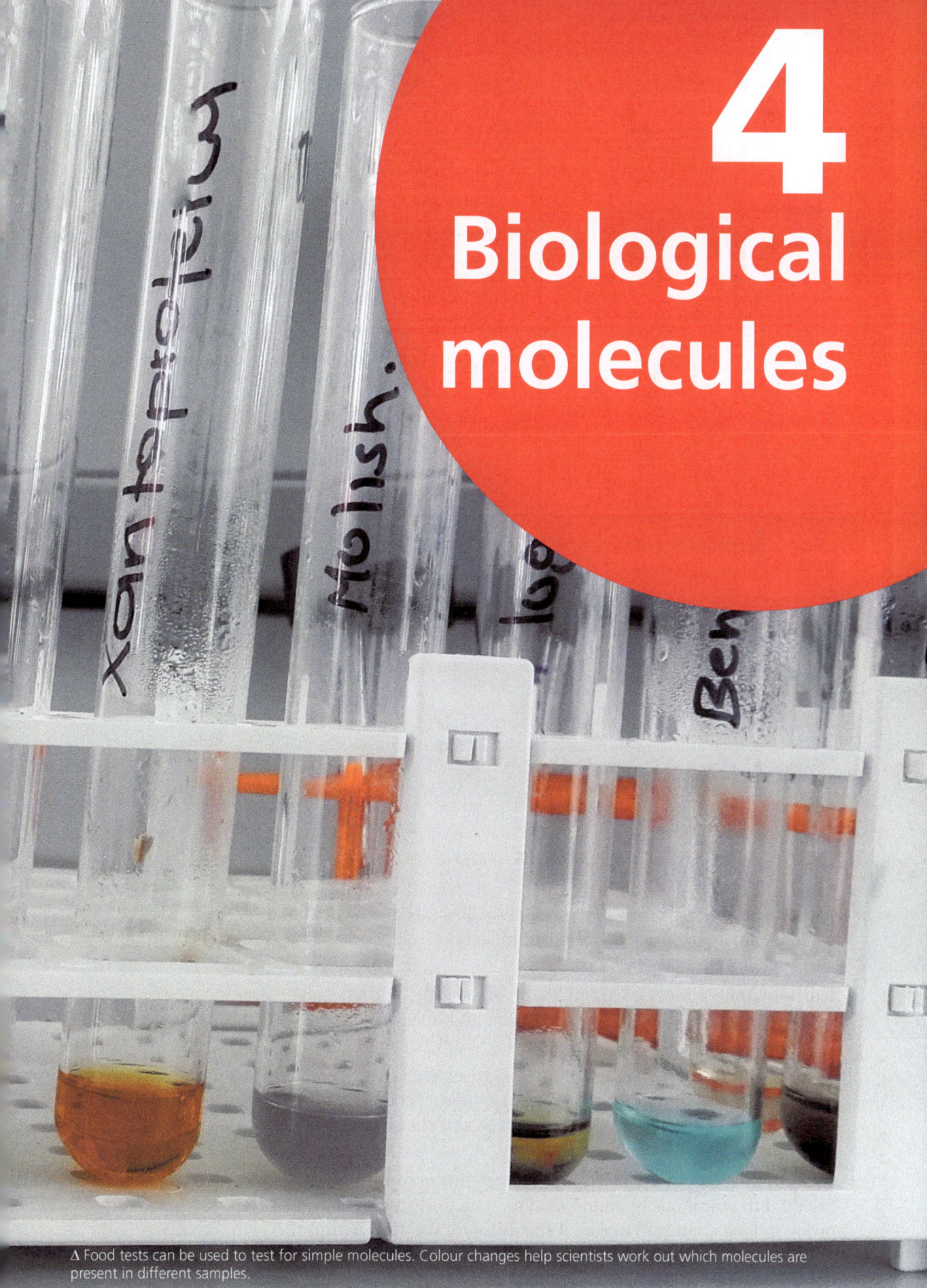

4
Biological molecules

△ Food tests can be used to test for simple molecules. Colour changes help scientists work out which molecules are present in different samples.

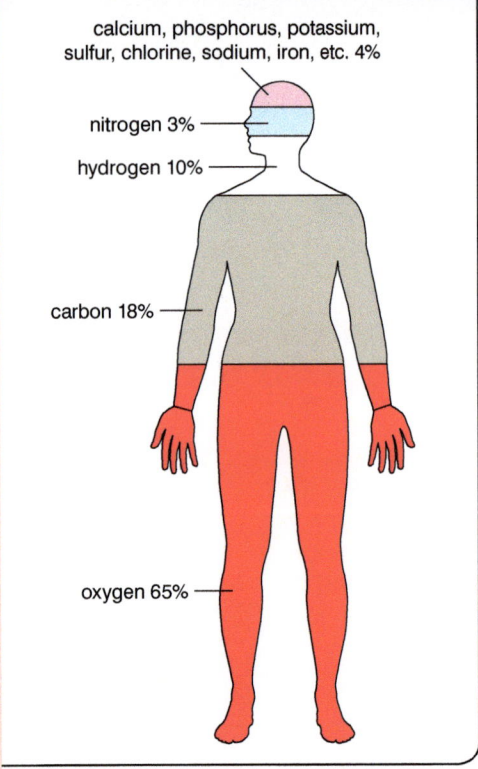

calcium, phosphorus, potassium, sulfur, chlorine, sodium, iron, etc. 4%

nitrogen 3%

hydrogen 10%

carbon 18%

oxygen 65%

Δ Fig. 4.1 The proportions of elements in the human body.

Biological molecules

INTRODUCTION

Around 65% of your body mass is oxygen, another 18% is carbon and 10% is hydrogen. The remainder of your mass is made up of a large range of other elements, including nitrogen, sulfur, calcium and iron. These elements are combined in different ways to form all the compounds in your body.

KNOWLEDGE CHECK

✓ Most of the foods that we eat can be grouped into carbohydrates, proteins or fats.
✓ Carbohydrates, proteins, and fats are formed from smaller molecules.

LEARNING OBJECTIVES

✓ List the chemical elements that make up: carbohydrates, fats, and proteins.
✓ State that large molecules are made from smaller molecules, limited to: starch, glycogen and cellulose from glucose; proteins from amino acids; fats and oils from fatty acids and glycerol.
✓ Describe the use of: iodine solution test for starch; Benedict's solution test for reducing sugars; biuret test for proteins; ethanol emulsion test for fats and oils.

CARBOHYDRATES, PROTEINS AND LIPIDS

Most of the molecules found in living organisms fall into three main groups: carbohydrates, proteins, and lipids, which are commonly called fats and oils. All of these molecules contain carbon, hydrogen and oxygen. In addition, all proteins contain nitrogen and some also contain sulfur.

Carbohydrate molecules are made up of small basic units called **simple sugars**. These are formed from carbon, hydrogen and oxygen atoms, sometimes arranged in a ring-shaped molecule. One example of a simple sugar is **glucose**.

Simple sugar molecules can link together to form larger molecules. They can join in pairs, such as **sucrose** (the 'sugar' we use in our food). They can also form much larger molecules called polysaccharides, such as **starch**, **glycogen** and **cellulose**, which are long chains of glucose molecules.

Protein molecules are made up of long chains of **amino acids** linked together. There are 20 different kinds of amino acid in plant and animal cells, and they can join in any order, in long chains, to make all the different proteins within the plant or animal body. Examples include the structural proteins in muscle, as well as enzymes that help to control cell reactions.

A **lipid** is what we commonly call a fat or oil. At room temperature **fats** are solid and **oils** are liquid, but they have a similar chemical structure. Both fats and oils are made from basic units called **fatty acids** and **glycerol**. There are three fatty acids in each lipid, and the fatty acids vary in different lipids. Lipids are important in forming cell membranes, and many other molecules in the body such as fats in storage cells.

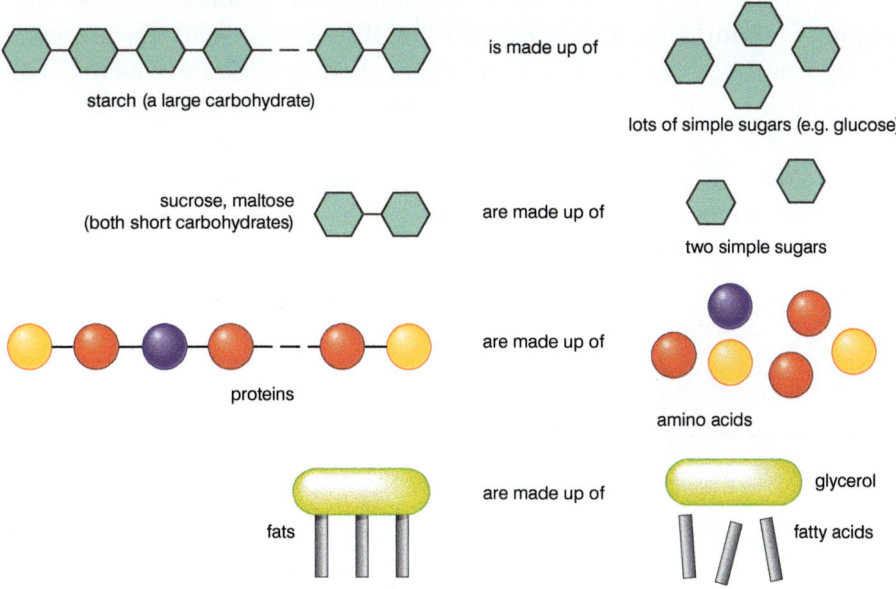

△ Fig. 4.2 Large biological molecules are formed from smaller molecules.

QUESTIONS

1. What are the basic units of:

   a) fats

   b) carbohydrates

   c) proteins?

2. Using the diagram of food molecules in Fig. 4.2, give two differences between the structures of a protein and a carbohydrate.

Tests for food molecules

We can use simple tests to indicate whether or not a food contains particular food molecules, such as starch, glucose, proteins, lipids, or vitamin C.

Test for starch

Starch is the storage molecule of plants, and is found in many foods that are made from plant tissue. When **iodine solution** (sometimes called iodine/potassium iodide solution) is mixed with a solution of food containing starch, or is dropped

△ Fig. 4.3 The blue-black colour shows there is starch in the biscuit.

onto food containing starch (Fig. 4.3), it changes from brown to dark blue (blue-black). This happens when even small amounts of starch are present and can be used as a simple test for the presence of starch. The colour change is easiest to see if the test is examined against a white background, such as on a white spotting tile.

Test for glucose

Glucose is a 'reducing sugar' that is important in respiration and photosynthesis. So it is commonly found in plant and animal tissues, and therefore in our food. Its presence can be detected using **Benedict's solution** (sometimes called Benedict's reagent). The pale blue Benedict's solution is added to a prepared sample that contains glucose and heated to around 95 °C. If it changes colour or forms a precipitate, this indicates the presence of glucose. A green colour means there is only a small amount of glucose in the solution. A medium amount of glucose produces a yellow colour. A significant amount of glucose produces a precipitate that is an orange-red colour.

◁ Fig. 4.4 Benedict's solution with a range of concentrations of reducing sugars (none in the tube on the left, getting more concentrated towards the right).

Test for protein

The **biuret test** is used to check for the presence of protein. A small sample of the food under test is placed in a test tube.

An approximately equal volume of biuret solution is carefully poured down the side of the tube. If the sample contains protein, a blue ring forms at the surface. If the sample is then shaken the blue ring disappears and the solution turns a light purple (lilac).

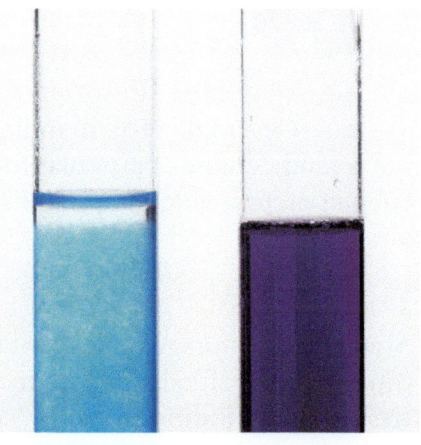

△ Fig. 4.5 A positive biuret test for protein.

Test for lipid

This test (called the **ethanol emulsion test**) depends upon the fact that lipids (fats and oils) do not dissolve in water but do dissolve in ethanol. The test sample is mixed with ethanol. If lipid is present it will be dissolved in the ethanol to form a solution. This solution is poured into a test tube of water, leaving behind any solid that has not dissolved. If there is any lipid dissolved in the ethanol it will form a cloudy white emulsion when mixed with the water.

Δ Fig. 4.6 Lipids have formed an emulsion, making it appear cloudy.

QUESTIONS

1. Describe what you would see if you tested samples of the following separately with **(i)** Benedict's solution and **(ii)** iodine solution:

   a) glucose syrup

   b) a cake made with wheat flour, table sugar (sucrose), fat and eggs.

   Explain your answers.

2. Explain how you would test seeds to see if they contained stores of:

   a) fat

   b) protein.

End of topic checklist

Key terms

amino acid, Benedict's solution, biuret test, carbohydrate, cellulose, ethanol emulsion test, fat, fatty acid, glucose, glycerol, glycogen, iodine solution, lipid, oil, protein, reducing sugar, simple sugar, starch, sucrose

During your study of this section you should have learned:

○ That carbohydrates, proteins, fats and oils all contain the elements carbon, hydrogen and oxygen.

○ That proteins also contain the element nitrogen and some may contain sulfur.

○ That large carbohydrate molecules, such as starch, glycogen and cellulose, are made from smaller molecules, such as glucose.

○ That proteins are made from smaller molecules called amino acids.

○ That fats and oils are made from smaller molecules called fatty acids and glycerol.

○ How to use iodine solution to test for starch.

○ How to use Benedict's solution to test for reducing sugars.

○ How to carry out the biuret test for proteins.

○ How to carry out the ethanol emulsion test for fats and oils.

End of topic questions

1. **a)** Which of the following elements is found in proteins but **not** in carbohydrates or lipids?

    A carbon

    B hydrogen

    C nitrogen

    D oxygen

    b) Explain why the body needs other elements, in addition to those listed in part **a)**.

2. A sample of bread was ground up. Some of the breadcrumbs were tested with Benedict's solution and some with iodine solution. The rest of the crumbs were mixed with Substance A. After 20 minutes, some of the mixture was tested with Benedict's reagent and some with iodine solution. The results of the tests are shown in the table.

	Test with Benedict's solution	Test with iodine solution
Before adding Substance A	no precipitate	change to blue-black colour
After 20 min with Substance A	orange-red precipitate	no colour change

    a) Describe what the results show.

    b) Suggest what Substance A was. Explain your answer.

3. Before solid foods are tested for the presence of different food substances it is usually best to cut or grind them into small pieces. Suggest an explanation for this.

Many useful enzymes come from bacteria and other microorganisms, and these organisms are harnessed both in industry and in the home, for uses as wide-ranging as washing detergents, to baking leavened bread. Yeast is a single-celled organism that produces enzymes that break down the sugars in flour, and in the process tiny bubbles of carbon dioxide gas are released, which cause the bread to rise.

The first enzyme used commercially in washing products was introduced in the 1960s. It was a protease that broke down protein-based stains such as blood, and it was extracted from a bacterium. Since then a much wider range of enzymes has been added to washing products, to digest fats, starches, and other molecules.

STARTING POINTS

1. What are enzymes?

2. Why are enzymes important in all living organisms?

3. What factors affect enzyme activity?

SYLLABUS SECTIONS COVERED

5.1 Enzymes

5
Enzymes

Δ Enzymes released into the gut, and attached to the gut surface (shown here), digest food so that nutrients can be absorbed.

Enzymes

INTRODUCTION

Δ Fig. 5.1 Enzymes in the mouth, stomach and small intestine will break down this food into much smaller molecules.

Many of our staple foods, such as rice, potato, pasta or bread, contain large quantities of starch. Take a mouthful of one of these, without anything else, and you won't taste a lot to start with. But continue chewing on it for a few minutes, to mix it with saliva and reduce it to a slush, and you will find it starts to taste sweeter. This is because there are enzymes in saliva that start to break down the starch into smaller sugar molecules that taste sweet. Enzymes are essential in digestion, to break down the large molecules in our food into molecules small enough for diffusion through the cells of the gut wall and into our bodies.

KNOWLEDGE CHECK

✓ Food is digested in the gut into smaller molecules.

LEARNING OBJECTIVES

✓ Describe enzymes as proteins that are involved in all metabolic reactions, where they function as biological catalysts.
✓ **SUPPLEMENT** Describe and explain enzyme action with reference to the active site, enzyme-substrate complex, substrate and product.
✓ **SUPPLEMENT** Describe and explain the specificity of enzymes in terms of the complementary shape and fit of the active site with the substrate.
✓ Investigate and describe the effect of changes in temperature and pH on enzyme activity.
✓ **SUPPLEMENT** Explain the effect of changes in temperature on enzyme activity in terms of kinetic energy, shape and fit, frequency of effective collisions, and denaturation.
✓ **SUPPLEMENT** Explain the effect of changes in pH on enzyme activity in terms of shape and fit, and denaturation.

ENZYMES AS CATALYSTS

A **catalyst** is a substance that increases the rate of a chemical reaction but is not itself changed by the reaction. Living cells use catalysts to change the rate of reactions that happen inside them. These are known as *metabolic reactions* because they are the reactions of the metabolism (all the processes that keep a living organism alive). This makes enzymes very important to all living organisms.

Catalysts that control metabolic reactions are **enzymes,** and because they work in living cells they are called **biological catalysts.** Enzymes are proteins. They help cells carry out all the life processes quickly. Without them, most metabolic reactions would happen too slowly for life to carry on.

Some enzymes help two or more small molecules join together, such as when the polysaccharides starch and glycogen are built from glucose. Other enzymes help large molecules break down into smaller ones, such as when proteins are broken down into separate amino acids.

1. Define the term *catalyst*.
2. Explain what is meant by *biological catalyst*.
3. Explain why cells need enzymes.

SUPPLEMENT

How enzymes work

Enzymes are proteins and, like all proteins, they have a three-dimensional (3D) shape produced by the way the molecule folds up. A molecule that an enzyme joins with at the start of a reaction is called a substrate, and the molecule that is formed by the end of the reaction is called a product. So, during a reaction, substrate molecules are changed to product molecules. Enzymes have a part of the molecule with a particular 3D shape. This space is called the enzyme's **active site**. The active site matches the shape of the substrate molecule. We say the shapes are **complementary**, because the substrate fits neatly into the active site in the enzyme, like fitting two jigsaw pieces together.

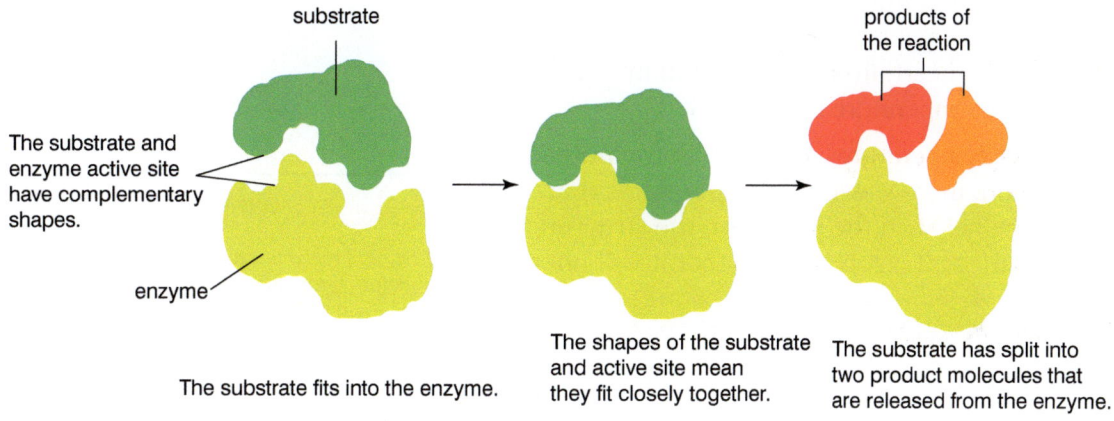

substrate

products of the reaction

The substrate and enzyme active site have complementary shapes.

enzyme

The substrate fits into the enzyme.

The shapes of the substrate and active site mean they fit closely together.

The substrate has split into two product molecules that are released from the enzyme.

Δ Fig. 5.2 In this reaction, the enzyme helps a substrate molecule split into two product molecules.

The substrate fits tightly into the enzyme's active site, forming an **enzyme–substrate complex**. This makes it easier for the bonds inside the substrate to be rearranged to form the products. Once the products are formed, they no longer fit the active site, so they are released, leaving the active site free and the enzyme unchanged. This means the enzyme molecule is able to bind with another substrate molecule.

Explaining enzyme specificity

Enzymes are **specific**, which means that each enzyme only works with one substrate or a group of similar-shaped substrates. For example:

- amylase is a type of carbohydrase enzyme produced in the mouth, which starts the digestion of starch in food into simple sugars
- proteases are digestive enzymes that break down proteins into smaller units
- lipases are digestive enzymes that break down lipids in foods.

(You will find out more about digestive enzymes in Section 7.)

The complementary shapes of the enzyme and substrate helps to explain the fact that enzymes are specific, because only a substrate with the correct shape can fit into the active site and so be affected by the enzyme.

QUESTIONS

1. Define the terms *substrate* and *product*.

2. Describe how an enzyme causes a substrate molecule to change into product molecules.

3. Explain what is meant by the active site of an enzyme.

4. SUPPLEMENT Explain how the shape of the active site is related to the specificity of an enzyme.

Enzymes and temperature

Enzymes work best at a particular temperature, called their **optimum temperature**. For many enzymes in the human body, particularly those that work in the organs in the core (centre) of the body, such as the heart, liver, kidneys, and lungs, the optimum temperature is around 37 °C.

At lower temperatures, enzymes in the human body work more slowly. At temperatures that are much higher than the optimum, the structure of an enzyme will be changed so that it will not work.

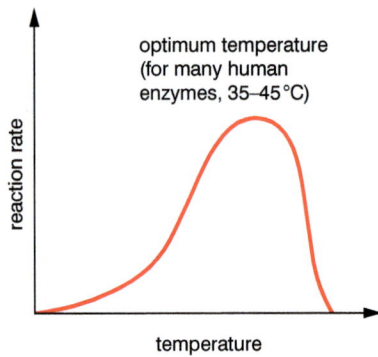

Δ Fig. 5.3 Many enzymes work best at an optimum temperature.

SUPPLEMENT

Remember the relationship between enzyme activity, temperature, and pH, particularly when learning about topics like homeostasis (Section 13), as this helps to explain why maintaining particular conditions in the body, such as a constant internal body temperature, is so important for health.

Investigating the effect of temperature on enzymes

The effect of temperature on an enzyme can be tested by measuring the rate of action of the enzyme at different temperatures. One method is shown in the Developing practical skills box on page 63. Alternatively, you could use the following method to investigate the optimum temperature of amylase.

Background knowledge:

- Amylase breaks down starch to simple reducing sugars.
- Starch reacts with iodine solution by turning it blue-black.
- Reducing sugars do not react with iodine solution, so leave it orange-brown.

Outline method:

- Mix starch solution with amylase solution and place different tubes of the mixture in water baths of different temperatures.
- For each tube in turn, take a sample every minute for testing with iodine solution.
- For the first samples, any starch present will turn the iodine solution blue-black.
- However, when the starch has all been broken down it will stop reacting with the iodine solution, leaving it orange-brown.
- Identify the tube from which the samples stop reacting with the iodine solution quickest. The temperature that this tube had been kept at is the one that is closest to the optimum temperature for amylase.

Developing practical skills

Before the widespread use of digital photography, most photographs were taken using photographic film. Photographic film consists of a celluloid plastic backing covered with a layer of gelatin. Where the film has been exposed to light the gelatin layer contains tiny particles of silver, which make that area black. Gelatin is a protein and is easily digested by proteases.

Strips of exposed film were soaked in protease solution at different temperatures. When the gelatin had been digested, the silver grains fell away from the celluloid backing, leaving transparent film. The table shows the results.

Tube	Temperature in °C	Time to clear
1	10	6 min 34 s
2	20	3 min 15 s
3	30	2 min 43 s
4	40	3 min 55 s
5	50	8 min 33 s

Devise and plan investigations

1. Describe how you would set up this investigation to get results like those shown in the table.

Analyse and interpret data

2. Draw a graph using the data in the table.

3. Describe the shape of the graph.

4. Explain the shape of the graph.

Evaluate data and methods

5. Suggest how you could modify this experiment to get a more accurate estimate of the optimum temperature for this enzyme.

The effect of temperature

Kinetic energy is the energy of moving particles. Particles that have a greater kinetic energy move faster. The kinetic energy of molecules that are free to move will cause them to bump into surrounding molecules. The kinetic energy of atoms held within larger molecules by bonds will cause them to vibrate.

An enzyme molecule and substrate molecule can only form an enzyme–substrate complex when they collide into each other with sufficient energy, and the substrate fits into the active site. These are known as **effective collisions**.

- At a low temperature the enzyme and substrate molecules move slowly, so effective collisions are relatively rare.
- As the temperature increases, the molecules gain more energy and move faster, so the *frequency* (or likelihood) of effective collisions increases. The rate of reaction increases up to the optimum temperature.
- Beyond the optimum temperature, the atoms in the enzyme molecule are vibrating so much that they start to change the shape of the active site. This means the substrate doesn't fit as well, so the chances of an enzyme–substrate molecule forming decreases. The rate of reaction decreases.
- If the temperature increases too much, the bonds between atoms in the enzyme molecule start to break, changing the shape of the active site. This is a permanent change, and when it happens the enzyme is said to be **denatured**.

Enzymes and pH

Enzymes also often work best at a particular pH, called their **optimum pH**. Extremes of (very high or very low) pH can slow down the rate of action of the enzyme and even stop it from working completely.

Different enzymes have different optimum pHs, depending on where they are normally found in the body. Pepsin digests proteins in the stomach, which is a highly acidic environment. Trypsin digests proteins in the small intestine, where conditions are more alkaline.

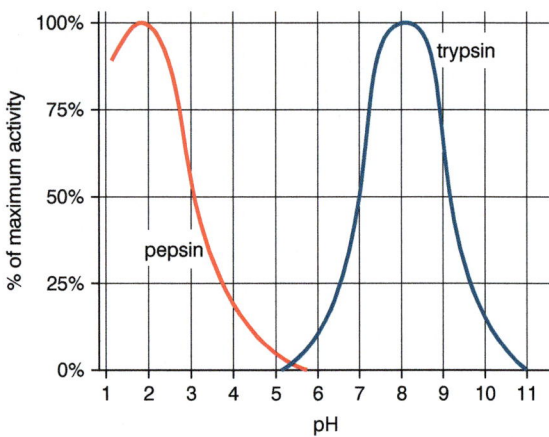

△ Fig. 5.4 Pepsin (an enzyme found in the stomach) and trypsin (an enzyme released into the small intestine) have different optimum pHs.

Investigating the effect of pH on enzymes

You can investigate the effect of pH on amylase enzyme using a similar method to the one above for temperature.

Outline method:

- Set up one tube for each pH to be investigated and add buffer solution, which will keep the contents at a particular pH.
- Add starch solution to each tube.
- For each tube in turn, add amylase solution and then take a sample every minute and test for starch using iodine solution.
- The sample that is the quickest to stop turning iodine solution blue-black comes from the tube where digestion of starch was fastest. The pH that this tube was kept at is therefore the closest to the optimum pH for amylase.

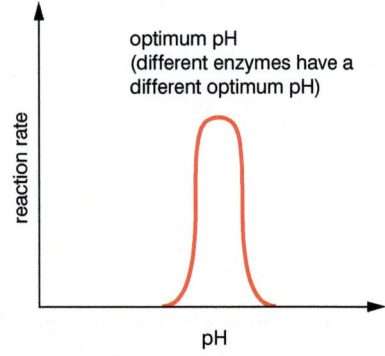

△ Fig. 5.5 Many enzymes work best at an optimum pH.

Δ Fig. 5.6 The tubes show the results of an experiment on the digestion of meat. Pepsin is a protease enzyme that is released in the stomach where it starts the digestion of proteins in food. Acid is also secreted into the stomach contents, reducing the pH and providing the optimum pH for pepsin. The left tube shows that acid has no effect on the meat. The middle tube shows that pepsin on its own digests the meat slowly. Only when the pepsin is mixed with acid can the enzyme work quickly to digest the protein.

REMEMBER

pH can be measured or monitored in different ways, for example, by using indicators such as hydrogencarbonate, litmus, or universal indicator.

QUESTIONS

1. Describe the effect of temperature on the rate of an enzyme-controlled reaction.

2. Compare the optimum pHs for pepsin and trypsin, shown in the graph in Fig. 5.4, and explain the differences.

The effect of pH

Proteins are made of amino acids, joined together in a chain. The amino acids then interact with nearby amino acids, which causes the chain to fold up into the 3D shape of the enzyme.

Some of the interactions between amino acids in the enzyme molecule depend on the pH of the surrounding solvent. So the shape of the enzyme will depend on the surrounding pH. If the pH changes too much from the optimum pH, the shape of the enzyme, and particularly its active site, will change. So the substrate will not fit as well and the rate of reaction will decrease.

QUESTIONS

1. Explain the effect of temperature on enzyme activity:

   a) at temperatures below the optimum

   b) at temperatures above the optimum.

2. Explain the effect of pH on pepsin (see graph in Fig. 5.4) in terms of the active site of the enzyme.

End of topic checklist

Key terms

biological catalyst, catalyst, enzyme, optimum pH, optimum temperature

SUPPLEMENT active site, complementary, denature, effective collision, enzyme–substrate complex, kinetic energy, product, specific, substrate

During your study of this topic you should have learned:

○ That enzymes are proteins which act as biological catalysts and control the rate of metabolic reactions.

○ **SUPPLEMENT** How to describe enzyme action in terms of the shape of the enzyme's active site being complementary to its substrate, leading to the formation of products.

○ **SUPPLEMENT** How to explain enzyme action in terms of the active site, enzyme–substrate complex, substrate and product.

○ **SUPPLEMENT** How to explain the specificity of enzymes in terms of the complementary shape and fit of the active site with the substrate.

○ How to investigate the effect of changes in temperature and pH on enzyme activity and to describe the role of optimum temperature where relevant.

○ **SUPPLEMENT** How to explain the effect of changes in temperature on enzyme activity in terms of kinetic energy, shape and fit, frequency of effective collisions and denaturation.

○ **SUPPLEMENT** How to explain the effect of changes in pH on enzyme activity in terms of shape and fit and denaturation.

End of topic questions

1. **SUPPLEMENT** What is the name of the part of an enzyme that binds to the substrate?

   A active site

   B complementary site

   C metabolic site

   D product site

2. **SUPPLEMENT** There are around 75 000 different enzymes in the human body. Explain why we need so many.

3. Describe how you would investigate the optimum temperature for a particular enzyme.

4. Sketch a graph to show the effect of temperature on the rate of reaction for an enzyme from humans. Label the value of the optimum temperature on your graph.

5. The body has many mechanisms for keeping internal conditions within limits. One of the internal conditions that is controlled is the concentration of carbon dioxide in the blood. Carbon dioxide gas is acidic and highly soluble.

   a) State which process in cells produces carbon dioxide.

   b) State how this gas is removed from the body.

   c) Explain what you would expect to happen to the amount of carbon dioxide in the body during exercise.

   d) Describe what effect this would have on conditions inside cells if the carbon dioxide was not removed.

   e) Explain what problem this would cause for enzymes and the cell processes that they control.

6. **SUPPLEMENT** Explain fully the shape of the graph you drew for Question 4.

7. Some students investigated the effect of temperature on activity of the enzyme amylase. They added amylase to starch solution in three test tubes placed in water baths at different temperatures. For five minutes, at 30 second intervals they removed a drop from each test tube and tested it for the presence of starch. The results are shown in the table.

	Temperature (°C)	Time when no starch was detected (s)
Tube 1	10	180
Tube 2	20	120
Tube 3	30	60

At what temperature was the rate of breakdown of starch highest? How do you know?

The world's tallest known living tree, which has been named Hyperion, is a coast redwood tree growing in Northern California in the US. When it was measured in 2006 it was found to be 115.61 m tall (379.3 ft). Like all plants, from the smallest seedling to the tallest redwood giant, Hyperion makes its own food from three simple ingredients: sunlight, water and carbon dioxide. This process is known as photosynthesis, and is one of the features of plants that distinguishes them from animals.

STARTING POINTS

1. How do plants get their food?

2. Why do plants need mineral ions?

3. Why do plants need leaves?

SYLLABUS SECTIONS COVERED

6.1 Photosynthesis

6.2 Leaf structure

6
Plant nutrition

△ Saplings growing towards a controlled light source.

Plant nutrition

INTRODUCTION

Even from space, we can distinguish different environments by looking at where the land is green, brown or white. The green areas are a result of chlorophyll in photosynthesising plants. We can also see where land use is changing, by looking at how the green areas of rainforests are slowly becoming brown as a result of deforestation.

Δ Fig. 6.1 The green on this satellite image shows plant growth on Earth.

KNOWLEDGE CHECK

✓ Plants make their own food in their leaves using photosynthesis.
✓ Plant structures, such as leaf and root cells, are specialised for their functions in nutrition.

LEARNING OBJECTIVES

✓ Describe photosynthesis as the process by which plants synthesise carbohydrates from raw materials using energy from light.
✓ State the word equation for photosynthesis as: carbon dioxide + water → glucose + oxygen, in the presence of light and chlorophyll.
✓ **SUPPLEMENT** State the balanced symbol equation for photosynthesis as:
  $6CO_2 + 6H_2O \rightarrow C_6H_{12}O_6 + 6O_2$.
✓ State that chlorophyll is a green pigment that is found in chloroplasts.
✓ **SUPPLEMENT** State that chlorophyll transfers energy from light into energy in chemicals, for the synthesis of carbohydrates.
✓ **SUPPLEMENT** Outline the subsequent use and storage of the carbohydrates made in photosynthesis, limited to: starch as an energy store; cellulose to build cell walls; glucose used in respiration to provide energy; sucrose for transport in the phloem; nectar to attract insects for pollination.
✓ **SUPPLEMENT** Explain the importance of: nitrate ions for making amino acids and magnesium ions for making chlorophyll.
✓ Investigate and understand the need for chlorophyll, light, and carbon dioxide for photosynthesis.
✓ **SUPPLEMENT** Understand and describe the effects of varying light intensity, carbon dioxide concentration, and temperature on the rate of photosynthesis.
✓ **SUPPLEMENT** Understand and describe the effect of light and dark conditions on gas exchange in an aquatic plant using hydrogencarbonate indicator solution.
✓ State that most leaves have a large surface area and are thin, and explain how these features are adaptations for photosynthesis.
✓ Identify in diagrams and images the following structures in the leaf of a dicotyledonous plant: chloroplasts, cuticle, guard cells and stomata, upper and lower epidermis, palisade mesophyll, spongy mesophyll, air spaces, vascular bundles, xylem and phloem.
✓ **SUPPLEMENT** Explain how the structures listed above adapt leaves for photosynthesis.

PHOTOSYNTHESIS

Plant tissue contains the same types of chemical molecules (carbohydrates, proteins and lipids) as animal tissue. However, whereas animals eat other organisms to get the nutrients they need to make these molecules, plants make these molecules from basic building blocks, beginning with the process of **photosynthesis**. ('Photo' means using light, and '**synthesis**' means making larger molecules from smaller ones.)

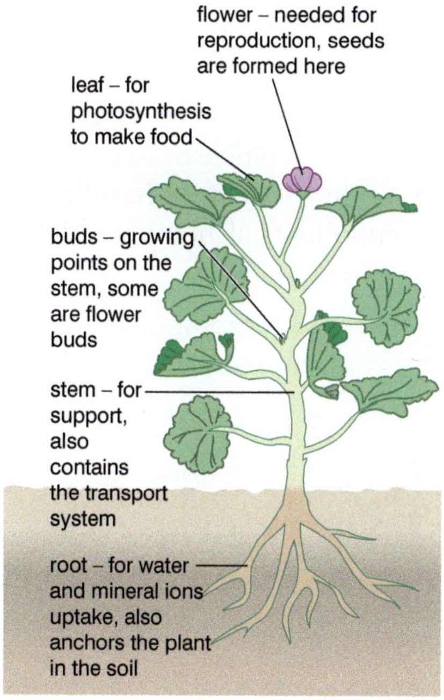

In photosynthesis, plants combine the raw materials carbon dioxide (from the air) and water (from the soil) to form **glucose**, a simple sugar which is also a **carbohydrate**. This process transfers energy from light (usually from sunlight) into chemical energy in the bonds of the glucose. The light is absorbed by the green pigment **chlorophyll** in the plants' **chloroplasts**.

Photosynthesis is fundamental to almost all life on Earth, because most organisms other than plants get their energy from the chemical energy in the food that they eat, whether that is herbivores getting energy directly from plants or carnivores consuming the herbivores.

flower – needed for reproduction, seeds are formed here

leaf – for photosynthesis to make food

buds – growing points on the stem, some are flower buds

stem – for support, also contains the transport system

root – for water and mineral ions uptake, also anchors the plant in the soil

△ Fig. 6.2 Anatomy of a plant.

Oxygen is also produced in photosynthesis. Although some is used inside the plant for respiration (releasing energy from food), most is not needed and is given out as a **waste product**.

The process of photosynthesis can be summarised in a word equation:

$$\text{carbon dioxide + water} \xrightarrow[\text{light energy}]{\text{chlorophyll}} \text{glucose + oxygen}$$

Photosynthesis can also be summarised as a balanced symbol equation:

$$6CO_2 + 6H_2O \xrightarrow[\text{light energy}]{\text{chlorophyll}} C_6H_{12}O_6 + 6O_2$$

Some of the glucose formed by photosynthesis is used in respiration to release energy (see Section 12). The rest of the glucose is converted into other substances, such as other carbohydrates. One of these is starch, which the plant uses as an energy store. Starch molecules are large carbohydrate molecules made of lots of simple sugar molecules joined together. As starch is insoluble it can be stored in cells without affecting water movement into and out of the cells by osmosis. Some plants, such as potato and rice, store large amounts of starch in particular parts of the plant (tubers or seeds). We use these parts as sources of starch in our food.

Some glucose is converted to cellulose, another large insoluble carbohydrate molecule, to build cell walls. The energy needed to join simple sugars together to make larger carbohydrates comes from respiration.

Some glucose is converted to a different sugar, **sucrose**. This is still soluble, but not as reactive as glucose, so can easily be transported around the plant in the phloem (see Section 8). Sucrose has different uses in plants, one example being that it is one of the components of **nectar**, which attracts insects for pollination.

QUESTIONS

1. Write the word equation for photosynthesis.

2. Explain the importance of light in photosynthesis.

3. SUPPLEMENT **a)** Write the balanced symbol equation for photosynthesis.

   b) Annotate your equation to show where each of the reactants come from, and where each of the products go to.

4. SUPPLEMENT Explain why the transfer of energy from light to energy in chemicals in plant cells is essential for life on Earth.

SUPPLEMENT

Mineral requirements of plants

Photosynthesis produces carbohydrates, but plants contain many other types of chemical. Carbohydrates contain just the elements carbon, hydrogen, and oxygen, but the amino acids that make up proteins also contain nitrogen. So plants need a source of nitrogen, which they get from nitrate ions. Other chemicals in plants contain different elements; for example, chlorophyll molecules contain magnesium and nitrogen. Without a source of magnesium ions and nitrate ions, a plant cannot produce chlorophyll and so cannot photosynthesise.

These additional elements are dissolved in water in the soil as **mineral ions**. The plant absorbs the mineral ions through their roots, using active transport because the concentration of the ions in the soil is lower than in the plant cells (see Section 8).

QUESTIONS

1. Explain why plants need a supply of mineral ions.

2. Outline what plants use the following mineral ions for:

   a) nitrate ions

   b) magnesium ions.

Investigating and understanding photosynthesis

We can use the iodine test to show that photosynthesising parts of a plant produce starch. Before carrying out this test, though, we must start by leaving the plant in a dark place for 24 hours. This will make sure that the plant uses up its stores of starch (this is known as de-starching) and means that any starch identified by the test is the result of photosynthesis during the investigation.

- The production of starch after photosynthesis can be shown simply by placing a de-starched plant in light for an hour. A single leaf is removed from the plant and placed in boiling water followed by hot ethanol to remove the chlorophyll. After rinsing in cold water the leaf can be tested for starch using iodine solution. The leaf should turn blue-black, indicating the presence of starch.

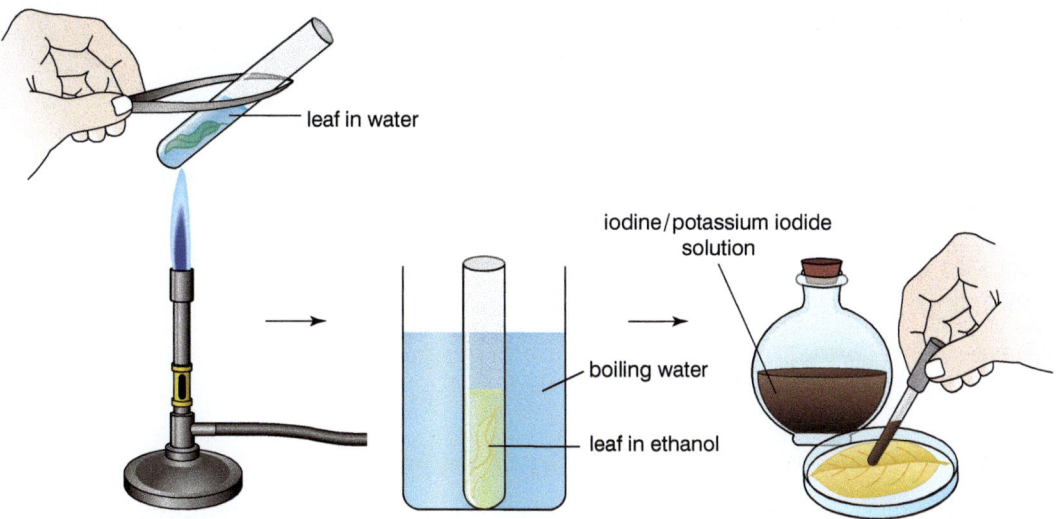

△ Fig. 6.3 Steps for preparing and testing a leaf for starch. The Bunsen burner must be turned off before ethanol is used.

- The investigation above can be adjusted to show the need for light by covering part of the leaf before the de-starched plant is brought into the light. Only the part of the leaf that received light should test positive for the presence of starch, showing that photosynthesis is linked to the production of starch.
- This investigation can also be adjusted to show the need for chlorophyll by using variegated leaves. Variegated leaves are partly green (where the cells contain chlorophyll) and partly white (where there is no chlorophyll). A variegated leaf after this investigation will show the presence of starch where there was chlorophyll but not in the parts of the leaf that had no chlorophyll.
- A simple test to show the need for carbon dioxide can be carried out by setting up two bell jars on glass sheets.

△ Fig. 6.4 Light was excluded from all of the lower leaf except an L-shaped window. After exposure to light, only the L shape tests positive for starch.

△ Fig. 6.5 Only the green parts of a variegated leaf can photosynthesise, as shown by the leaf on the right, which has been tested for starch.

Sodium or potassium hydroxide reacts with the carbon dioxide, removing it from the air. So a dish of one of the hydroxides is placed in one bell jar. Carbon dioxide is added to the other bell jar by burning a candle in it, which also removes some of the oxygen. Similar de-starched plants are placed in each bell jar, and the base of the jar sealed to the glass sheet, for example with petroleum jelly. After a few hours in light, a leaf from each plant is tested for starch, which should show that the plant with the least carbon dioxide produces little starch.

QUESTIONS

1. Describe a test, using an appropriate control, that would show the need for chlorophyll in photosynthesis.

2. Explain the precautions that should be taken when boiling ethanol to remove chlorophyll in a leaf.

3. Describe how you could show that plants need carbon dioxide for photosynthesis.

Factors affecting the rate of photosynthesis

The rate at which a process can occur depends on how quickly the required materials can be supplied.

Photosynthesis needs light, water, and carbon dioxide.

- Light provides the energy needed for the process of photosynthesis. So, as light intensity increases, so does the rate of photosynthesis.
- Carbon dioxide provides some of the molecules needed to make sugar. So as carbon dioxide concentration increases, so does the rate of photosynthesis.
- Temperature is needed to maintain the optimum temperature of the enzymes that are used in the process of photosynthesis. So as temperature increases, so does the rate of photosynthesis, unless it gets too hot and the enzymes denature.

Investigating gas exchange in aquatic plants

All organisms use cellular respiration to release the chemical energy in food molecules such as glucose. Cellular (usually aerobic) respiration must take place all the time because organisms need energy continually for other life processes. So animal and plant cells are always taking in oxygen and releasing carbon dioxide for respiration.

Plants photosynthesise as well as respire. For photosynthesis, plant cells need to take in carbon dioxide and release oxygen. However, photosynthesis can only take place when there is sufficient light.

At night, plants do not photosynthesise, but they do continue respiring. So plants give out carbon dioxide and take in oxygen. At daybreak, as light intensity increases, the rate of photosynthesis increases. At a particular light intensity, the amount of oxygen produced by photosynthesis will balance the amount used by the plant in respiration, and the *net production of oxygen* will be zero. This is known as a compensation point.

During daylight, oxygen production from photosynthesis exceeds its use in respiration, and the opposite is true for the use (by photosynthesis) and production (by respiration) of carbon dioxide. This continues until light intensity decreases, when the Sun sets and a second compensation point is reached.

You can investigate the effect of light on net **gas exchange** in a plant using a pH indicator, because carbon dioxide is acidic when dissolved in water. Hydrogencarbonate solution is often used because it is non-toxic and can be used with living organisms. Before use in an investigation it needs to be *equilibrated*, so that the concentration of carbon dioxide in the solution is the same as the concentration of carbon dioxide in the surrounding air. This is done by drawing air through the solution using a vacuum pump for a few minutes.

Discs can be cut from leaves using a core borer of large diameter. The discs are placed in Petri dishes containing equilibrated **hydrogencarbonate indicator**. Placing one dish in bright light and covering the other with dark paper shows a difference in colour of the indicator after 10–15 minutes as a result of the net release or net uptake of carbon dioxide. A similar investigation using an aquatic plant is shown in the Developing practical skills box below.

Developing practical skills

Hydrogencarbonate indicator can be used to indicate the acidity or alkalinity of a solution. At neutral pH it is a red-orange colour. In acidic solutions it is yellow, and in alkaline solutions it is purple.

Devise and plan investigations

- Add equal-sized measures of pondweed to boiling tubes and fill two-thirds full with hydrogen carbonate indicator solution.
- Seal with stoppers and clamp one next to a desk lamp. Cover the other with foil, or leave in a dark cupboard.
- Leave both for 30 minutes and write down observations every 2 minutes.

1. Using the information above, and that from the previous page, write a plan for testing the effect of light on the net gas exchange from an aquatic plant such as Elodea (pondweed).

Analyse and interpret data

The following results were obtained in a similar experiment to the one you have described.

Time in minutes	Tube in the light	Tube in the dark
0	red-orange	red-orange
2	red	light orange
4	reddish-purple	yellowish-orange
6	purple	yellow
8	purple	yellow
10	purple	yellow

2. Describe the results shown in the table for:
   a) the light dish
   b) the dark dish.
3. Explain what caused the colour change in the dark dish. What does this suggest is happening in the plant cells?
4. Explain what caused the colour change in the light dish. What does this suggest is happening in the plant cells?

5. Explain what other process is happening in the plant cells in the light dish that we cannot see because its effects are masked.

Evaluate data and methods

6. Explain what no change in the colour of the indicator would mean.

7. Explain how you would adapt your method to find the compensation point for this plant.

8. Light is not the only factor that can affect the rate of photosynthesis.

   a) State one other factor that might have had an effect on these measurements.

   b) Suggest how the method could be changed to avoid this problem.

QUESTION

1. **SUPPLEMENT** Explain why a pH indicator can be used to investigate the net exchange of gases in aquatic plants.

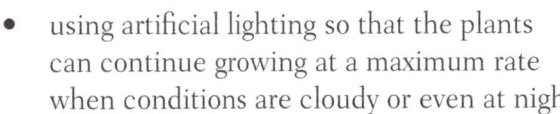

SCIENCE IN CONTEXT CHANGING GLASSHOUSE CONDITIONS

Farmers and plant growers want their crops to grow well, but in open fields it is not usually possible to control the amount of carbon dioxide or light the plants receive, or the temperature at which they are growing. However, if the plants are grown in sheltered conditions, such as in glasshouses, then it may be possible to change conditions, for example by:

Δ Fig. 6.6 Plants can be kept under the best conditions for growth in a greenhouse.

- using artificial lighting so that the plants can continue growing at a maximum rate when conditions are cloudy or even at night
- enriching the atmosphere around the plants with carbon dioxide by burning coal or oil
- using a heating system to increase the temperature to an optimum for photosynthesis.

Remember that enzymes have an optimum temperature at which they work, so glasshouses and polytunnels may also need to be ventilated to release hot air if the temperature rises too high, otherwise the rate of photosynthesis will decrease.

Challenge Question: Controlling conditions in glasshouses can increase the rate of photosynthesis, and so increase the rate of growth, of crop plants. However, this may not always be economically viable for the plant grower. Explain why.

1. Describe how each of the following factors affects the rate of photosynthesis:

   a) light intensity

   b) carbon dioxide concentration

   c) temperature.

2. **SUPPLEMENT** Explain why the factors have the effects you described in Question 1.

LEAF STRUCTURE

Photosynthesis takes place mainly in the leaves, although it can occur in any cells that contain green chlorophyll. Leaves are adapted to make them very efficient as sites for photosynthesis, gas exchange, transport, and support. Fig. 6.7 shows the external adaptations of the leaf that help to maximise the rate of photosynthesis.

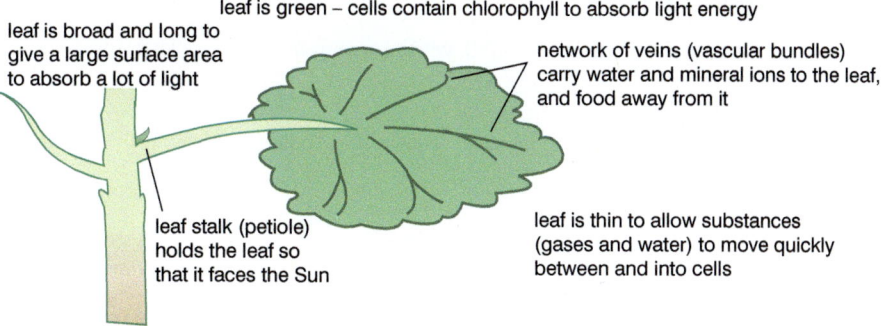

leaf is green – cells contain chlorophyll to absorb light energy

leaf is broad and long to give a large surface area to absorb a lot of light

network of veins (vascular bundles) carry water and mineral ions to the leaf, and food away from it

leaf stalk (petiole) holds the leaf so that it faces the Sun

leaf is thin to allow substances (gases and water) to move quickly between and into cells

Δ Fig. 6.7 Adaptations of a leaf.

◁ Fig. 6.8 The leaves of trees are often arranged so that they do not overlap each other, which makes it possible for the tree to capture as much light energy as possible.

Fig. 6.9 shows the arrangement of cells and tissues inside a leaf.

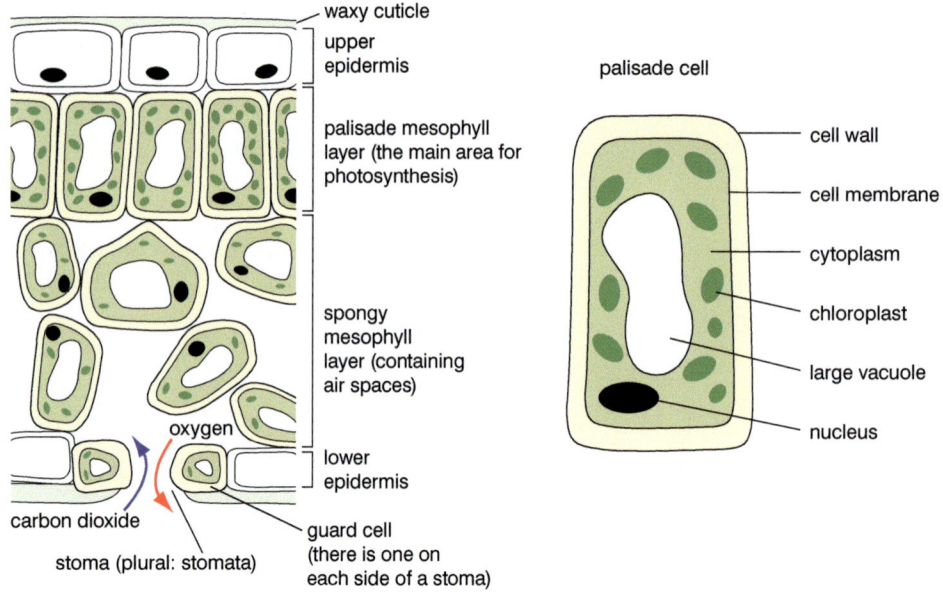

Δ Fig. 6.9 Cells in a section of a leaf (left), and a palisade cell (right), which contains many chloroplasts, for photosynthesis.

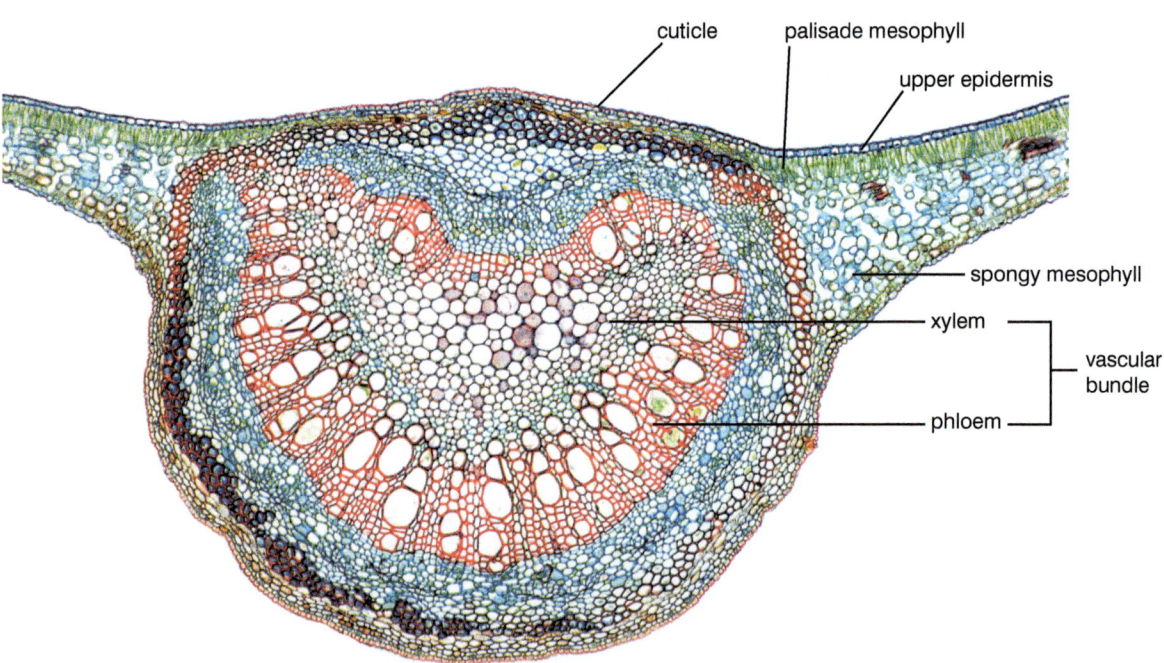

Δ Fig. 6.10 Photomicrograph of a section of a dicotyledonous plant leaf. Stomata are not easily seen in this section.

Adaptations for photosynthesis

Many structures in a leaf are adapted so that photosynthesis can be carried out as efficiently as possible.

- The waxy **cuticle** that covers the leaf, particularly the upper surface, prevents the loss of water from epidermal cells and helps to stop the plant from drying out too quickly.

- The transparent upper **epidermis** allows as much light as possible to reach the photosynthesising cells within the leaf. The lower epidermis is also transparent.
- The **palisade mesophyll cells**, where most photosynthesis takes place, are tightly packed together in the uppermost half of the leaf so that as many as possible can receive sunlight.
- Chloroplasts containing chlorophyll are concentrated in the palisade cells in the uppermost half of the leaf to absorb as much sunlight as possible.
- The **spongy mesophyll** cells and air spaces in the lower part of the leaf provide a large internal surface area to volume ratio to allow the efficient exchange of the gases carbon dioxide and oxygen between the cells and the air in the leaf.
- Many pores or **stomata** (singular: **stoma**) allow the movement of gases into and out of the leaf, to allow efficient gas exchange between the leaf and the air surrounding it. Each stoma is surrounded by two **guard cells** which open the stoma to allow gas exchange for photosynthesis. They can close the stomata at night to help conserve water.
- The **vascular bundles** form the veins in the stem and leaf. The thick cell walls of the tissue in the bundles help to support the stem and leaf.
- **Phloem** tissue transports sucrose, formed from glucose in photosynthesising cells, away from the leaf. **Xylem** tissue transports water and minerals to the leaf from the roots.

QUESTIONS

1. State four tissues in a leaf.

2. Give as many adaptations of a plant leaf for photosynthesis as you can.

3. **SUPPLEMENT** Explain why a large surface area inside the leaf is essential for photosynthesis.

4. **SUPPLEMENT** Explain why a transparent epidermis is an adaptation for photosynthesis.

End of topic checklist

Key terms

carbohydrate, chlorophyll, chloroplast, cuticle, epidermis, glucose, guard cell, hydrogencarbonate indicator, nectar, palisade mesophyll, phloem, photosynthesis, spongy mesophyll, stomata, sucrose, synthesis, vascular bundle, xylem, waste product

SUPPLEMENT gas exchange, mineral ion

During your study of this topic you should have learned:

○ How to describe photosynthesis as the process by which plants make carbohydrates from raw materials using energy from light.

○ That the word equation for photosynthesis is:

- carbon dioxide + water → glucose + oxygen
  in the presence of light and chlorophyll.

○ That chlorophyll is a green pigment found in chloroplasts.

○ **SUPPLEMENT** That chlorophyll transfers energy from light into energy in chemicals, for the synthesis of carbohydrates.

○ **SUPPLEMENT** That the balanced symbol equation for photosynthesis is:

- $6CO_2 + 6H_2O \rightarrow C_6H_{12}O_6 + 6O_2$

○ **SUPPLEMENT** How to explain the importance of: nitrate ions for making amino acids and magnesium ions for making chlorophyll.

○ How to understand the need for chlorophyll, light, and carbon dioxide for photosynthesis, using appropriate controls.

○ **SUPPLEMENT** How to investigate and describe the effects of varying light intensity, carbon dioxide concentration, and temperature on the rate of photosynthesis.

○ **SUPPLEMENT** How to investigate and describe the effect of light and dark conditions on gas exchange in an aquatic plant using hydrogencarbonate indicator solution.

○ That most leaves have a large surface area and are thin, and how to explain how these features are adaptations for photosynthesis.

○ How to identify the following structures in diagrams and images of the leaf of a dicotyledonous plant: chloroplasts, cuticle, guard cells and stomata, upper and lower epidermis, palisade mesophyll, spongy mesophyll, air spaces, vascular bundles, xylem and phloem.

○ How to explain how the structures listed above adapt leaves for photosynthesis.

End of topic questions

1. Which is a product of photosynthesis?

   A carbon dioxide

   B chlorophyll

   C oxygen

   D water

2. Some students used an oxygen sensor to measure the amount of dissolved oxygen there was in pond water containing pondweed, when a light was on and when it was switched off. The graph shows their results.

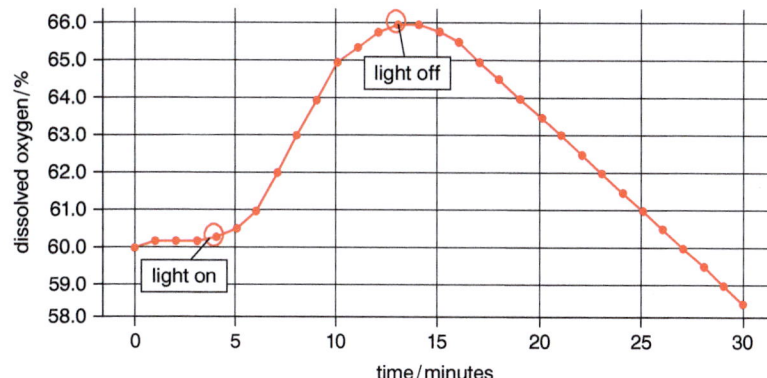

a) The light was switched on after 4 minutes. Describe and explain what happened until the light was switched off again.

b) Describe and explain what happened after the light was switched off.

3. SUPPLEMENT Explain why gardeners may add a liquid feed containing nitrate ions and magnesium ions to the water for the plants that they are growing.

4. a) What is needed for photosynthesis to take place?

   b) Describe and explain how leaf structures are adapted to photosynthesis.

5. SUPPLEMENT a) Sketch the axes of a graph with time of day along the *x*-axis and rate of photosynthesis on the *y*-axis. The units on the *x*-axis should start at midnight on one day and end at midnight on the following day. Add an arrowhead at the top of the *y*-axis to show that the units are arbitrary (have no values) but increase as you go up the axis.

   b) Draw a line on your graph to show how the rate of photosynthesis might change during the day for a large tree.

6. SUPPLEMENT In a greenhouse a grower is growing tomato plants. Explain why she might do the following:

a) leave lights on in the greenhouse all night

b) close the greenhouse windows at night but open them during the day.

7. SUPPLEMENT It is sometimes stated that 'Plants produce oxygen during the day and carbon dioxide at night.' Evaluate this statement.

8. A student measured the surface area of 10 nettle leaves grown in the shade and 10 leaves grown in full sun. They drew round the leaves on millimetre squared graph paper and then estimated the surface area by counting the grid squares inside the outline of the leaf.

	Surface area of a leaf in shade (mm²)	Surface area of a leaf in sun (mm²)
1	6902	2800
2	5150	1998
3	4940	2325
4	4437	2352
5	5717	1587
6	6159	1810
7	6726	2268
8	4900	2592
9	7020	2109
10	8880	2688

a) Calculate the mean surface area for the leaves grown in the shade and in full sun.

b) Describe the relationship between the surface area of the leaf and the amount of sunlight.

c) Suggest an explanation for this relationship.

Nutrition is one of the seven characteristics of living organisms. For humans and other animals, nutrition is the taking in of nutrients (including organic substances and mineral ions) that contain the raw materials needed by the body to make essential molecules, such as proteins, which are used as the building blocks to maintain healthy growth and tissue repair.

We think of malnutrition as not having enough food, but many people may be malnourished because they don't eat enough of the kinds of foods that contain all the nutrients that their bodies need, or they eat too much of the types of foods that will supply more energy than the body needs, so the excess is stored as fat.

STARTING POINTS

1. Why do we need food?

2. What is digestion?

3. What happens to food in the different parts of the digestive system?

SYLLABUS SECTIONS COVERED

7.1 Diet

7.2 Digestive system

7.3 Digestion

7

Human nutrition

△ Green beans contain vital nutrients essential to maintaining growth and tissue repair.

△ Fig 7.1 Honey is still a highly prized source of sugar today.

Human nutrition

INTRODUCTION

Human taste buds have evolved to give us useful information about what we are putting into our mouth. Sour or bitter tastes can indicate food that is decaying or poisonous, and so is dangerous to eat. Sweet (presence of sugars), salty and savoury (presence of proteins) tastes indicate nutrients that are essential for healthy growth. These were particularly important in our hunter-gatherer past, when it could be difficult to find foods containing these nutrients. They are not so useful to us now because, for many of us, foods containing large quantities of these are easily available. The urge to eat foods containing high levels of sugars and salt has led to problems of obesity and disease, particularly heart disease in people who have increasingly sedentary lifestyles.

KNOWLEDGE CHECK

✓ Animals eat other organisms to get the food they need for their life processes.
✓ The organs, tissues and cells of the digestive system are adapted to digest and absorb nutrients from food.
✓ Food may be chemically or physically digested before absorption.
✓ Different groups of people need different diets.

LEARNING OBJECTIVES

✓ Describe what is meant by a balanced diet.
✓ State the principal dietary sources and describe the importance of: carbohydrates, fats and oils, proteins, vitamins (limited to C and D), mineral ions (limited to calcium and iron), fibre (roughage), water.
✓ State the causes of scurvy and rickets.
✓ Identify in diagrams and images the main organs of the digestive system, limited to: alimentary canal: mouth, oesophagus, stomach, small intestine (duodenum and ileum) and large intestine (colon, rectum, anus); associated organs: salivary glands, pancreas, liver, and gall bladder.
✓ Describe the functions of the organs of the digestive system listed above, in relation to: ingestion – the taking of substances, e.g. food and drink, into the body; digestion – the breakdown of food; absorption – the movement of nutrients from the intestines into the blood; assimilation – uptake and use of nutrients by cells; egestion – the removal of undigested food from the body as faeces.

✓ Describe physical digestion as the breakdown of food into smaller pieces without chemical change to the food molecules.

✓ State that physical digestion increases the surface area of food for the action of enzymes in chemical digestion.

✓ Describe chemical digestion as the breakdown of large insoluble molecules into small soluble molecules.

✓ State the role of chemical digestion in producing small soluble molecules that can be absorbed.

✓ **SUPPLEMENT** Describe the functions of enzymes as follows: amylase breaks down starch to simple reducing sugars; proteases break down protein to amino acids; lipases break down fats and oils to fatty acids and glycerol.

✓ **SUPPLEMENT** State where, in the digestive system, amylase, proteases, and lipases are secreted and where they act.

✓ **SUPPLEMENT** Describe the functions of hydrochloric acid in gastric juice, limited to killing harmful microorganisms in food and providing an acidic pH for optimum enzyme activity of proteases in the stomach.

✓ **SUPPLEMENT** Explain that bile is an alkaline mixture that neutralises the acidic mixture of food and gastric juices entering the duodenum from the stomach, to provide a suitable pH for enzyme action in the small intestine.

✓ **SUPPLEMENT** Outline the role of bile in emulsifying fats and oils to increase the surface area for chemical digestion.

DIET

Essential nutrients

To keep healthy, humans need a diet that includes all the nutrients that our cells and tissues use, such as:

- **proteins** – which are broken down to make amino acids. The amino acids are used to form other proteins needed by cells, including enzymes. Protein sources include eggs, milk and milk products (cheese, yoghurt, etc.), meat, fish, legumes (peas and beans), nuts and seeds.
- **carbohydrates** – which are broken down to simple sugars for use in respiration. This releases energy in our cells and enables all the life processes to take place. Good sources of carbohydrate include rice, bread, potatoes, pasta and yams.
- **fats** and **oils** – which are deposited in many parts of the body, including just below the skin. Some fat helps to maintain body temperature. Fat is also a store of energy to supply molecules for respiration if the diet does not contain enough energy for daily needs. Fat is present in meat and can also come in the form of oils, milk products (butter, cheese), nuts, avocados and oily fish.
- **vitamins** and **mineral ions** – which are needed in tiny amounts for the correct functioning of the body. Vitamins and minerals cannot be produced by the body, and cooking food destroys some vitamins. For example, **vitamin C** is best supplied by eating raw fruit and vegetables. A lack of specific vitamins or mineral ions can cause different deficiency diseases, such as **scurvy** or **rickets** (see Table 7.1).

Essential vitamins and minerals	Role	Good food source	Deficiency disease
Vitamin C	for healthy skin, teeth and gums, and keeps lining of blood vessels healthy	citrus fruit, green vegetables, potatoes	scurvy (bleeding gums and wounds do not heal properly)
Vitamin D	for strong bones and teeth	fish, eggs, liver, cheese and milk	rickets (softening of the bones)
Calcium	needed for strong teeth and bones, and involved in the clotting of blood	milk and eggs	rickets (softening of the bones)
Iron	needed to make haemoglobin in red blood cells	red meats, liver and kidneys, leafy green vegetables such as spinach	anaemia (reduction in number of red blood cells, person soon becomes tired and short of breath)

Δ Table 7.1 Vitamins and mineral ions, their roles, sources and effects of deficiency.

- fibre (roughage) – which is made up of the cell walls of plants. Good sources are leafy vegetables, such as cabbage, and unrefined grains such as brown rice and wholegrain wheat. It adds bulk to food so that it can be easily moved along the digestive system by peristalsis. This is important in preventing constipation. Fibre is thought to help prevent bowel cancer.
- water – which is the major constituent of the body of living organisms and is necessary for all

Δ Fig 7.2 A healthy meal contains a good balance of the foods your body needs and nothing in too large an amount.

life processes. Water is continually being lost through excretion and sweating, and must be replaced regularly through food and drink in order to maintain health. Most foods contain some water, but most fruit and vegetables contain a lot of water.

The right balance

A **balanced diet** contains all of these nutrients in the right proportions to stay healthy because we need more of some nutrients than of others. As most foods contain more than one kind of nutrient, trying to work out what a balanced diet looks like can be difficult. Governments use images like the ones in Figs. 7.3 and 7.4, of food on a plate, to guide people on what proportions of food to eat.

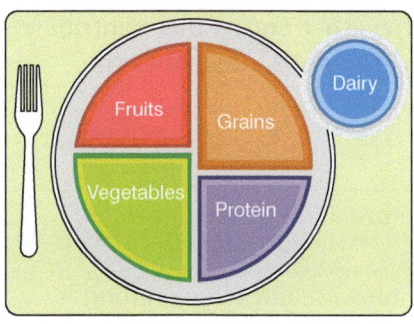

△ Fig 7.3 Guidance from the USDA (United States Department of Agriculture) on the proportions of different nutrients in a balanced diet.

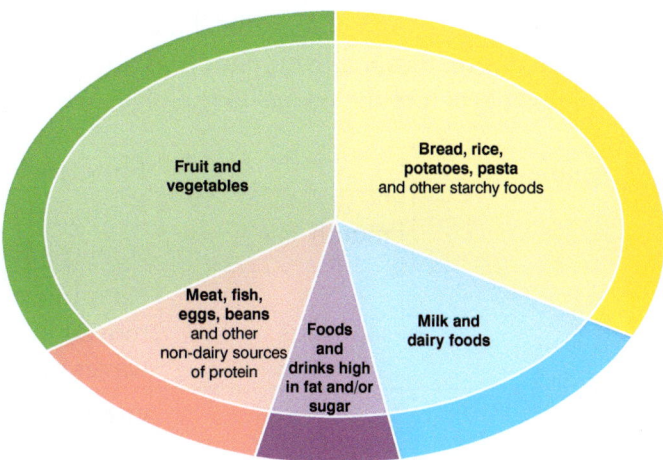

△ Fig 7.4 Guidance from the UK Government on the proportions of different nutrients in a balanced diet.

QUESTIONS

1. State which three groups of food molecules we need most of in a healthy diet.

2. Give examples of foods that are good sources of each group of food molecules.

3. State which other substances are needed in our diet.

4. Explain the role of each of these substances in our diet.

5. Explain what is meant by a *balanced diet*.

DIGESTIVE SYSTEM

Eating food involves several different processes:

- **ingestion** – taking substances, e.g. food and drink, into the body (through the mouth in humans)
- **digestion** – the breakdown of food
- **absorption** – the movement of nutrients from the intestines into the blood
- **assimilation** – the uptake and use of nutrients by cells
- **egestion** – the removal of undigested food from the body as **faeces**.

All these different processes take place in different parts of the **alimentary canal**.

The alimentary canal is a continuous tube through the body, from the mouth where food is ingested, through the **oesophagus**, **stomach**, **small intestine** and **large intestine**, to the anus where faeces are egested. You could say that materials in the alimentary canal aren't truly in the body. Not until food molecules are absorbed do they cross cell membranes into body tissue. Then they can be assimilated and **waste products** *excreted* through other organs.

The digestive system includes the alimentary canal and the other organs that contribute to digestion, such as the **liver, pancreas**, and **gall bladder**. Table 7.2 describes the functions of each of the organs in the digestive system.

Organs			Functions
Alimentary canal	mouth		teeth and tongue break down food into smaller pieces
	oesophagus		each lump of swallowed and chewed food is moved from the mouth to the stomach by waves of muscle contraction called peristalsis
	stomach		enzymes are secreted to start protein digestion; movements of the muscular wall churn up food into a liquid
	small intestine	duodenum	secretions from the gall bladder and pancreas enter to complete the process of digestion
		ileum	digested food molecules and water are absorbed in the ileum
	large intestine	colon	water is absorbed from the remaining material
		rectum	the remaining, unabsorbed, material (faeces), plus dead cells from the lining of the alimentary canal and bacteria, are compacted and stored
		anus	faeces are egested through a sphincter
Associated organs	salivary glands		produce liquid saliva, which moistens food so it is easily swallowed and contains the enzymes to begin breakdown of starch
	pancreas		secretes digestive enzymes in an alkaline fluid into the duodenum
	gall bladder		stores bile to be released into the small intestine

Δ Table 7.2 The functions of parts of the human digestive system.

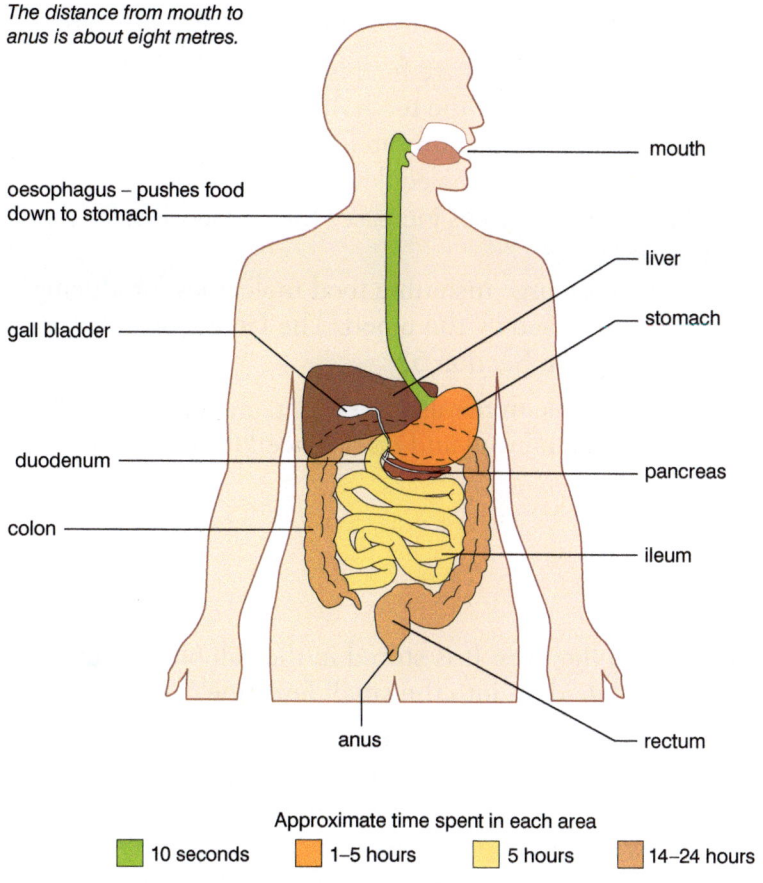

The distance from mouth to anus is about eight metres.

mouth

oesophagus – pushes food down to stomach

liver

gall bladder

stomach

duodenum

pancreas

colon

ileum

anus

rectum

Approximate time spent in each area

10 seconds	1–5 hours	5 hours	14–24 hours

△ Fig. 7.5 The human digestive system.

Food moves along the alimentary canal because of the contractions of the muscles in the walls of the alimentary canal. This is called **peristalsis**. Fibre in the food keeps the bolus bulky and soft, making peristalsis easier.

QUESTIONS

1. Sketch the diagram of the digestive system shown in Fig. 7.5. Label the organs, and add notes to each organ to explain its function in the system.

2. Explain the difference between egestion and excretion.

Different types of digestion

If food is to be of any use to us, the food molecules must enter the blood so that they can travel to every part of the body. Many of the foods we eat are made up of large, **insoluble** molecules that cannot cross the wall of the alimentary canal and the cell membranes of cells lining the blood vessels. This means the food molecules have to be broken down into small, **soluble** molecules that can easily cross cell membranes and enter the blood. Breaking down food is called digestion.

There are two types of digestion.

- **Physical digestion** occurs mainly in the mouth, where food is broken down into smaller pieces by the biting and chewing action of the teeth. It also happens in the stomach, where movements of the muscular wall churn up food. Physical digestion does not involve any chemical change to the food molecules themselves, but is important because breaking food into smaller pieces increases the surface area for the action of enzymes in chemical digestion.
- **Chemical digestion** is the breakdown of large insoluble food molecules, producing small soluble molecules that can be absorbed by the blood. The large molecules are broken down by the action of enzymes (see Section 5).

Some molecules, such as glucose, vitamins, minerals, and water, are already small enough to pass through the alimentary canal wall and do not need to be digested.

SUPPLEMENT

Bile in digestion

Bile is a substance produced by cells in the liver. It is stored in the gall bladder until it is needed and then passes along the bile duct into the small intestine.

Bile is important in the physical digestion of fats and oils. Fats and oils do not mix well with aqueous (water-based) mixtures such as the digesting food, and so remain as large droplets. This produces a small surface area for lipase enzymes to work on, which slows down the rate of digestion. Bile **emulsifies** fats and oils, breaking them up into much smaller droplets, increasing the surface area so that the rate of chemical digestion is much faster.

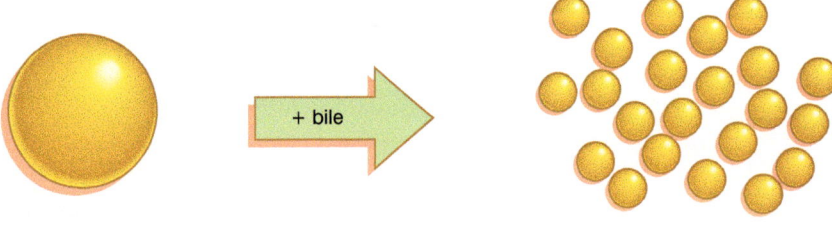

large fat or oil droplet   small fat or oil droplets

Δ Fig 7.6 Bile lowers the surface tension of large droplets of fat or oil so that they break up. This part of the digestive process is called emulsification.

QUESTIONS

1. Explain the difference between chemical and physical digestion.

2. **SUPPLEMENT** Explain the role of bile in digestion.

CHEMICAL DIGESTION

Chemical digestion in the alimentary canal is the result of enzymes. **Digestive enzymes** are a group of enzymes that are produced in the cells lining parts of the digestive system and are **secreted** (produced) into the alimentary canal to mix with the food.

The digestive enzymes include:

- carbohydrases that break down carbohydrates, one example of which is **amylase**
- **proteases**
- **lipases**.

REMEMBER

The -ase at the end of the name means it is an enzyme, and the first part usually names the substrate that the enzyme works on.

Each of the food groups (carbohydrates, proteins, and fats and oils) contains many different molecules. As each enzyme is specific to its substrate, this means that in each group of digestive enzymes there are many different enzymes.

Different enzymes are made in different parts of the digestive system, as shown in Table 7.3.

Enzyme	Where produced	Where it acts	Substrate	Products
Amylase	salivary glands (mouth) pancreas	mouth and small intestine (duodenum)	starch	simple reducing sugars
Proteases (many types)	stomach wall pancreas	stomach Small intestine	proteins	amino acids
Lipases (many types)	pancreas	small intestine (duodenum)	fats and oils (lipids)	fatty acids and glycerol

Δ Table 7.3 Digestive enzyme sites and functions.

SCIENCE IN CONTEXT

LACTOSE

Lactose is a sugar found in milk (from *lactis*, meaning 'milk'), which can be broken down in the alimentary canal by the enzyme lactase to the simple sugars glucose and galactose.

Like all young mammals, human babies produce lactase, which helps them to digest the lactose in breast milk. In most mammals the production of lactase decreases as the young mature, because the adult diet does not include milk. This also happens in adults from many human cultures in which adults generally do not drink milk, such as in South-East Asia. However, there are human cultures where mammals such as sheep or goats are kept to supply meat and milk for food. In these human groups the adults continue to produce lactase and are able to digest the lactose in milk. Adults who cannot do this are *lactose intolerant*. In these people bacteria in the alimentary canal break down the lactose, producing gas, which causes great discomfort.

Challenge Question: Explain why it is usually not advisable to give milk to adult cats to drink.

Stomach acid

Different enzymes work better in different conditions (see Section 5). Those enzymes that digest food in the stomach work best in acid conditions. Special cells in the lining of the stomach secrete **hydrochloric acid** into the gastric juices inside the stomach providing an acidic pH for optimum enzyme activity. The acid is also helpful in killing any harmful microorganisms taken in with food (see Section 10).

QUESTIONS

1. Explain why enzymes are needed in the digestive system.

2. a) State which enzyme has starch as its substrate.

   b) State which products are formed by the digestion of fats and oils.

3. Describe the roles of

   a) stomach acid

   b) bile in digestion.

End of topic checklist

Key terms

absorption, alimentary canal, anus, assimilation, balanced diet, carbohydrate, chemical digestion, colon, digestion, digestive enzymes, duodenum, egestion, faeces, fat, fibre (roughage), gall bladder, glycogen, ileum, ingestion, insoluble, large intestine, liver, mineral ion, oesophagus, oil, pancreas, peristalsis, physical digestion, protein, rectum, rickets, salivary gland, scurvy, secretion, small intestine, soluble, stomach, urea, vitamin, vitamin C, waste product

SUPPLEMENT amylase, bile, emulsify, hydrochloric acid, lipase, protease

During your study of this topic you should have learned:

◯ What is meant by a balanced diet.

◯ How to state the principal dietary sources and to describe the importance of: carbohydrates; fats and oils; proteins; vitamins C and D; the mineral ions calcium and iron; fibre (roughage); water.

◯ How to describe the causes of scurvy and rickets.

◯ How to identify the main organs of the digestive system: mouth and salivary glands, oesophagus, stomach, small intestine (duodenum and ileum), pancreas, liver and gall bladder, large intestine (colon, rectum, anus).

◯ How to describe the functions of the organs of the digestive system, in relation to:

- ingestion – the taking of food and drink into the body;
- digestion – the breakdown of food;
- absorption – the movement of nutrients from the intestines into the blood;
- assimilation – uptake and use of nutrients by cells;
- egestion – the removal of undigested food from the body as faeces.

◯ That physical digestion is the breakdown of food into smaller pieces without chemical change to the food molecules.

◯ That physical digestion increases the surface area of food for the action of enzymes in chemical digestion.

◯ **SUPPLEMENT** That the role of bile in emulsifying fats and oils is to increase the surface area for chemical digestion.

◯ That chemical digestion is the breakdown of large insoluble molecules into small soluble molecules so that they can be absorbed.

○ **SUPPLEMENT** How to describe the functions of enzymes:

- amylase breaks down starch to simple reducing sugars;
- proteases break down protein to amino acids;
- lipases break down fats and oils to fatty acids and glycerol.

○ **SUPPLEMENT** How to state where in the digestive system amylase, proteases, and lipases are secreted and where they act.

○ **SUPPLEMENT** That the functions of hydrochloric acid in gastric juice are killing harmful microorganisms in food and providing an acidic pH for optimum enzyme activity of proteases.

○ **SUPPLEMENT** That bile is an alkaline mixture that neutralises the acidic mixture of food and gastric juices entering the duodenum from the stomach, providing a suitable pH for enzyme action in the small intestine.

End of topic questions

1. Explain the importance of the following in a healthy diet:

   a) vitamins C and D

   b) the minerals calcium and iron

   c) water

   d) fibre.

2. There is an old saying that you should chew your food 100 times before swallowing to help look after your stomach. Explain why chewing food well helps digestion.

3. Outline the organs of the digestive system involved, and their roles, in each of the following processes:

   a) ingestion

   b) digestion

   c) absorption.

4. This is the diet schedule for a male Olympic athlete training for a competition, not including drinks during training.

Breakfast	large bowl of cereal, such as porridge or muesli
	half pint semi-skimmed milk plus chopped banana
	1–2 thick slices wholegrain bread with olive oil or sunflower spread and honey or jam
	glass of fruit juice + 1 litre fruit squash
Post-training 2nd breakfast	portion of scrambled eggs
	portion of baked beans
	1–2 pieces of grilled tofu
	portion of grilled mushrooms or tomatoes
	2 thick slices wholegrain bread with olive oil spread
	1 litre fruit squash
Lunch	pasta with Bolognese or chicken and mushroom sauce
	mixed side salad
	fruit
	1 litre fruit squash

Post-training snack	4 slices toast with olive oil or sunflower spread and jam
	large glass of semi-skimmed milk
	fruit
	500 ml water
Dinner	grilled lean meat or fish
	6–7 boiled new potatoes, large sweet potato or boiled rice
	large portion of vegetables, e.g. broccoli, carrots, corn or peas
	1 bagel
	1 low-fat yoghurt and 1 banana or other fruit
	750 ml water and squash
Bedtime snack	low-fat hot chocolate with 1 cereal bar

a) Identify the foods that contribute to each of these food types:

   i) carbohydrates

   ii) proteins

   iii) fats and oils

   iv) vitamins and mineral ions

   v) fibre.

b) Identify which food type is most represented in this diet.

c) Explain why this food type is so important in this diet.

d) Suggest which food group you would expect to be more represented in an athlete's diet in the early stages of training. Explain your answer.

e) Suggest an explanation why this diet is not suitable for everyone.

5. **SUPPLEMENT** Describe and explain the roles of bile in digestion.

6. **SUPPLEMENT** For each substance say where it is acts in the digestive system and describe its role in the process of digestion:

   a) hydrochloric acid

   b) amylase

   c) protease

   d) lipase.

Plants take in substances they need from their surroundings. Their roots take in water and dissolved mineral ions from the soil. But these then need to be transported to all the other parts of the plant, including growing shoots and leaves at the top of the plant. The leaves use energy from sunlight to produce molecules such as sugars, which are then used to make other molecules such as proteins. The water and mineral ions provide raw materials for different parts of these processes. Later, many of the products will be transported to other parts of the plant, including back to the roots.

STARTING POINT

1. What substances are transported around a plant?

2. How are these substances transported around a plant?

SYLLABUS SECTIONS COVERED

3.1 Xylem and phloem

3.2 Water uptake

3.3 Transpiration

3.4 Translocation

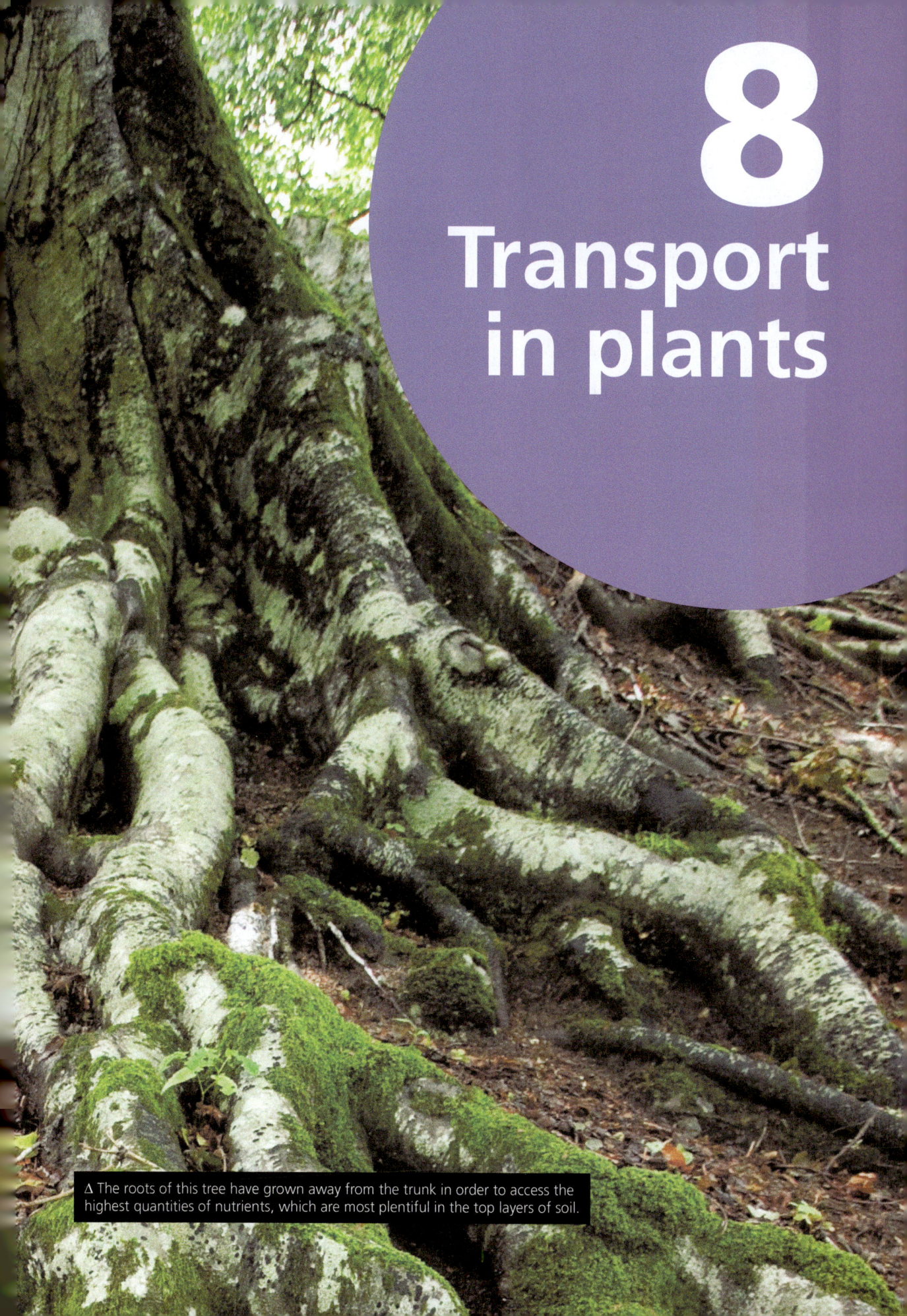

8
Transport in plants

△ The roots of this tree have grown away from the trunk in order to access the highest quantities of nutrients, which are most plentiful in the top layers of soil.

Transport in plants

INTRODUCTION

In one day, hundreds of litres of water will be absorbed from the soil and transpired through a fully grown tree in the Amazonian rainforest. This has a major impact on the environment in the rainforests. It reduces the amount of water in the soil. It also cools the air around the trees as the water evaporates into the air. The increase in water vapour in the air also affects where rainfall occurs. So the trees are effectively controlling the climate.

Δ Fig 8.1 These fine clouds over the Amazon rainforest are formed from water moving through the trees and into the atmosphere over the day.

KNOWLEDGE CHECK

✓ Cells in a plant leaf make glucose by photosynthesis.
✓ **SUPPLEMENT** Glucose is converted to sucrose and transported to other parts of the plant in phloem cells.
✓ Xylem vessels transport water and mineral ions.
✓ **SUPPLEMENT** Xylem vessels transport water and mineral ions from the roots of a plant thorough the stem to the leaves.

LEARNING OBJECTIVES

✓ State the functions of xylem and phloem: xylem – transport of water and mineral ions, and support; phloem – transport of sucrose and amino acids.
✓ Identify in diagrams and images the position of xylem and phloem as seen in sections of roots, stems, and leaves of non-woody dicotyledonous plants.
✓ Identify in diagrams and images root hair cells and state their functions.
✓ State that the large surface area of root hairs increases the uptake of water and mineral ions.
✓ Outline the pathway taken by water through the root, stem and leaf as: root hair cells, root cortex cells, xylem, mesophyll cells.
✓ Describe transpiration as the loss of water vapour from leaves.
✓ State that water evaporates from the surfaces of the mesophyll cells into the air spaces and then diffuses out of the leaves through the stomata as water vapour.
✓ Investigate and describe the effects of variation of temperature and wind speed on transpiration rate.
✓ **SUPPLEMENT** Explain the effects on the rate of transpiration of varying the following: temperature, wind speed, and humidity.
✓ **SUPPLEMENT** Explain how and why wilting occurs.
✓ **SUPPLEMENT** Describe translocation as the movement of sucrose and amino acids in phloem from sources to sinks.
✓ **SUPPLEMENT** Describe: sources as the parts of plants that release sucrose or amino acids; sinks as the parts of plants that use or store sucrose or amino acids.

XYLEM AND PHLOEM

In plants, water and dissolved substances are transported throughout the plant in a series of tubes or vessels. There are two types of transport vessel in plants, called **xylem** and **phloem**.

- Xylem tissue contains xylem vessels that form long tubes through the plant. Their thick strong cell walls help to support the plant. Xylem vessels are important for carrying water and dissolved mineral ions, which have entered the plant through the roots, to all the parts of the plant that need them. They are particularly important for supplying the water that the leaf cells need for photosynthesis.
- Phloem cells are living cells that are linked together to form continuous phloem tissue. Dissolved food materials, particularly sucrose and amino acids that have been formed in the leaf, are transported all over the plant from the leaves. For example, sucrose will be carried to any cell that needs glucose for respiration. Sucrose is less reactive than glucose and therefore is easier to transport without causing problems for other cells. Sucrose may also be carried to parts of the plants where it will be stored, often as another carbohydrate such as starch which is stored in seeds and root tubers.

In roots the xylem and phloem vessels are usually grouped together separately, but in the stem and leaves they are found together as **vascular bundles** or **veins**.

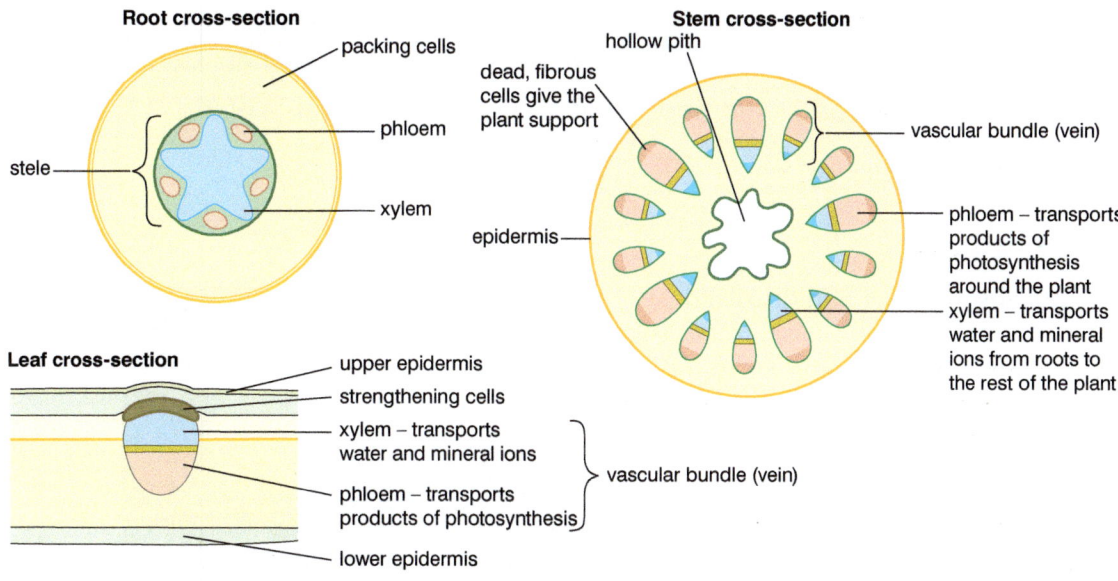

Δ Fig 8.2 The positions of xylem and phloem tissue in a root, stem, and leaf.

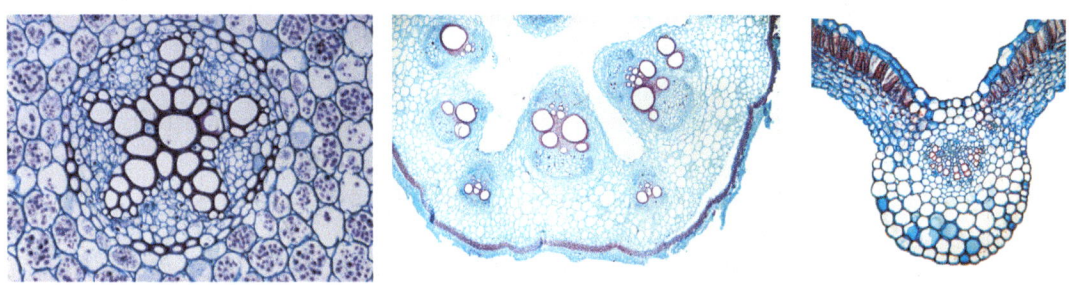

Δ Fig 8.3 Photomicrographs of (left) a cross-section of the middle of a buttercup root, (middle) a cross-section of part of a pumpkin stem, (right) a cross-section of a vein in a meadow beauty leaf.

SCIENCE IN CONTEXT — TREE RINGS

The wood of a tree is mostly xylem tissue. Every year, new xylem cells are produced from a ring of cells just inside the bark of the tree. When the tree is growing rapidly, the new xylem cells are large. In temperate regions, such as the UK, the rate of growth and the size of new cells decrease as autumn approaches, and growth stops during winter. The difference in size of cells produced over one year gives the tree its 'rings' and makes it possible to estimate the age of the tree.

Challenge Question: Scientists have used data from growth rings from very old wood to provide evidence for the climate conditions at the time the tree was growing. Suggest how they can do this.

Δ Fig. 8.4 Growth rings occur in temperate climates when new xylem cells alternately grow (in spring and summer) and stop growing (in winter).

QUESTIONS

1. Describe where you would find xylem and phloem tissue in a plant.

2. Outline the function of phloem tissue.

WATER UPTAKE

Plants absorb water and dissolved mineral ions from the soil through **root hair cells**. Root hair cells are found in a short region just behind the growing tip of every root. They are very delicate, and easily damaged. As the root grows, the hairs of the cells are lost, and new root hair cells are produced near the tip of the root.

Root hair cells are specially adapted for absorption of substances, because they have a fine extension that sticks out into

Δ Fig 8.5 The root of this germinating seed has many fine root hair cells that greatly increase its surface area.

the soil. This greatly increases their surface area for absorption. Root hair cells absorb water by osmosis and mineral ions by active transport because the concentration of the ions in the soil is lower than in the plant cells (see Section 3).

Water enters the root hair cells (Fig. 8.6), then passes across the root through the **root cortex cells** by osmosis. It then enters the xylem tissue in the root and can move from there to all other parts of the plant, including the stem and leaves.

In the leaves, water moves out of the xylem cells in the vascular bundle, into the **palisade mesophyll cells** and **spongy mesophyll cells** by osmosis.

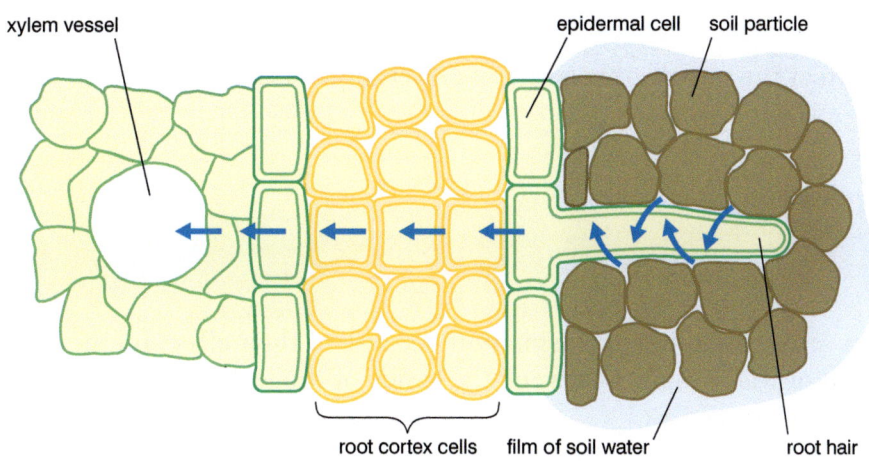

xylem vessel   epidermal cell   soil particle

root cortex cells   film of soil water   root hair

△ Fig 8.6 The movement of water through a root.

QUESTIONS

1. Describe the route that water takes as it moves through a plant.

2. State which process is used in a root to absorb

   a) water

   b) mineral ions from soil water.

TRANSPIRATION

Water is a small molecule that easily crosses cell membranes. Inside the leaf, water molecules cross the cell membranes of the spongy mesophyll cells into the air spaces. This process is called **evaporation** because the liquid water in the cells becomes water vapour in the air spaces. Whenever the **stomata** in a leaf are open, water molecules **diffuse** from the air spaces out into the air (where there are usually fewer water molecules). So, in addition to using water in the process of photosynthesis, plants lose water by evaporation from the leaf. This loss of water from the leaves is called **transpiration**.

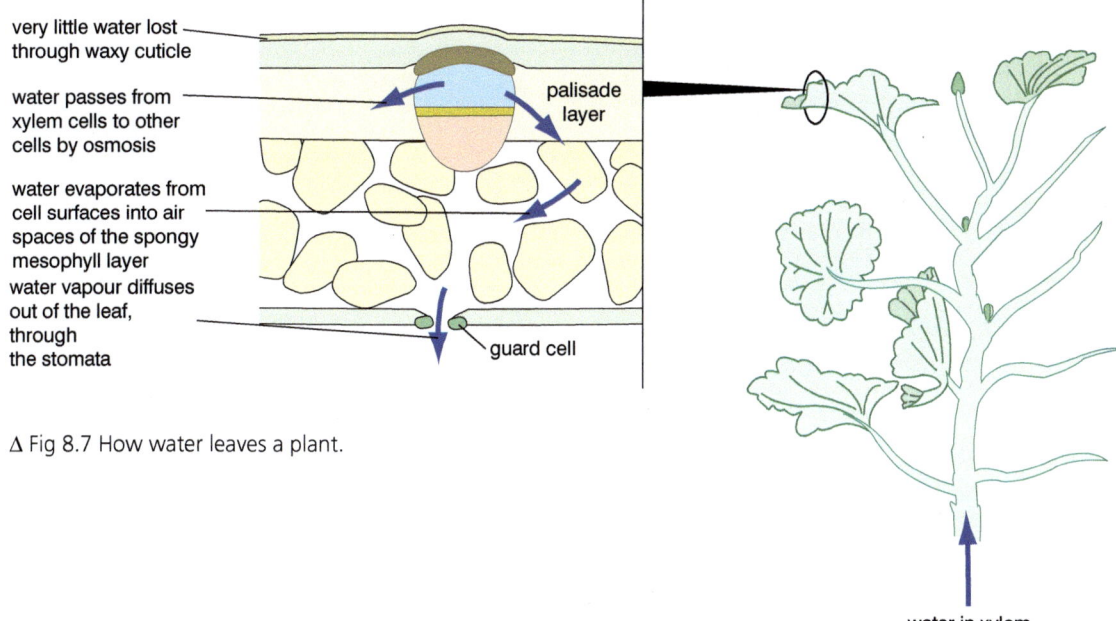

very little water lost through waxy cuticle

water passes from xylem cells to other cells by osmosis

palisade layer

water evaporates from cell surfaces into air spaces of the spongy mesophyll layer

water vapour diffuses out of the leaf, through the stomata

guard cell

water in xylem from the roots

△ Fig 8.7 How water leaves a plant.

Factors that affect the rate of transpiration

Several factors can affect the rate of transpiration. For example, transpiration is faster when:

- the temperature is higher
- the wind speed is greater.

The following factors affect the rate of transpiration because of their effect on the concentration gradient of water vapour between the inside of the leaf and the air outside.

- **Temperature** – Increased temperature means particles have more energy, which results in faster movement of the particles. The faster particles move, the easier it is for them to evaporate from cell surfaces into the air spaces, diffuse out of the leaf and move away. Another way of thinking about this is that with increased temperature and increased evaporation of water vapour into the internal air spaces of the leaf, the concentration of water vapour there also increases. This increases the concentration gradient between here and the outside air, so increasing the rate of transpiration from the leaf.
- **Wind speed** – In windy conditions, water vapour that has left the leaf will be carried away, so reducing the concentration of water vapour in the air around the leaf. This increases the concentration gradient between the air outside and inside the leaf. Therefore, the greater the wind speed, the greater the concentration gradient and the greater the rate of transpiration.
- **Humidity** – This is a measure of the concentration of water vapour in the air. When the air is very humid, it feels damp because there is a high concentration of water vapour in the air. When the air feels dry, the humidity is low. The concentration of water molecules inside the air spaces in the leaf is high. The higher the humidity of the air outside, the lower the concentration gradient between the air outside and inside the leaf, and so the lower the rate of transpiration.

Developing practical skills

Fig. 8.8 shows apparatus called a potometer that can be used to investigate the effect of a range of factors on the rate of transpiration. As water evaporates from the leaf surface, the bubble of air in the potometer moves nearer to the leafy twig.

Devise and plan investigations

1. Suggest how you could use a potometer to measure the effects of the following factors on transpiration:

    a) temperature

    b) wind.

Analyse and interpret data

The table below shows the results of an investigation, using a potometer in five different sets of conditions.

Conditions	Time for water bubble to move 5 cm, in seconds
still air, sunlight	135
moving air, sunlight	75
still air, dark cupboard	257
moving air, dark cupboard	122
hot, moving air, sunlight	54

2. For each of the following factors, identify which data in the table should be compared to show the effect of the factor, and explain why those are the right data to compare:

  a) temperature

  b) wind speed.

3. Using the data you have identified, make a conclusion about the effects of the following on the rate of transpiration:

  a) temperature

  b) wind speed.

Evaluate data and methods

4. Explain why the time taken for the bubble to move 5 cm is a measure of the rate of transpiration.

5. Suggest what else could make the bubble move.

6. Explain how could you improve the reliability of the conclusions you drew in Question 3.

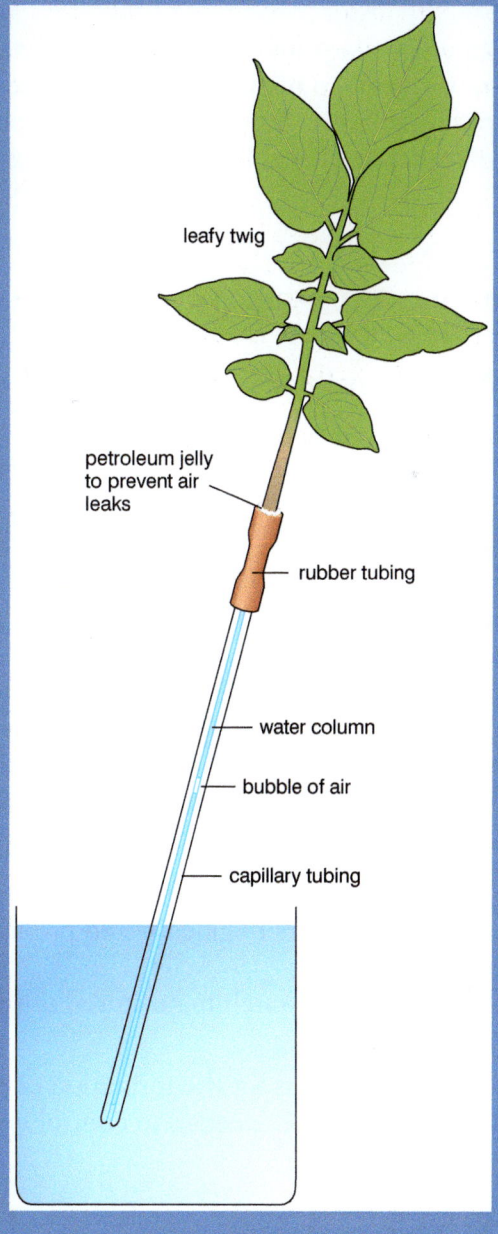

leafy twig

petroleum jelly to prevent air leaks

rubber tubing

water column

bubble of air

capillary tubing

△ Fig 8.8 A potometer.

Wilting

If the rate of transpiration of water from the leaves of a plant is greater than the rate at which water is supplied to the plant through absorption from soil water by root hair cells, the plant will start to **wilt**. This is when all the cells of the plant are not full of water, so the strength of the cell walls cannot support the plant and it starts to collapse (see Section 3).

QUESTIONS

1. Define the word *transpiration*.

2. Copy the diagram of the plant and leaf section in Fig. 8.7 and add your own annotations to explain how water moves through the plant. Include the following words in your labels: evaporation, diffusion, transpiration, root hair cells, root cortex cells, xylem, and mesophyll cells.

3. SUPPLEMENT Suggest an explanation of the advantage to plants of closing their stomata at night in terms of water loss.

4. SUPPLEMENT Explain, in terms of the movement of water molecules, why the transpiration rate is faster when:

   a) the temperature is higher

   b) the humidity of the air is lower.

TRANSLOCATION

Translocation is the transport or movement of materials, such as sucrose and amino acids, in the phloem of a plant. Sucrose is made directly from glucose that is produced in photosynthesis. Sucrose is a less reactive molecule than glucose, and so is more straightforward to transport around the plant. Amino acids are produced by cells using glucose from photosynthesis and nitrate ions taken from the soil.

Sucrose is transported in the phloem to other parts of the plant:

- for conversion back to glucose and use in respiration
- for conversion back to glucose and production of other molecules in the cell for cell growth
- for conversion to starch to be stored until needed.

Amino acids are used:
- to produce proteins for the formation of new plant tissue during growth
- to produce enzymes needed to control cell reactions.

Sources and sinks

We can describe translocation as the movement of sucrose and amino acids from sources to sinks.

- **Sources** are the parts of plants that release sucrose or amino acids.
- **Sinks** are the parts of plants that use or store sucrose or amino acids.

QUESTIONS

1. State in which tissue translocation occurs in a plant.

2. Give two examples of substances that are translocated.

3. Explain, with examples, what is meant by *sources* and *sinks* in translocation.

End of topic checklist

Key terms

diffusion, evaporation, palisade mesophyll cells, phloem, root cortex cell, root hair cell, spongy mesophyll cells, stomata, transpiration, vascular bundle, vein, xylem

SUPPLEMENT humidity, sink, source, translocation, wilt

During your study of this topic you should have learned:

○ That xylem transports water and mineral ions, and provides support; that phloem transports sucrose and amino acids.

○ How to identify the position of xylem and phloem in diagrams and images of root, stem, and leaf sections of non-woody dicotyledonous plants.

○ How to identify root hair cells in diagrams and images and state their functions.

○ That the large surface area of root hairs increases the uptake of water and mineral ions.

○ That the pathway taken by water through the root, stem, and leaf is: root hair cells, root cortex cells, xylem, mesophyll cells.

○ That transpiration is the loss of water vapour from leaves.

○ That water evaporates from the surfaces of the mesophyll cells into the air spaces and then diffuses out of the leaves through the stomata as water vapour.

○ How to investigate and describe the effects of variation of temperature and wind speed on transpiration rate.

○ **SUPPLEMENT** How to explain the effects on the rate of transpiration of varying temperature, wind speed, and humidity.

○ **SUPPLEMENT** How and why wilting occurs.

○ **SUPPLEMENT** That translocation is the movement of sucrose and amino acids in phloem from sources to sinks.

○ **SUPPLEMENT** That sources are the parts of plants that release sucrose or amino acids, and that sinks are the parts of plants that use or store sucrose or amino acids.

End of topic questions

1. Which is **not** a function of xylem tissue?

   A support

   B transport mineral ions

   C transport sucrose

   D transport water

2. Look at the photos in Fig. 8.3 of sections through a plant leaf, stem, and root.

   a) Using a sharp pencil, make careful diagrams that show the positions of all the main tissues in the sections. (Only draw a few cells in each tissue, to show their form. It will take too long to draw all the cells.)

   b) Use the labelled diagrams in Fig 8.2 to help you to clearly label your drawings to show the position of xylem.

   c) Clearly label the position of phloem in each of your drawings.

3. **SUPPLEMENT** A potted plant sitting on a sunny windowsill wilts more quickly than an identical plant placed on a shaded shelf in the room. Suggest an explanation why.

4. **SUPPLEMENT** Cactus plants have many adaptations to help them survive in a dry desert. One adaptation is that they close their stomata during the day and open them at night. Suggest an explanation for the advantage of this adaptation.

5. A researcher wanted to investigate the effect of temperature on the transpiration rate of a plant. After inserting the cut end of a plant stem into a potometer they placed the apparatus in an incubator and measured the how far the bubble in the potometer moved in 30 minutes. They repeated the investigation in different temperatures. The graph shows the data the researcher collected:

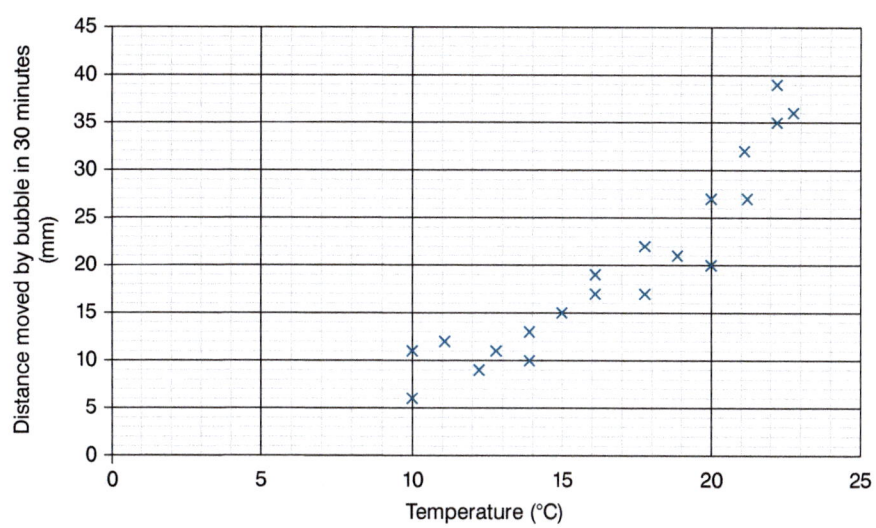

a) Describe the relationship shown on the graph between the distance the bubble moved in the potometer over 30 minutes and the temperature of the plant.

b) Use the graph to estimate the distance the bubble of air would move in the potometer at 19°C.

c) Predict, with a reason, how the results would change if a fan was placed inside the incubator.

Practice questions for Sections 1–8

Note: Practice questions, sample answers and comments have been written by the authors. The marks awarded for these questions indicate the level of detail required in the answers. In examinations, the way marks are awarded may be different. References to assessment and/or assessment preparation are the publisher's interpretation of the syllabus requirements and may not fully reflect the approach of Cambridge Assessment International Education.

Example answer

Question 1

The table lists some large biological molecules.

a) Give the smaller molecules that each of the large biological molecules is made up of. (5)

Large biological molecule	Smaller molecules that make up the molecule
Glycogen	*Glucose* ✓ ①
Fat	*Fatty acids* ✓ ①
Protein	*Amino acids* ✓ ①
Starch	*Glucose* ✓ ①

b) Describe a test that can be carried out in the laboratory for each of the following carbohydrates:

i) glucose (a reducing sugar) (4)

add Benedict's solution, to the sample to be tested ✓ ① *and heat it* ✗

Colour changes to red if glucose is present ✓ ①

ii) starch. (3)

add iodine solution, to a food sample on a spotting tile ✓ ①

colour changes from orange-brown to blue-black ✓ ①, *indicates*

that starch is present ✓ ①

(Total 12 marks)

> ### COMMENTS
>
> **a)** Fatty acids is correct, but glycerol should also be included.
>
> **b) i)** The first sentence is good, but the second part is too vague, and should include detail on the temperature to heat it to (95°C). This is correct, but should also say that the colour started off as blue. It would be more accurate to say a reducing sugar rather than glucose.
>
> **ii)** This is a good answer, giving the correct name of the chemical, with a clear description of the colour change as well as what the positive test indicates.

2. In which process do particles move from a region of their lower concentration to a region of their higher concentration? (1)

   A active transport

   B diffusion

   C osmosis

   D respiration

(Total 1 mark)

3. What is the cause of scurvy? (1)

   A lack of calcium

   B lack of iron

   C lack of vitamin C

   D lack of vitamin D

(Total 1 mark)

Question 4

This question is about the characteristics of living things.

a) The boxes on the left show characteristics of living organisms. The boxes on the right show some definitions of characteristics.

Draw one straight line from each characteristic to its correct definition. (3)

Characteristic		Definition
nutrition		a permanent increase in size and dry mass
sensitivity		the taking in of materials for energy, growth and development
growth		the ability to detect and respond to changes in the internal or external environment

b) Define the term *excretion*. (2)

c) Explain why a motor car is not a living thing. (3)

(Total 8 marks)

Question 5

Most living organisms are made up of cells.

a) Name the cell structure that holds the cell together and controls the substances entering and leaving the cell. (1)

b) State two differences between the structures found in plant cells but not in animal cells. (2)

c) Describe the role of the nucleus. (2)

d) Which cell structures are found in both animal and plant cells but not bacterial cells? (2)

(Total 7 marks)

Question 6

The light micrograph shows cells from the liver of a human.

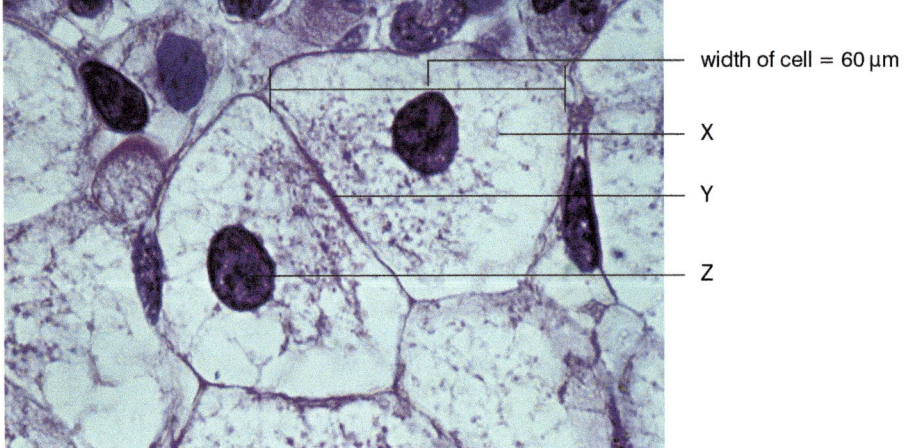

a) **i)** Identify the cell structures labelled X and Y in the image. (2)

ii) State the function of structure Z. (1)

b) **SUPPLEMENT** Use a ruler to measure the width of the image of the cell shown by the line on the micrograph. Use this to calculate the magnification of the image. Show your working. (4)

c) **SUPPLEMENT** Calculate how many liver cells of width 60 μm, laid side-to-side, there would be in a piece of liver tissue 1 cm across. Show your working. (3)

(Total 10 marks)

Question 7

A student set up an investigation in which three cubes of different sizes were cut from an agar jelly block. The agar jelly contained a pink indicator that turns colourless in the presence of acid.

The cubes were placed in an acid. The student measured the time taken for the cubes to turn completely colourless.

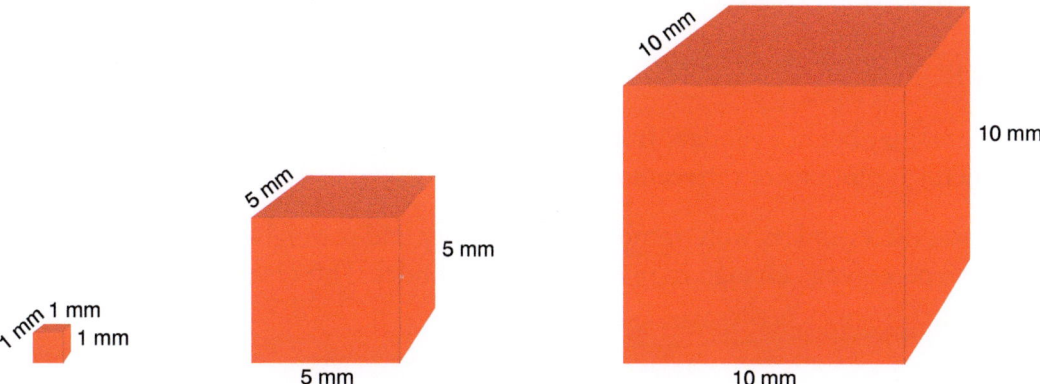

a) State the name of the process that causes the acid to penetrate the agar jelly. (1)

b) The student used cubes of three different dimensions:

Dimensions of cube, mm
$1 \times 1 \times 1$
$5 \times 5 \times 5$
$10 \times 10 \times 10$

For each cube, calculate its:

 i) volume (length × width × depth) (3)

ii) surface area : volume ratio (surface area/volume). (3)

Give the surface area : volume ratio in the form 'n : 1'.

Dimensions of cube/mm	Surface area of cube/mm²	i) Volume of cube/mm³	ii) Surface area : volume ratio
$1 \times 1 \times 1$	6		
$5 \times 5 \times 5$	150		
$10 \times 10 \times 10$	600		

c) SUPPLEMENT Describe the relationship between surface area and volume as the cube increases in size. (1)

d) SUPPLEMENT During the lesson, the acid completely penetrated the two smaller cubes, the smallest one first, but by the end of the lesson, the 10 mm × 10 mm × 10 mm cube had still not turned completely colourless.

    i) Explain these results. (2)

    ii) Suggest what implications this has for organisms of increasing size. (3)

(Total 13 marks)

Question 8

A student cut a number of cylinders from a potato and weighed them. These were placed in sucrose solutions of different concentrations.

After one hour, the cylinders were removed, blotted dry and reweighed. The student calculated the percentage change in mass for each cylinder. The results are shown in the table.

Concentration of sucrose / g per cm³	Percentage change in mass of potato cylinders				Average percentage change in mass
	Experiment 1	Experiment 2	Experiment 3	Experiment 4	
0.0	+31.4	+33.7	+31.2	+32.5	
0.2	+20.9	+22.2	+22.8	+21.3	21.8
0.4	−2.7	−1.8	−1.9	−2.4	
0.6	−13.9	−12.8	−13.7	−13.6	−13.5
0.8	−20.2	−19.7	−19.3	−20.4	
1.0	−19.9	−20.3	−21.1	−20.3	−20.4

a) State the process involved in these changes in the potato cylinders. (1)

b) Calculate the average percentage changes in mass for each of the sucrose concentrations that are missing from the table. (3)

c) **i)** Plot a graph of the average percentage changes in mass against the concentration of sucrose. (3)

    ii) Draw the line of best fit curve. (1)

    iii) Determine at what concentration of sucrose there was no net movement of water. (2)

    iv) Describe the changes in mass over a range of sucrose concentrations. (3)

(Total 13 marks)

Question 9

A company that produces enzymes publishes information sheets of their performance. The graphs show the performance of an enzyme, pectinase, at different pHs and temperatures.

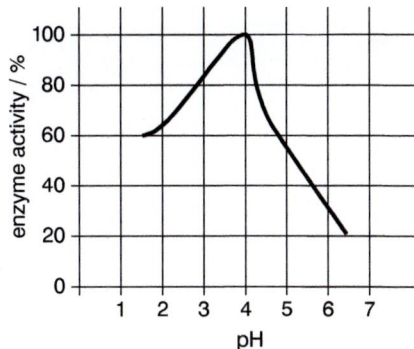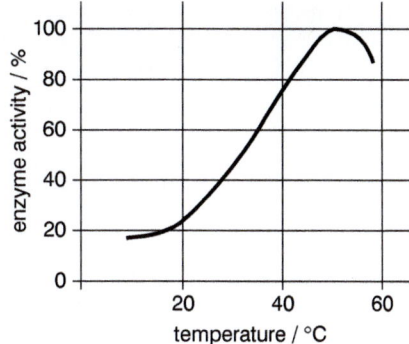

a) Describe and explain the effect on the activity of pectinase of:

   i) pH   (6)

   ii) temperature.   (5)

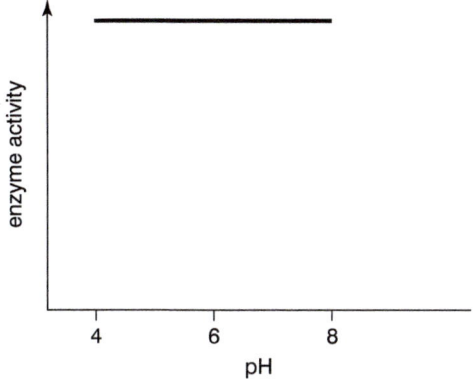

b) A student says that the graph above shows that papain is unaffected by pH. Evaluate whether the student is correct. Explain your answer.   (2)

(Total 13 marks)

Question 10

a) In humans, proteases are produced by the stomach, pancreas and small intestine.

   State the letters on the diagram to show the location of:

   i) the small intestine

   ii) the stomach

   iii) the pancreas.   (3)

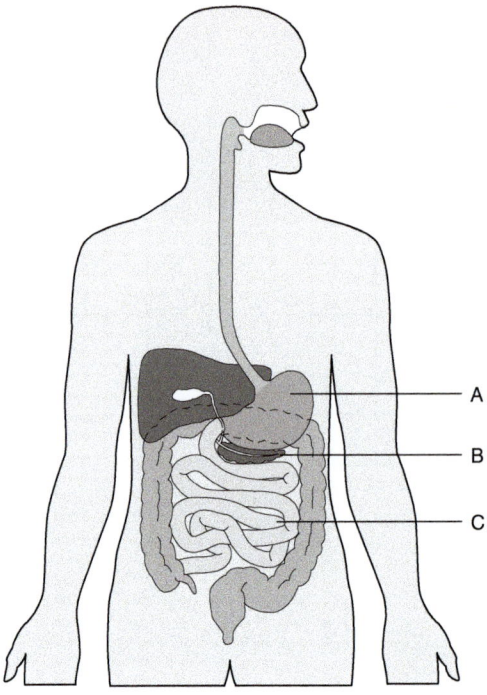

b) In the laboratory, proteases can be added to egg whites which become more transparent as the protease activity increases.

Plan an investigation to find out how pH affects the activity of one of these proteases. (5)

c) Starch is broken down by amylase, an enzyme found in saliva. Maltose is produced by the reaction.

Plan an investigation to find out how temperature affects the activity of amylase.

i) What apparatus could you use to change the temperature? (1)

ii) What would be a suitable range of temperatures? (1)

iii) What is the test to show the presence for starch? (1)

iv) What could be a possible dependent variable? (2)

(Total 13 marks)

Question 11

Plants respond to light in different ways.

The graph shows the effect of light intensity on the rate of photosynthesis in a plant.

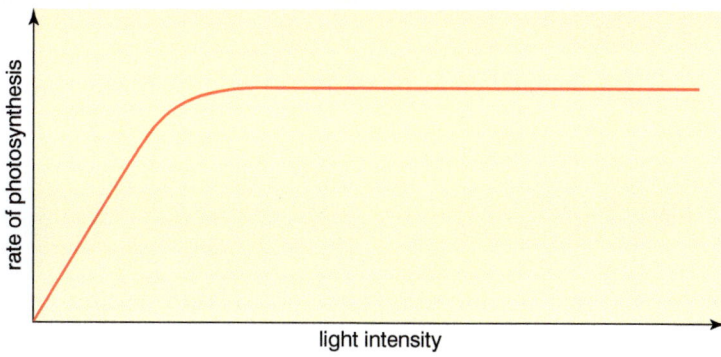

a) Describe and explain the effect of light intensity on the rate of photosynthesis. (3)

b) State one other variable that could affect the rate of photosynthesis. (1)

c) Give the name of the cell that is specialised to carry out photosynthesis. (1)

(**Total 5 marks**)

SUPPLEMENT Question 12

The leaf is the main organ of photosynthesis.

a) Write a balanced symbol equation for photosynthesis. (3)

b) Explain how a leaf is adapted to exchanging gases required for photosynthesis. (5)

(**Total 8 marks**)

Question 13

The diagram shows a section through a leaf.

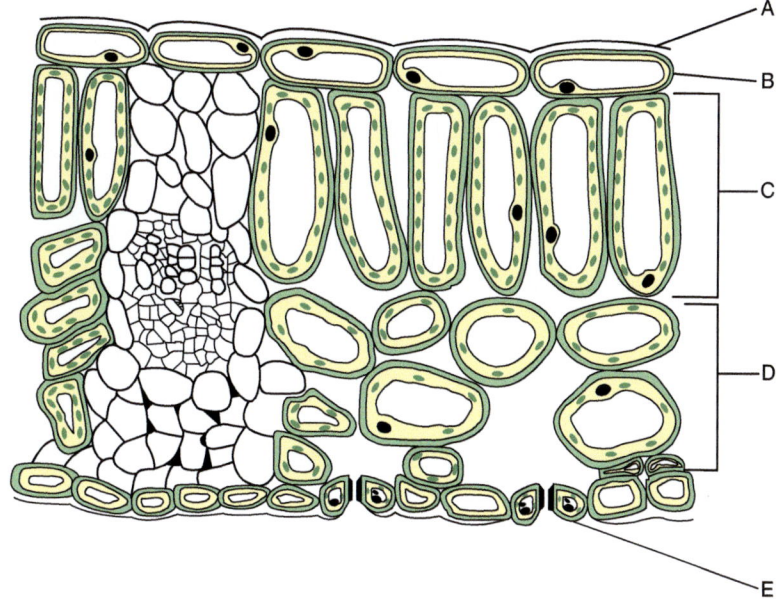

a) Give the names of parts A–E shown in the diagram. (5)

b) The leaf is the main organ of photosynthesis. Write a word equation for photosynthesis. (3)

c) Define the term *transpiration*. (3)

d) Describe an experiment used to investigate the effect of temperature on transpiration rate. (7)

(Total 18 marks)

Question 14

a) The diagram below shows a section of a plant root surrounded by soil particles. The blue arrows show the movement of water.

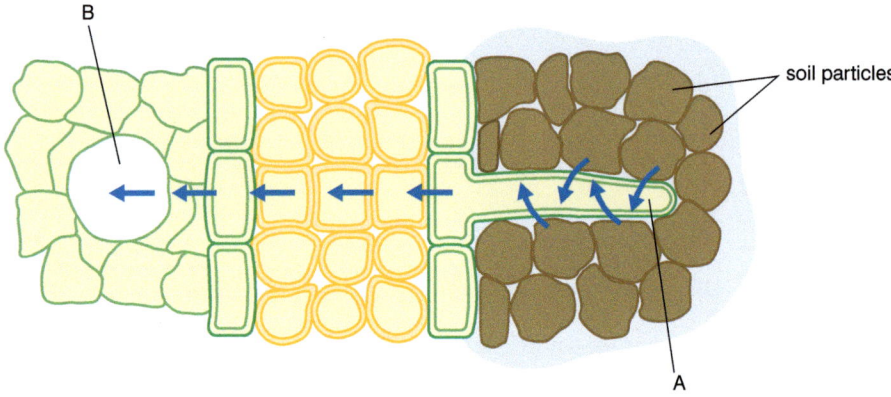

i) Identify the parts of the plant root, A and B. (2)

ii) **SUPPLEMENT** Explain the role of water potential in the uptake and transport of water in plant roots. (6)

b) **SUPPLEMENT** Explain what is meant by the term *translocation*. In your answer, include references to *sources* and *sinks*. (6)

(Total 14 marks)

An average-sized human adult carries about 4.5–5.0 litres of blood. Blood is pumped around the bodies of mammals by the heart to carry oxygen, nutrients and water to every organ, tissue and cell, and transport waste products away from cells. The human heart is about the size of an adult fist, located just about in the centre of the chest. It beats over 100 000 times a day, and the characteristic double sound of a beat corresponds to the two sets of heart valves opening and closing in order.

STARTING POINTS

1. Why do we need a circulatory system?

2. What are the different parts of the circulatory system?

3. What are the jobs of the different parts of the circulatory system?

SYLLABUS SECTIONS COVERED

9.1 Circulatory systems

9.2 Heart

9.3 Blood vessels

9.4 Blood

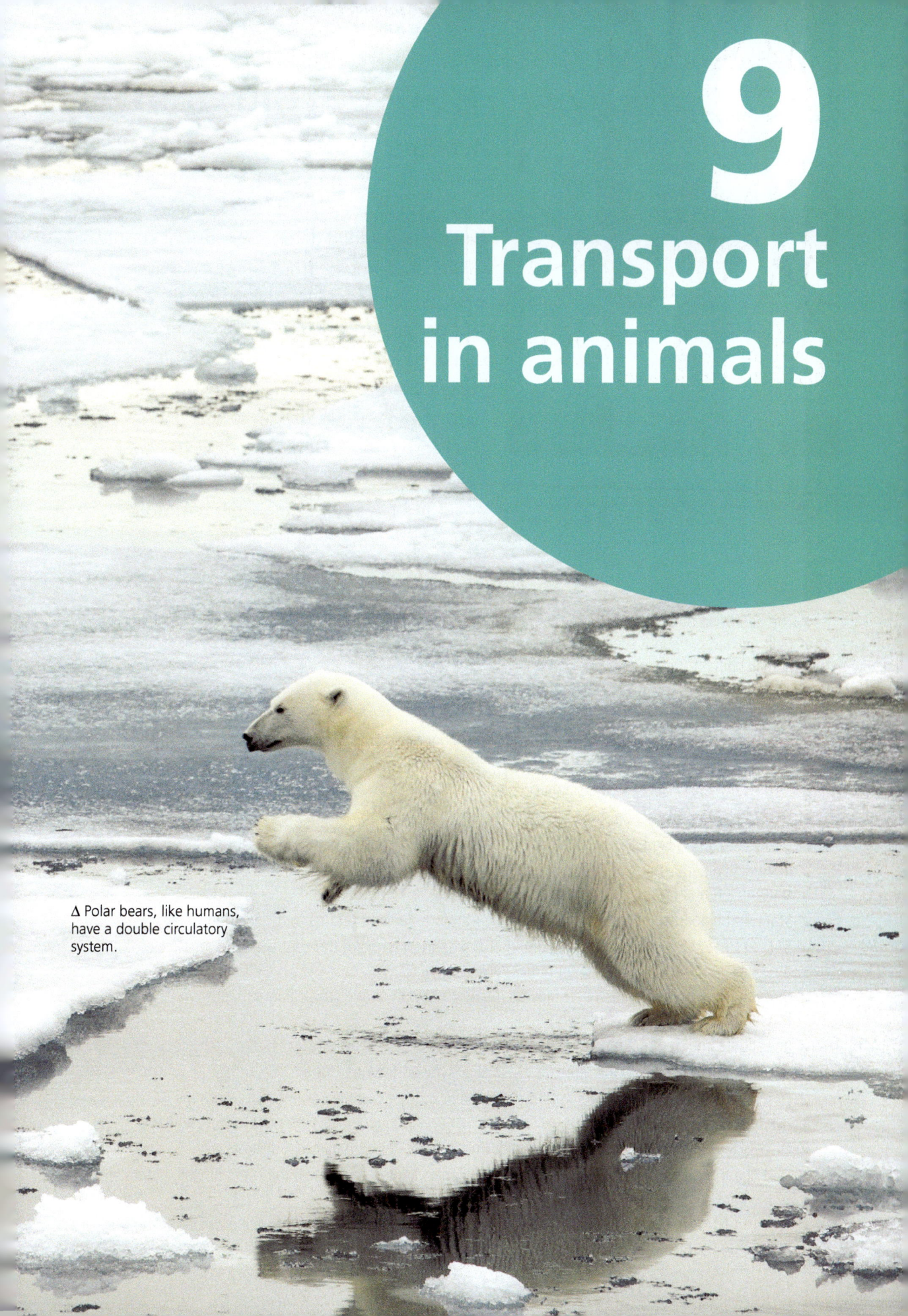

9

Transport in animals

△ Polar bears, like humans, have a double circulatory system.

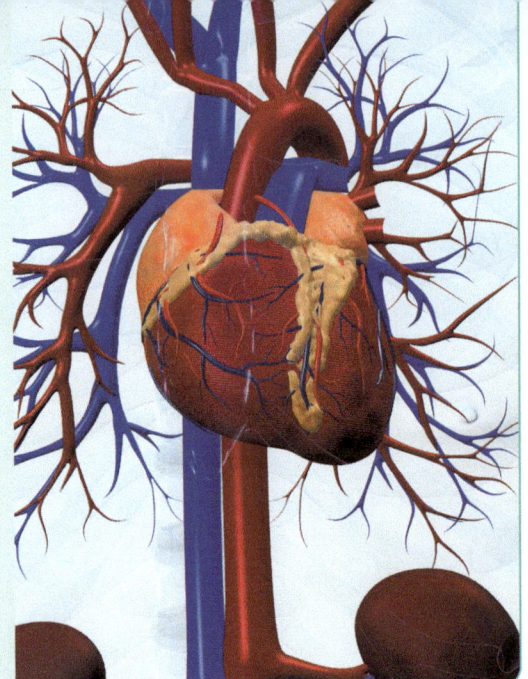

Δ Fig. 9.1 This image of the heart shows it surrounded by the larger blood vessels that are found in a lung. The millions of capillaries are not visible.

Transport in animals

INTRODUCTION

Almost no cell in your body is more than 20 μm (0.02 mm) from a blood vessel. This is necessary because the blood delivers a constant supply of oxygen and glucose, for respiration, without which the cells will rapidly die. So it's not surprising that, no matter where your skin is damaged, you will bleed. Many of the blood vessels that penetrate the tissues are extremely narrow – about 5 to 10 μm wide, which is about the width of one red blood cell. It has been calculated that if you placed the blood vessels of an adult in a line it would wrap four times around the equator of the Earth.

KNOWLEDGE CHECK

✓ The heart and blood vessels form the human circulatory system.
✓ Cells need a continuous supply of oxygen and glucose for respiration, which are supplied by the blood in a human body.

LEARNING OBJECTIVES

✓ Describe the circulatory system as a system of blood vessels with a pump and valves to ensure one-way flow of blood.
✓ **SUPPLEMENT** Describe the single circulation of a fish.
✓ **SUPPLEMENT** Describe the double circulation of a mammal.
✓ **SUPPLEMENT** Explain the advantages of a double circulation.
✓ Identify in diagrams and images the structures of the mammalian heart, limited to: muscular wall, septum, left and right ventricles, left and right atria, one-way valves, and coronary arteries.
✓ State that blood is pumped away from the heart in arteries and returns to the heart in veins.
✓ **SUPPLEMENT** Describe the functioning of the heart in terms of the contraction of muscles of the atria and ventricles, and the action of the valves.
✓ State that the activity of the heart may be monitored by: ECG (electrocardiogram), pulse rate, and listening to sounds of valves closing.
✓ Investigate and describe the effect of physical activity on the heart rate.
✓ **SUPPLEMENT** Explain the effect of physical activity on the heart rate.
✓ Describe coronary heart disease in terms of the blockage of coronary arteries and state the possible risk factors including: diet, lack of exercise, stress, smoking, genetic predisposition, age, and sex.
✓ Discuss the roles of diet and exercise in reducing the risk of coronary heart disease.
✓ Describe the structure of arteries, veins and capillaries, limited to: relative thickness of wall, diameter of the lumen, and the presence of valves in veins.
✓ **SUPPLEMENT** Explain how the structure of arteries and veins is related to the pressure of the blood that they transport.

- ✓ State the functions of capillaries.
- ✓ **SUPPLEMENT** Explain how the structure of capillaries is related to their functions.
- ✓ **SUPPLEMENT** Identify in diagrams and images the main blood vessels to and from the: heart, limited to: vena cava, aorta, pulmonary artery and pulmonary vein; lungs, limited to: pulmonary artery and pulmonary vein.
- ✓ List the components of blood as: red blood cells, white blood cells, platelets, and plasma.
- ✓ Identify red and white blood cells in photomicrographs and diagrams.
- ✓ **SUPPLEMENT** Identify lymphocytes and phagocytes in photomicrographs and diagrams.
- ✓ State the functions of the following components of blood: red blood cells in transporting oxygen, including the role of haemoglobin; white blood cells in phagocytosis and antibody production; platelets in clotting (details are not required); plasma in the transport of blood cells, ions, nutrients, urea, hormones, and carbon dioxide.
- ✓ **SUPPLEMENT** State the functions of: lymphocytes – antibody production; phagocytes – engulfing pathogens by phagocytosis.
- ✓ **SUPPLEMENT** State the roles of blood clotting as preventing blood loss and the entry of pathogens.

CIRCULATORY SYSTEMS

Transport in animals usually takes place inside a **circulatory system**. A circulatory system is formed from a system of continuous tubes (blood vessels) that carry blood around the body. The tubes are connected to a pump, the **heart**, which forces the blood through the circulation. **Valves** in the heart and in some of the blood vessels make sure that blood circulates in only one direction.

Different circulatory systems

Fish have a **single circulation**, which means that blood passes all parts of the body in one circuit before returning to the heart.

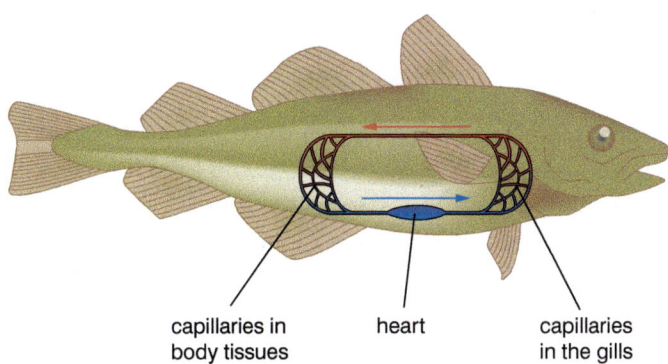

capillaries in body tissues   heart   capillaries in the gills

Δ Fig. 9.2 A fish has a single circulatory system.

The heart pumps deoxygenated blood through the gills, where oxygen from the surrounding water is absorbed into the blood. The oxygenated blood then passes to the body tissues. Here, oxygen is taken from the blood for respiration in cells, and carbon dioxide is added to the blood. The deoxygenated blood passes back to the heart to be pumped back to the gills, where carbon dioxide diffuses out into the water.

The mammalian circulatory system (as in humans) is more complex than that of a fish. It is described as a **double circulation** because the blood passes through the heart twice for each time that it passes through the body tissues. This is because when blood leaves the right side of the heart it passes through the tissues of the lungs before returning to the left side of the heart for pumping around the rest of the body.

One advantage of a double circulation over a single circulation is that, as blood is pumped twice on each complete journey around the body, a higher overall blood pressure can be maintained which means more efficient transport of the blood and the materials it contains. Separating the circulation to the lungs from the circulation to the body also means that the blood in the two circulations can be at different pressures.

- Blood leaving the right side of the heart is normally below 4 kPa. Blood does not travel far to the lung tissue, so there is little loss of pressure before it reaches the capillaries surrounding the alveoli. This lower pressure prevents damage to the delicate capillaries that pass through lung tissue.
- Blood leaving the left side of the heart has to travel all round the body and back to the heart. So it needs to start at a much higher pressure, at about 16 kPa as it leaves the heart. By the time it reaches the capillaries within body tissues, the pressure has dropped to below 3 kPa and so will not damage them.

QUESTIONS

1. State the role of the heart in the circulatory system.

2. Identify what prevents blood flowing the wrong way round the circulatory system.

3. SUPPLEMENT Explain what is meant by a *double circulatory system*.

4. SUPPLEMENT Explain the advantages of a double circulatory system compared with the single circulatory system of a fish.

HEART

The heart is a muscular organ that pumps blood by expanding in size as it fills with blood, and then contracting, forcing the blood on its way through the blood vessels. Blood is pumped away from the heart in arteries and returns to the heart through veins. The heart is two pumps in one. The right side pumps blood to the lungs to collect oxygen. The left side then pumps this blood around the rest of the body before it returns to the right side to be sent to the lungs again. Between the right and left sides of the heart is a layer of tissue called the **septum**.

SUPPLEMENT

The importance of the septum is that it keeps the **deoxygenated** (without oxygen) blood on the right side of the heart separate from the **oxygenated** blood on the left.

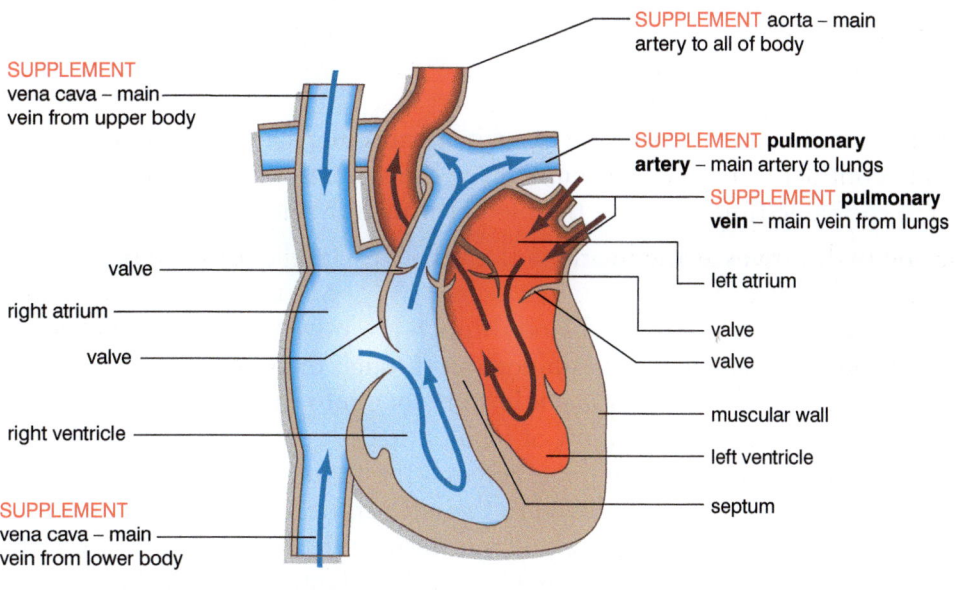

aorta – main artery to all of body

vena cava – main vein from upper body

– main artery to lungs

– main vein from lungs

valve

right atrium

valve

right ventricle

vena cava – main vein from lower body

left atrium

valve

valve

muscular wall

left ventricle

septum

Δ Fig. 9.3 Structure of the human heart. Oxygenated blood is shown in red and deoxygenated blood is shown blue. (Knowledge of the names of the major blood vessels is required at Supplement level only.)

REMEMBER

We always draw diagrams of the circulatory system and heart as if we are facing another person. So in the diagram the left side of the heart/circulation in a body is drawn on the right side of the diagram.

The heart consists of four chambers: two **atria** (single: **atrium**) and two **ventricles**. The walls of the chambers are formed from thick muscle. Blood that flows towards the heart passes through blood vessels called **veins**. Blood that leaves the heart passes through blood vessels called **arteries**. To make sure that blood only flows in one direction through the heart, there are one-way valves at the points where arteries leave the heart, and between the atria and ventricles. These close when the heart contracts, to prevent backflow of blood.

The heart has its own separate blood supply, to provide the muscle tissue with oxygen and glucose for respiration so that it can contract. These blood vessels are called the **coronary arteries** and coronary veins. You can see some of these on the outside of a whole heart (Fig. 9.4).

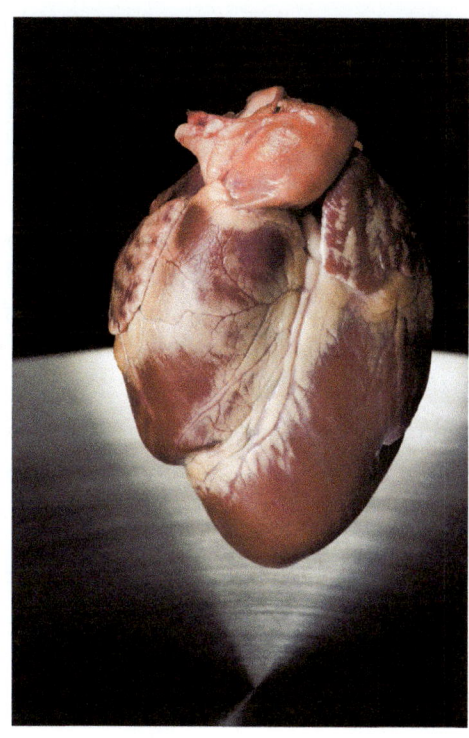

Δ Fig. 9.4 Whole human heart.

Blood flow through the heart

Blood passes through the chambers of the heart in a particular sequence as the walls of the chambers contract. First the atria contract at the same time, then the ventricles both contract at the same time, to move the blood through the heart.

- Blood from the body arrives at the heart via the vena cava, and enters the right atrium.
- Contraction of the right atrium forces blood through the valve to the right ventricle.
- Contraction of the right ventricle forces blood out through the pulmonary artery to the lungs. The valve closes to prevent backflow of blood into the right atrium, and the valve then closes to prevent backflow of blood into the ventricle.
- Blood enters the left atrium from the lungs through the pulmonary vein.
- Contraction of the left atrium forces blood to the left ventricle through the valve.
- Contraction of the left ventricle forces blood out through the aorta towards the rest of the body. The valve closes to prevent backflow of blood into the left atrium, and then the valve closes to prevent backflow of blood into the ventricle.

QUESTIONS

1. State the four chambers of the mammalian heart.

2. Describe how *arteries* and *veins* are different from each other.

3. SUPPLEMENT Starting in the vena cava, give the structures of the heart and blood vessels in the order that blood passes through them until it reaches the aorta.

Heart rate

Heart rate is the measure of how frequently the heart beats, generally given as beats per minute. We can take measurements of heart rate by feeling for a pulse point, where the blood flows through an artery near to the skin, such as in the wrist or at the temple.

Taking the pulse rate is actually measuring the expansion and relaxation of the artery wall as the blood passes through it. However, as each pulse of blood is created by one contraction of the ventricles, we say that we are measuring *heart beats*.

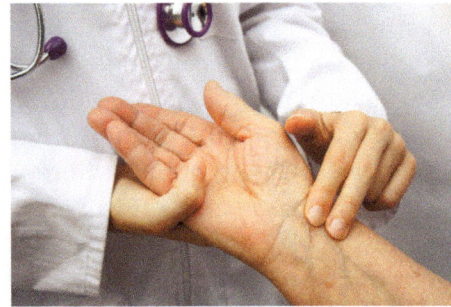

Δ Fig. 9.5 Taking the pulse at the wrist.

Heart rate can also be measured by listening to the heart. The 'lub, dub' sounds of one complete contraction are the sounds of the valves inside the heart as they open and shut.

The contraction of the heart is regulated by electrical activity in nerves in the heart. We can measure this electrical activity by attaching sensors to the skin. A normal heart beat produces a particular pattern of activity, which can be recorded as an **electrocardiogram (ECG).**

Resting heart rate is the rate at which the heart beats when the person is at rest. On average it is between 60 and 80 beats per minute for an adult human, but this range is very variable. Resting heart rate may vary as a result of:

- age – children usually have a faster average than adults
- fitness – a trained athlete may have a resting rate as low as 40 beats per minute because their heart contains more muscle and can pump out more blood with each contraction
- illness – infection can raise resting heart rate, but some diseases of the circulatory system can slow resting heart rate
- drugs – some drugs can change heart rate (see Section 14).

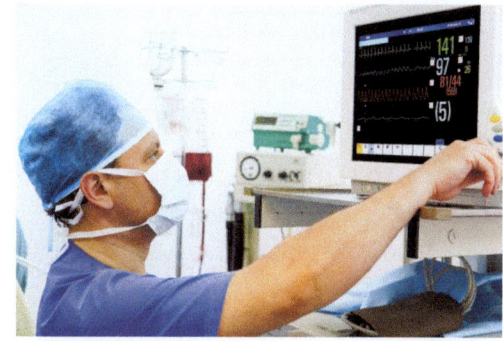

Δ Fig. 9.6 During an operation, a doctor continually checks the patient's ECG trace to make sure all is well.

Changing heart rate

Heart rate increases during physical activity. The harder you exercise, the faster the heart beats.

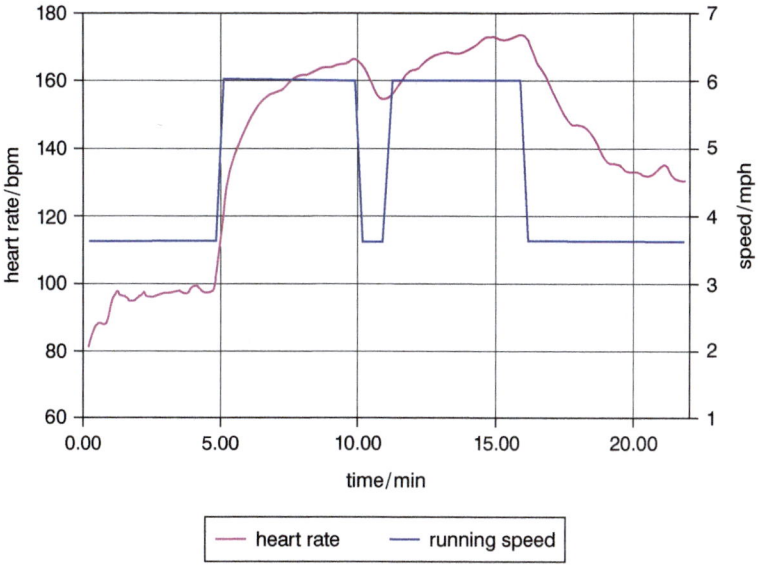

Δ Fig. 9.7 How heart rate changes with exercise.

During exercise, muscles contract more, therefore needing more energy, which is provided by increased respiration in the muscle tissue. Muscles therefore require oxygen and glucose to be supplied more rapidly, which is done by a faster heart rate pumping blood more quickly. Waste products of respiration are also removed more quickly.

Developing practical skills

The effect of exercise on heart rate can be measured by taking pulse measurements after different levels of exercise.

Devise and plan investigations

In an investigation of the effect of exercise on the heart rate, a student was asked to exercise at a particular level for 2 minutes, at which point the pulse rate was measured. The student was then allowed to rest for 5 minutes and then asked to exercise at the next level of activity.

1. a) Explain why the pulse rate was taken after 2 minutes of exercise and not sooner.

   b) Explain why the student rested for 5 minutes before starting the next level of exercise.

Analyse and interpret data

Table 9.1 shows the results of an investigation into the effect of exercise on the heart rate of one student.

	Resting	Walking	Jogging	Running
Heart rate in beats per minute	72	81	96	122

Δ Table 9.1 Heart rate with activity.

2. Describe the pattern shown in the data.

3. Use the data to give a conclusion for the investigation.

4. **SUPPLEMENT** Explain why heart rate responds like this to different levels of exercise.

Evaluate data and methods

5. Evaluate this conclusion, and suggest how the investigation could be improved.

QUESTIONS

1. Describe three ways of monitoring heart rate.

2. Explain why resting heart rate in an adult is given as a range of values and not a single value.

3. Describe the effect of physical activity on heart rate.

4. **SUPPLEMENT** Explain the effect of physical activity on heart rate.

Coronary heart disease

The muscle of the heart needs its own blood supply to provide the oxygen and glucose it needs for respiration. It cannot get these materials from the blood that is pumped through the heart, so there are coronary arteries and veins that pass through heart muscle. If the blood flow through these coronary blood vessels is reduced, it can reduce the amount of oxygen and glucose getting to the muscle cells, and so reduce the amount of energy that they can release through respiration.

Blockage of the coronary arteries can occur when layers of cholesterol are deposited on the inner lining of the blood vessel. This causes **coronary heart disease**. Even partial blockage can cause a health problem, such as angina (heart pains) or high blood pressure. A full blockage will cause the blood supply to the heart to stop, which we call a heart attack, and this may result in death.

Several factors can increase the risk of a blockage of the coronary arteries and so coronary heart disease.

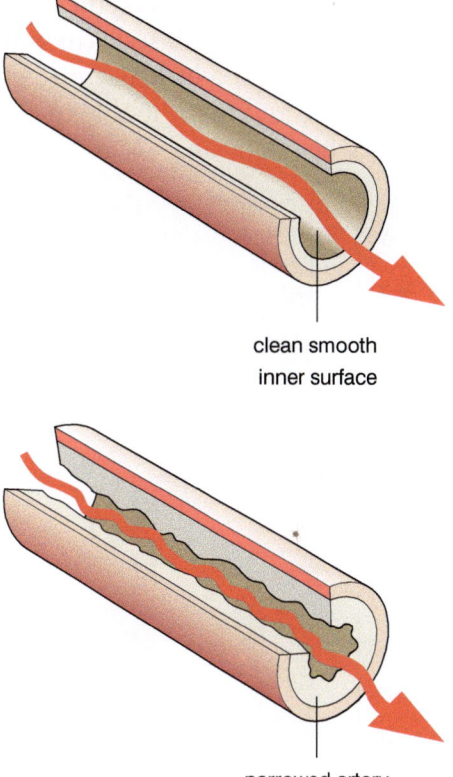

clean smooth
inner surface

narrowed artery
due to cholesterol deposits

Δ Fig. 9.8 Deposits of cholesterol inside arteries makes it more difficult for blood to flow through freely, increasing the risk of diseases of the circulatory system.

- **Diet**: high levels of saturated fats in the diet may cause increased deposits of cholesterol.
- **Lack of exercise**: high-energy foods that aren't used to provide energy for exercise are stored as fat, increasing the likelihood of cholesterol deposits.
- **Smoking**: chemicals in tobacco smoke that pass into the blood can damage the delicate lining of arteries, which increases the chance that deposits of cholesterol are laid down at these points.
- **Genetic predisposition**: genes can give some people an increased tendency towards developing coronary heart disease. As this tendency can be inherited, the disease may appear more commonly in some families than in others, and more frequently in some ethnic groups than others. For example, African-American people are at greater risk of developing coronary heart disease than White American people.
- **Stress**: stress itself may not be a direct cause of coronary heart disease, but responses to stress, such as smoking or eating for comfort, particularly over a long time, can increase the risk of heart disease.
- **Age**: the risk of coronary heart disease increases with age and is most common in people over 50.
- **Sex**: men are more likely to develop coronary heart disease at an earlier age than women.

Preventing coronary heart disease

Medical advice for preventing coronary heart disease includes changing the diet and amount of exercise taken.

Many studies of the relationship between saturated fat and heart disease have concluded that reducing the amount of saturated fat in the diet should reduce the risk of heart disease. However, it can sometimes be difficult to demonstrate this relationship because people don't just change their diet when they are advised to live more healthily. For example, they may also change how much they exercise.

HEART

The amount of exercise taken each day also affects the risk of heart disease. Someone who has a sedentary lifestyle (mainly sitting) can significantly reduce their risk of an early death by just a little exercise every day. This exercise seems to strengthen the heart muscle and make it able to cope with sudden increases in heart rate more easily, such as when you run.

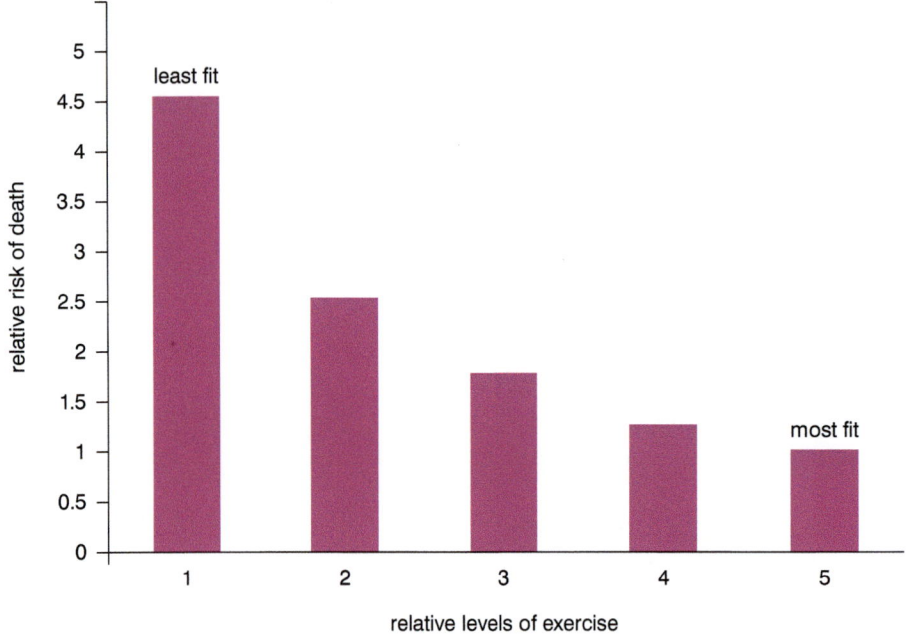

△ Fig. 9.9 The relationship between risk of death and level of exercise.

QUESTIONS

1. Explain why the heart needs its own blood supply.

2. Identify four possible risk factors for coronary heart disease.

3. Explain how a person can reduce their risk of coronary heart disease.

BLOOD VESSELS

The blood vessels are grouped into three different types: arteries, capillaries and veins.

- **Arteries** are large blood vessels that carry blood that is flowing away from the heart. Arteries have thick muscular and elastic walls, with a narrow central space (lumen) through which the blood flows.
- **Capillaries** are the tiny blood vessels that form a network throughout every tissue and connect arteries to veins. Capillaries have very thin walls. All the exchange of substances between the blood and tissues happens in the capillaries.
- **Veins** are large blood vessels that carry blood that is flowing back towards the heart. Veins have a large lumen through which blood flows. Valves in the veins prevent backflow.

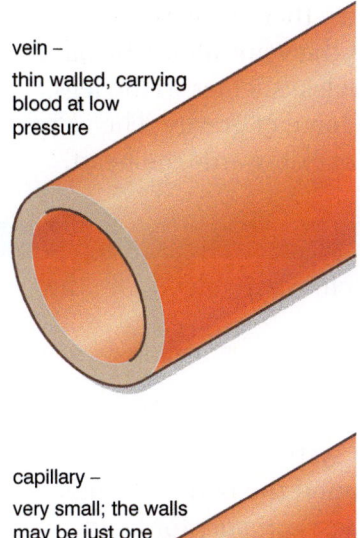

vein –

thin walled, carrying blood at low pressure

capillary –

very small; the walls may be just one cell thick

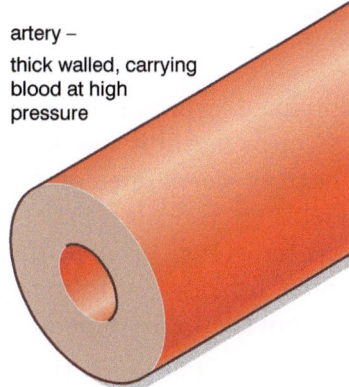

artery –

thick walled, carrying blood at high pressure

These diagrams are not to the same scale. Capillaries are much smaller than arteries and veins.

Δ Fig. 9.10 Veins vary in diameter from about 5 to 15 mm. Capillaries are very small, with a diameter of around 0.01 mm. Arteries vary in diameter from about 10 to 25 mm.

Structure related to function

The structure of different blood vessels enables them to carry out their function most efficiently.

- Blood carried in the arteries is at higher pressure than in the other vessels. The highest pressure is in the aorta, the blood vessel that leaves the left ventricle. The thick walls of arteries help to protect them from bursting when the pressure increases as the pulse of blood enters them. The recoil of the elastic wall after the pulse of blood has passed through helps to maintain the blood pressure and even out the pulses. By the time the blood enters the fine capillaries, the change in pressure during and after a pulse has been greatly reduced.
- The thin walls of capillaries helps to increase the rate of diffusion of substances by keeping the distance for diffusion between the blood and the cytoplasm of the surrounding cells to a minimum.

normal blood flow

veins have valves to stop the blood flowing backwards

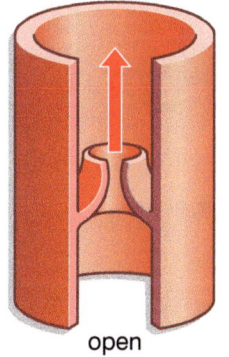
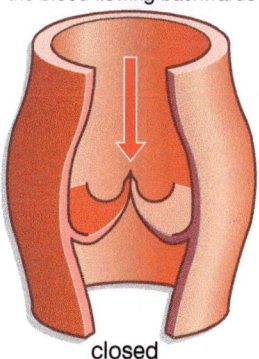

open   closed

Δ Fig. 9.11 Valves in the veins make sure that blood can only move in one direction, towards the heart.

Blood vessels	Blood pressure/kPa
aorta	>13
arteries	13–5.3
capillaries	3.3–1.6
veins	1.3–0.7
vena cava	0.3

Δ Table 9.2 Blood pressure in different blood vessels.

- By the time blood leaves the capillaries and enters the veins, there is no pulse and the blood pressure is very low. The large lumen (centre) of the veins allows blood to flow easily back to the heart. The contraction of body muscles, such as in the legs, helps to push the blood back toward the heart against the force of gravity. The valves make sure that blood can flow only in the right direction, back towards the heart.

REMEMBER

Remember, **a** for **a**rteries that travel **a**way from the heart. Veins carry blood into the heart and contain valves.

Fig. 9.12 shows a simplified layout of the human circulatory system, including the major blood vessels. Notice that this is a double circulation system because the blood travels through the heart twice on one complete journey around the body. The name of a major blood vessel is often related to the organ it supplies: *coronary* for heart (from the Latin *corona* for 'crown' because the blood vessels surround the top of the heart like a crown), *renal* for kidneys (from the Latin *renes* meaning 'the kidneys'), *pulmonary* for lungs (from the Latin *pulmonis* meaning 'lungs').

The largest vein and the largest artery in the body have special names. The **vena cava** is the vein that carries blood from the body to the heart, and the **aorta** is the artery that takes blood from the heart before it goes to the rest of the body.

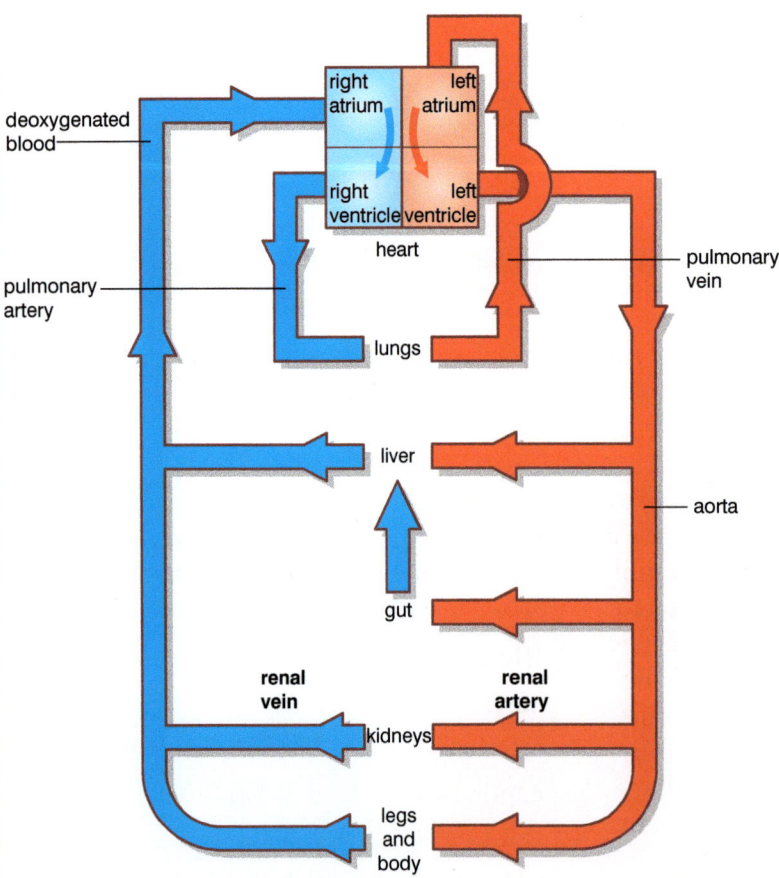

Δ Fig. 9.12 Plan of the human circulatory system.

1. Describe the differences in structure of arteries, capillaries, and veins.
2. **SUPPLEMENT** Give the names of the following blood vessels:
   a) the vessel that carries blood from the heart towards the body
   b) the vessels that carry blood from the lungs back towards the heart.
3. **SUPPLEMENT** Explain how the structure of arteries helps to reduce and even out the blood pulses from the heart.

BLOOD

The human circulatory system carries substances around the body. Table 9.3 shows some of the important substances transported around the human body. These substances are carried within the blood, in different forms.

Substance	Carried from	Carried to
Molecules absorbed from digested food, e.g. glucose, amino acids, fatty acids	small intestine	all parts of the body
Water	intestines	all parts of the body
Oxygen	lungs	all parts of the body
Carbon dioxide	all parts of the body	lungs
Urea (waste)	liver	kidneys
Hormones	glands	all parts of the body (different hormones affect different parts)

Δ Table 9.3 Substances carried by the blood around the body.

Blood is made from a liquid called plasma and the cells it transports: red blood cells, white blood cells, and platelets. Each of these has a particular function in the body.

Plasma

Plasma is the straw-coloured, liquid part of blood. It mainly consists of water, which makes it a good solvent for many substances. Nutrients and ions from digested food easily dissolve in plasma. **Urea**, which is formed by the liver from excess amino acids, is also soluble in plasma. Many hormones (see Section 13) are also soluble and are carried around the body dissolved in plasma. Carbon dioxide dissolves in water to form carbonic acid (H_2CO_3), and most carbon dioxide is carried in the blood in this form.

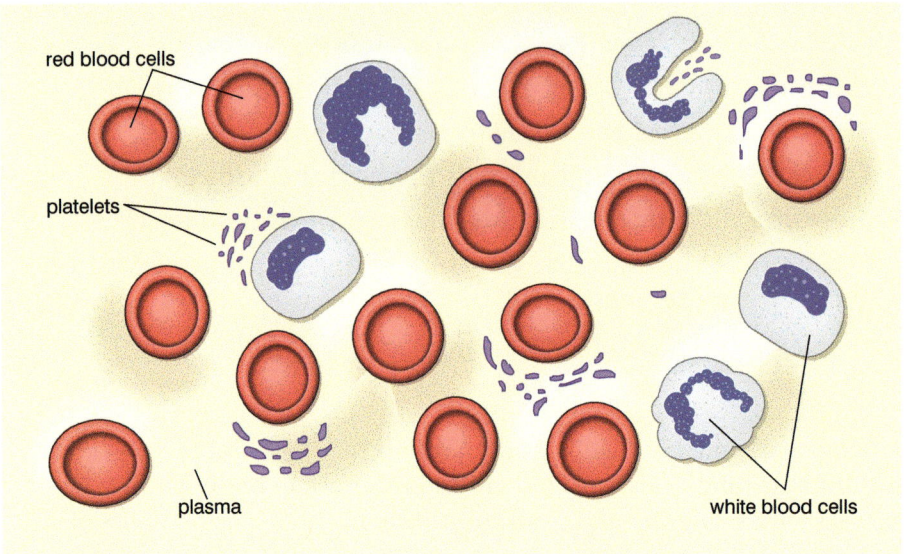

△ Fig. 9.13 Blood is mostly water, containing cells and many dissolved substances.

Red blood cells

Red blood cells are the most common cell in blood. Their main function is to carry oxygen around the body. The oxygen is attached to molecules of **haemoglobin** inside the cells, which give red blood cells their colour.

White blood cells

There are several different types of **white blood cell**, but they all play an important role in defending the body against disease. They are part of the **immune system** that responds to infection by trying to kill the **pathogens** (the disease-causing organisms). Some kinds of white blood cell kill pathogens by engulfing (flowing around the pathogen until it is completely enclosed), which is known as **phagocytosis**. Other kinds of white blood cell produce chemicals called **antibodies** that attack pathogens. (There is more on diseases and the immune system in Section 10.)

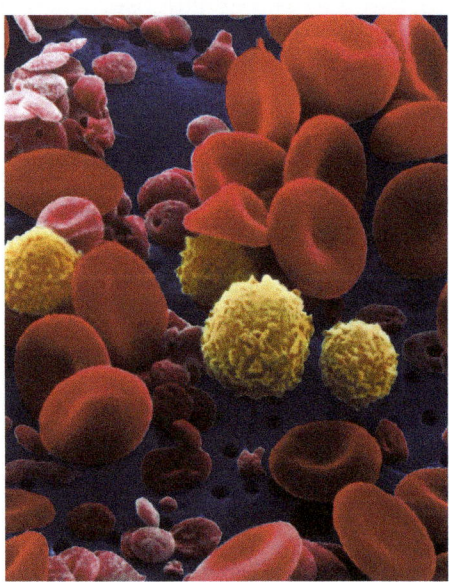

△ Fig. 9.14 Red blood cells (shown in red), white blood cells (yellow), and platelets (pink).

SUPPLEMENT

- **Phagocytes:** Several types of white blood cell belong to this group, but they all kill pathogens by phagocytosis (flowing round the pathogen and engulfing it). The phagocytes then digest the pathogens inside them. Different types of phagocytes target different pathogens, such as bacteria, fungi and protoctist parasites.

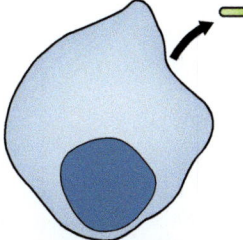
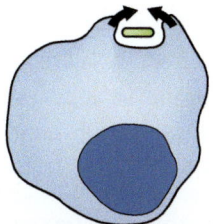
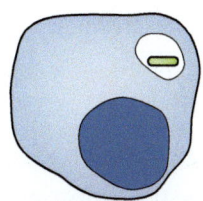

1 A phagocyte moves towards a bacterium.

2 The phagocyte pushes a sleeve of cytoplasm outwards to surround the bacterium.

3 The bacterium is now enclosed in a vacuole inside the cell. It is then killed and digested by enzymes.

△ Fig. 9.15 Phagocytosis of a bacterium by a phagocyte, a type of white blood cell.

- **Lymphocytes:** This type of white blood cell has a very large nucleus, and is responsible for producing antibodies.

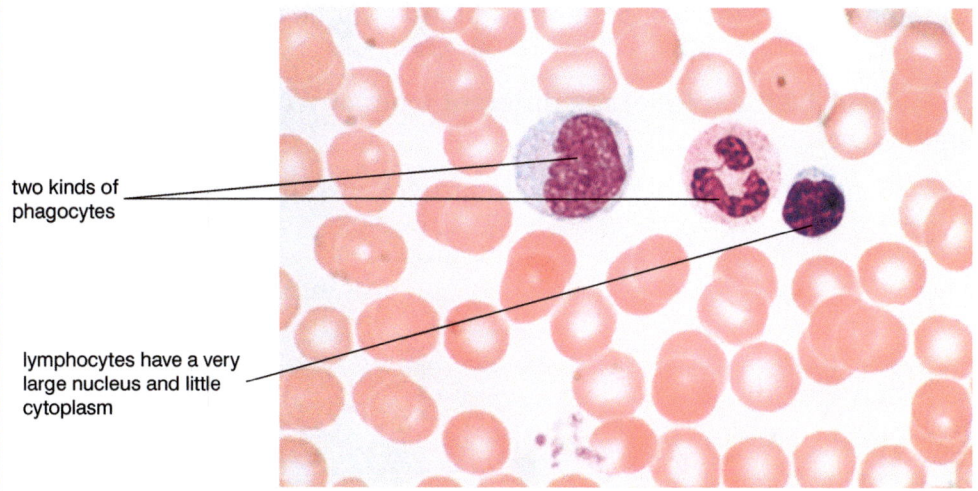

two kinds of phagocytes

lymphocytes have a very large nucleus and little cytoplasm

△ Fig. 9.16 Photomicrographs of (left) phagocytes and (right) a lymphocyte.

Platelets

Platelets are small fragments of much larger cells that are also important in protecting us from infection by causing blood to **clot** where there is damage to a blood vessel.

Clotting helps to seal the cut and prevent blood from leaking out and pathogens from getting in.

SCIENCE IN CONTEXT

BLOOD TESTS

Doctors often send blood off to be tested when a patient is unwell. These tests not only check the concentrations of different substances in the blood, such as glucose, they also count the number of different kinds of cells. Taking a blood sample can be quick and easy for a doctor to do, and the tests can help in diagnosing what is wrong with the patient.

- An abnormal concentration of glucose may indicate diabetes.
- Too few white blood cells may indicate liver or bone marrow disease.
- Too few red blood cells can make the blood look paler than normal. Low numbers of red blood cells cause anaemia, which usually results in the patient feeling more easily exhausted than usual.

Any abnormal results need following up with other tests to confirm diagnosis.

Challenge Question: Explain why having too few white blood cells could pose a risk to a patient.

QUESTIONS

1. Draw up a table to show the components of blood and the roles that they play in the body.

2. **SUPPLEMENT** Describe the roles of the different types of white blood cells in the immune system.

3. **SUPPLEMENT** Explain how platelets can help protect us from infection.

End of topic checklist

Key terms

antibody, artery, atrium (plural: atria), capillary, circulatory system, clot, coronary artery, coronary heart disease, electrocardiogram (ECG), haemoglobin, heart, heart rate, immune system, lumen, pathogen, phagocytosis, plasma, platelet, red blood cell, septum, urea, valve, vein, ventricle, white blood cell

SUPPLEMENT aorta, deoxygenated, double circulation, lymphocyte, oxygenated, phagocyte, pulmonary artery, pulmonary vein, single circulation, vena cava

During your study of this topic you should have learned:

◯ That the circulatory system is a system of blood vessels with a pump and valves to ensure one-way flow of blood.

◯ **SUPPLEMENT** How to describe the single circulation of a fish.

◯ **SUPPLEMENT** How to describe the double circulation of a mammal.

◯ **SUPPLEMENT** How to explain the advantages of a double circulation.

◯ How to identify the structures of the mammalian heart: muscular wall, septum, left and right ventricles, left and right atria, one-way valves, and coronary arteries.

◯ **SUPPLEMENT** How the heart functions in terms of the contraction of the muscles of the atria and ventricles, and the action of the valves.

◯ That blood is pumped away from the heart in arteries and returns to the heart in veins.

◯ That heart activity may be monitored by: ECG, pulse rate, and listening to sounds of valves closing.

◯ How to investigate and describe the effect of physical activity on the heart rate.

◯ **SUPPLEMENT** How to explain the effect of physical activity on the heart rate.

◯ How to describe coronary heart disease in terms of the blockage of coronary arteries and state the possible risk factors including: diet, lack of exercise, stress, smoking, genetic predisposition, age, and sex.

◯ How diet and exercise can reduce the risk of coronary heart disease.

◯ How to describe the structure of arteries, veins, and capillaries: relative thickness of wall, diameter of the lumen, and the presence of valves in veins.

◯ **SUPPLEMENT** That the structure of arteries and veins is related to the pressure of the blood that they transport.

End of topic checklist continued

○ How to state the functions of capillaries.

○ **SUPPLEMENT** That the structure of capillaries is related to their functions.

○ **SUPPLEMENT** How to identify the main blood vessels to and from the:

  ● heart: vena cava, aorta, pulmonary artery, and pulmonary vein

○ That the components of blood are: red blood cells, white blood cells, platelets, and plasma.

○ How to identify red and white blood cells.

○ **SUPPLEMENT** How to identify lymphocytes and phagocytes.

○ That the functions of the components of blood are:

  ● red blood cells (containing haemoglobin) to transport oxygen
  ● white blood cells for phagocytosis and antibody production
  ● platelets for clotting
  ● plasma for the transport of blood cells, ions, nutrients, urea, hormones, and carbon dioxide.

○ **SUPPLEMENT** That the function of:

  ● lymphocytes is antibody production
  ● phagocytes is engulfing pathogens by phagocytosis.

○ **SUPPLEMENT** That blood clotting prevents blood loss and the entry of pathogens.

End of topic questions

1. **SUPPLEMENT** Which blood vessel takes blood from the heart to the lungs?

   A aorta

   B pulmonary artery

   C pulmonary vein

   D vena cava

2. The diagram on the left shows a normal ECG, whereas the one on the right shows an ECG of a patient soon after a heart attack.

   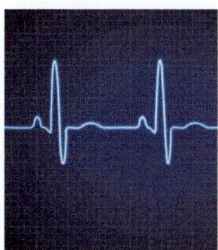 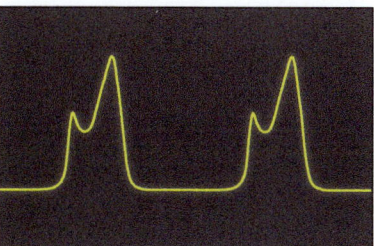

   patient A: healthy   patient B: just after a heart attack

   a) Explain what an ECG measures.

   b) Explain why the two ECGs look different.

   c) Describe how the ECG for patient A would change after 5 minutes of moderate exercise. Explain your answer.

   d) Describe one possible cause of patient B's heart attack.

3. People who suffer from anaemia often have a low red blood cell count (fewer blood cells per mm^3 blood) than usual. One of the symptoms of anaemia is becoming tired more easily than usual. Explain why this symptom occurs.

4. People who have suffered a thrombosis, in which the blood unnecessarily clots and blocks a blood vessel, may be given aspirin every day to help to protect them from it happening again. Aspirin interferes with the way platelets function.

   a) Describe the role of platelets in the blood.

   b) Explain why this normal function helps us to remain healthy.

   c) Suggest what damage might be caused if a blood vessel that supplies heart muscle is blocked by a blood clot.

   d) Explain why aspirin is effective in reducing the risk of another thrombosis.

5. Doctors look for risk factors in a person's lifestyle to help advise their patients on how to live more healthily.

   a) Describe four risk factors for coronary heart disease.

   b) For each of your answers to part **a)**, explain what the patient can or cannot do to reduce risk.

6. The graph shows the death rate from coronary heart disease (CHD) in some European countries compared with the proportion of the energy in the average diet in those countries that is provided by saturated fat.

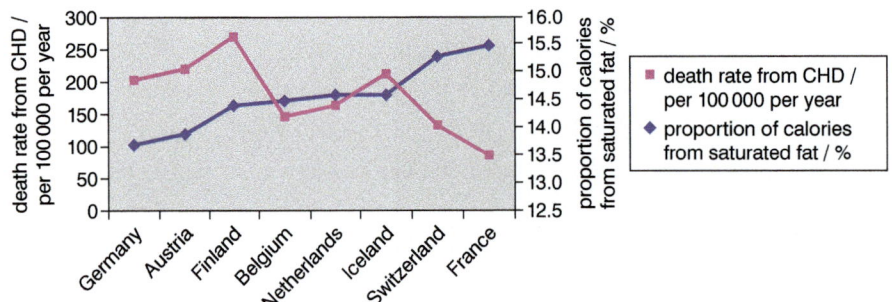

   a) Saturated fat consumption is a risk factor for coronary heart disease. Explain what is meant by a *risk factor*.

   b) Explain why saturated fat consumption is a risk factor for coronary heart disease.

   c) Describe what the graph shows.

   d) Some people think this graph indicates that there is no correlation between saturated fat consumption and CHD. Evaluate this idea.

7. **SUPPLEMENT** The blood pressure of blood leaving the right ventricle of the human heart is 3 kPa, and from the left ventricle is around 16 kPa. Explain how the heart can produce these different pressures, and why this difference is important for the body.

8. **SUPPLEMENT** **a)** Describe the differences between the circulation systems of a fish and a mammal.

   b) Explain the advantages of the circulation system of a mammal compared with that of a fish.

9. Look at the graph of how heart rate changes with exercise in Fig 9.7.

   a) Describe how the running speed changes during this experiment.

   b) What is the lowest heart rate, and when does this occur?

   c) What is the highest heart rate, and when does this occur?

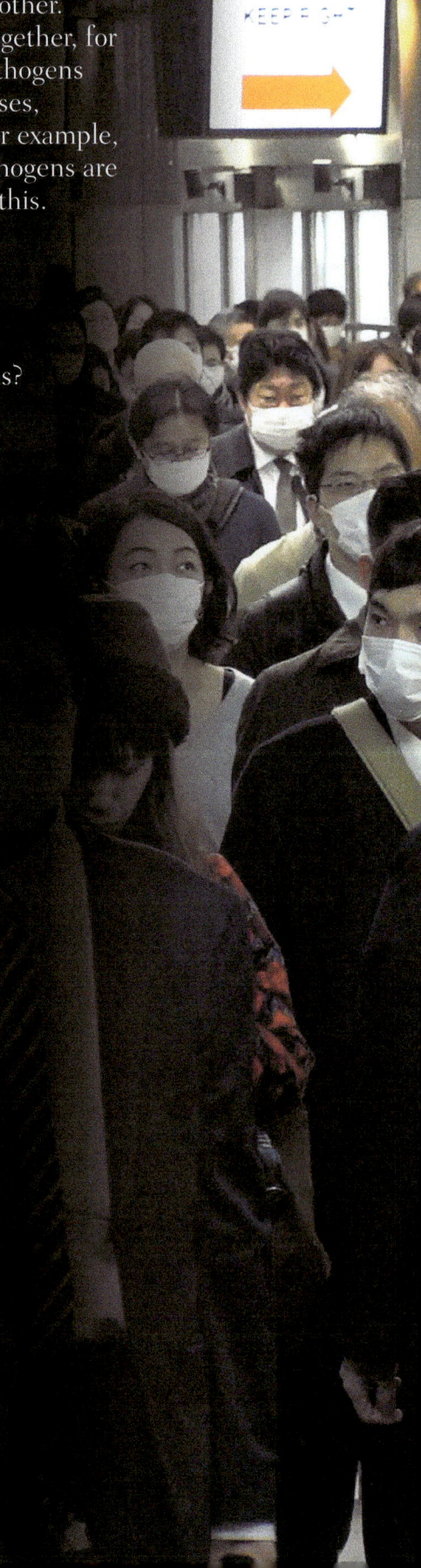

Many diseases are caused by microscopic organisms called pathogens. Examples of pathogens include many bacteria. Diseases spread when pathogens are passed from one person to another. Sometimes this can happen when two people are close together, for example, a healthy person may breathe in some of the pathogens that an infected person has just breathed out. In other cases, pathogens can be passed on without any close contact, for example, through contaminated food or water. If we know how pathogens are passed on then we can take precautions to try to prevent this.

STARTING POINTS

1. How do pathogens spread?

2. What are the body's natural defences against pathogens?

3. How do vaccinations make use of the body's natural defence systems?

SYLLABUS SECTIONS COVERED

10.1 Diseases and immunity

10
Diseases and immunity

Δ Face coverings can help prevent the transmission of airborne pathogens.

Diseases and immunity

△ Fig. 10.1 After the 2010 earthquake in Haiti many water sources were contaminated leading to a serious outbreak of cholera.

INTRODUCTION

Cholera is a disease that causes people to produce large amounts of watery diarrhoea, which can result in death through dehydration. In the 1800s it killed millions of people across the world. In 1854 John Snow, a doctor in London UK, identified a link between cholera and drinking water contaminated with human faeces. This led to massive efforts in cities to make sure that drinking water was kept separate from sewage. Today, good hygiene protects us from disease such as cholera. Although cholera is now a much rarer disease, it still occurs in disaster areas and refugee camps where the drinking water is not well protected.

KNOWLEDGE CHECK

✓ Disease may be caused by infection that can be passed from person to person.
✓ Good hygiene, such as washing hands, can prevent the spread of disease.
✓ Different infectious diseases may be spread in different ways.

LEARNING OBJECTIVES

✓ Describe a pathogen as a disease-causing organism.
✓ Describe a transmissible disease as a disease in which the pathogen can be passed from one host to another.
✓ State that a pathogen is transmitted: by direct contact, including through blood and other body fluids; indirectly, including from contaminated surfaces, food, animals, and air.
✓ Describe the body defences against pathogens, limited to: skin, hairs in the nose, mucus, stomach acid, and white blood cells.
✓ Explain the importance of the following in controlling the spread of disease: a clean water supply, hygienic food preparation, good personal hygiene, waste disposal, sewage treatment (details of the stages of treatment are **not** required).
✓ **SUPPLEMENT** Describe active immunity as defence against a pathogen by antibody production in the body.
✓ **SUPPLEMENT** State that each pathogen has its own antigens, which have specific shapes.
✓ **SUPPLEMENT** Describe antibodies as proteins that bind to antigens leading to direct destruction of pathogens or marking of pathogens for destruction by phagocytes.
✓ **SUPPLEMENT** State that specific antibodies have complementary shapes which fit specific antigens.
✓ **SUPPLEMENT** Explain that active immunity is gained after an infection by a pathogen or by vaccination.
✓ **SUPPLEMENT** Outline the process of vaccination: weakened pathogens or their antigens are put into the body; the antigens stimulate an immune response by lymphocytes which produce antibodies; memory cells are produced that give long-term immunity.
✓ **SUPPLEMENT** Explain the role of vaccination in controlling the spread of diseases.
✓ State that vaccinations are available for some pathogens to help control the spread of diseases.

Transmissible diseases

Many factors can cause disease, such as bad diet, smoking, and genetic tendency, but diseases that can be passed (transmitted) from one individual to another are called **transmissible diseases**, or infectious diseases. Common infectious diseases in humans include not only colds and flu, but also diseases such as food poisoning or sexually transmitted infections (STIs) (see Section 15).

Transmissible diseases are caused by **pathogens**, which are disease-causing organisms. Many of these organisms are microscopic, so it is only since the development of the microscope that we have been able to study them in detail and find out how to prevent them. For example, before cholera bacteria were seen in water, many people thought that the disease was transmitted through the air. However, treatment to clean the air, such as fumigation, had no effect on cholera.

The organism that is infected by the pathogen is called the **host**. Many groups of organisms include pathogens, such as bacteria, and fungi. Viruses that cause disease are also called pathogens, even though many scientists do not recognise them as being living organisms (see Section 1). All living organisms may be a host to an infection.

Disease	Pathogen	Host
Cholera	*Vibrio cholerae* (bacterium)	human
Tobacco mosaic disease	TMV virus	many plants including tobacco
Athlete's foot	*Trichophyton* species (fungus)	human

Δ Table 10.1 Transmissible diseases occur in every kind of living organism.

SUPPLEMENT

Viruses

Viruses are very simple structures, consisting of an outer protein coat that protects the genetic material inside.

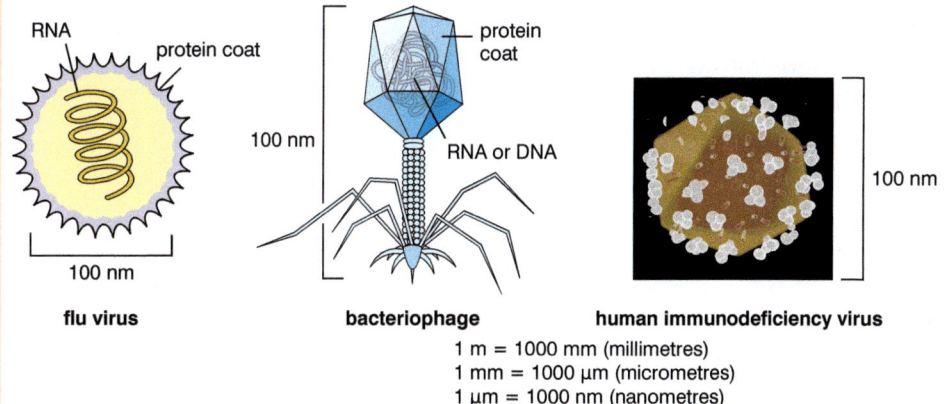

RNA   protein coat

100 nm

protein coat

RNA or DNA

100 nm

100 nm

flu virus   **bacteriophage**   **human immunodeficiency virus**

1 m = 1000 mm (millimetres)
1 mm = 1000 μm (micrometres)
1 μm = 1000 nm (nanometres)

Δ Fig. 10.2 Three kinds of virus. One thousand virus particles would fit across the width of one human hair.

QUESTIONS

1. Define the term *transmissible disease*.

2. Find out the names of three examples of bacteria and three examples of viruses that are pathogens.

Methods of transmission

Transmission is the passing of a disease from an infected person (host) to an uninfected person (new host). Inside the new host, the pathogen will reproduce rapidly to produce hundreds or thousands of new pathogens. These leave the host so that they can infect others.

There are many different ways that pathogens can be transmitted.

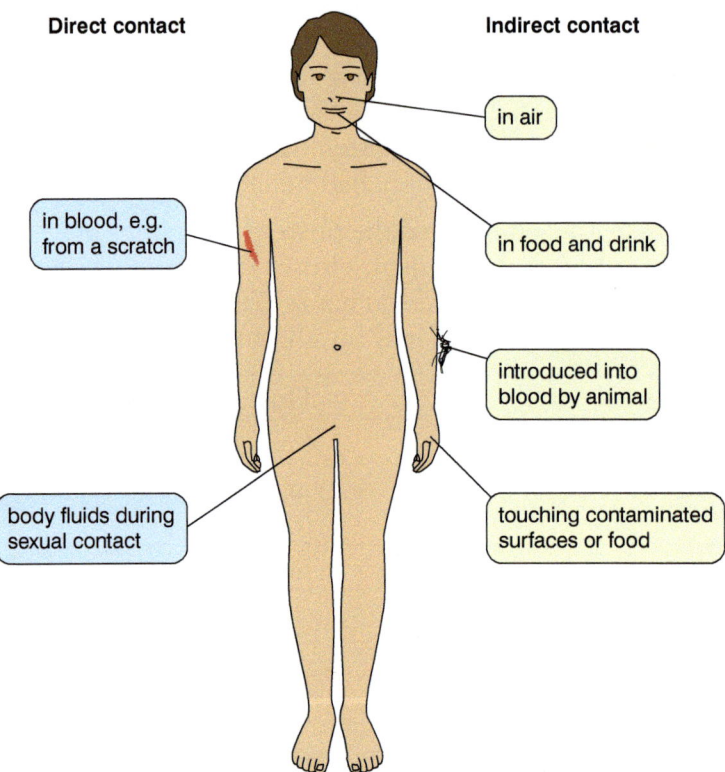

Direct contact

in blood, e.g. from a scratch

body fluids during sexual contact

Indirect contact

in air

in food and drink

introduced into blood by animal

touching contaminated surfaces or food

△ Fig. 10.3 Methods of transmission of disease.

Direct contact methods are those where the pathogen is transmitted by the transfer of body fluids from a host to an uninfected person. These fluids include blood, semen and other fluids made in the body. Body fluids can be exchanged during sexual contact, and this is the method of transmission of STIs such as syphilis, HIV and gonorrhoea. Blood is not usually exchanged between people, except by accident, during blood transfusions or if people share hypodermic needles. Diseases transmitted this way include HIV, hepatitis B and hepatitis C. Blood for transfusion is usually checked to make sure it is clear of pathogens before it is given to another person.

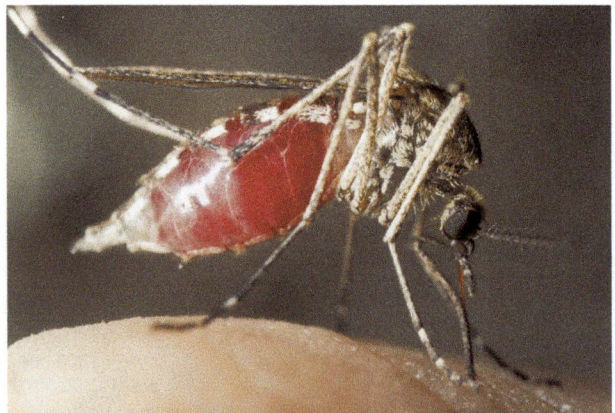

△ Fig. 10.4 A mosquito sucking blood from a human. The mosquito carries pathogens from an infected person to other people.

Indirect methods of transmission are where the pathogen leaves the host and is carried in some way to an uninfected individual. Examples include:

- transmission in water droplets in air – for example, colds and flu when the host coughs or sneezes
- transmission in infected drinking water – for example, cholera, typhoid, and dysentery
- transmission by touching contaminated surfaces – for example, athlete's foot fungus can be transmitted by walking on damp floors, and many food poisoning pathogens (e.g. *Salmonella* and *Clostridium*) are carried to food on the feet of flies that have recently been feeding on contaminated food or faeces
- transmission in insect bites – for example, insects, such as mosquitoes, that feed on human blood carry the pathogens in their body and introduce them into the blood when they feed, causing diseases such as malaria and dengue fever.

There are many animals that can transfer pathogens from one host to another, for example, mosquitoes and flies. These animals do not suffer from the disease that a pathogen causes in its host.

Controlling the spread of disease

We can control the spread of transmissible diseases by preventing pathogens in a host reaching uninfected individuals. One of the most important ways of preventing the spread of pathogens is to ensure a clean water supply, not only for drinking and cooking, but also for washing and cleaning.

Good **hygiene** means keeping things clean, which also reduces the numbers of pathogens on surfaces and on parts of the body. Good personal hygiene includes:

- washing hands thoroughly with warm water and detergent (soap) after going to the toilet; this removes pathogens that are found in faeces and makes it less likely that the pathogens can be transmitted by touch
- covering the mouth and nose with a tissue when coughing or sneezing; this will trap the pathogens and stop them passing into the air; disposing of the tissue properly and washing the hands afterwards will help reduce the risk of transmission by touch.

Δ Fig. 10.5 Large numbers of pathogens can be spread over a large area from a single sneeze.

Hygienic food preparation is important in preventing the transmission of pathogens that cause food poisoning. Food should be kept away from flies, and stored appropriately until it is prepared for eating. Meat, fish, and dairy products provide good growing environments for pathogens, so they should be kept in a fridge before use, and should be used within a limited time period after purchase. Foods that are cooked should be prepared separately from those that are uncooked, including using different utensils such as knives for chopping. The heat of cooking kills many pathogens, but uncooked food can easily transmit microorganisms.

Waste food and human waste (faeces) are attractive food sources for many insects that can carry the pathogens that cause transmissible diseases. The risk of transmission in this way can be reduced by:

- placing waste food into covered containers and then disposing of the waste away from human habitation in landfill tips, by burning or by composting it in sealed containers
- collecting human waste in sewage systems and keeping it completely separate from drinking water supplies until it has been treated to kill the pathogens and is safe to release into the environment.

1. Give one example of a direct method of transmission of disease, and one example of an indirect method. In each case identify a disease transmitted between humans in this way.

2. Using a particular example, suggest one way to reduce the risk of transmission of a disease. Explain why your suggestion would work.

Methods of defence

We are not completely at risk of infection by pathogens. In fact, the body has many defences that help to prevent us becoming infected.

Mechanical barriers, such as the skin, or hairs in the nose, physically prevent pathogens from getting into the body. Chemical barriers, such as lysozyme enzymes in **mucus** in the airways and acid in the stomach, are chemicals that kill pathogens and prevent them getting into body tissues.

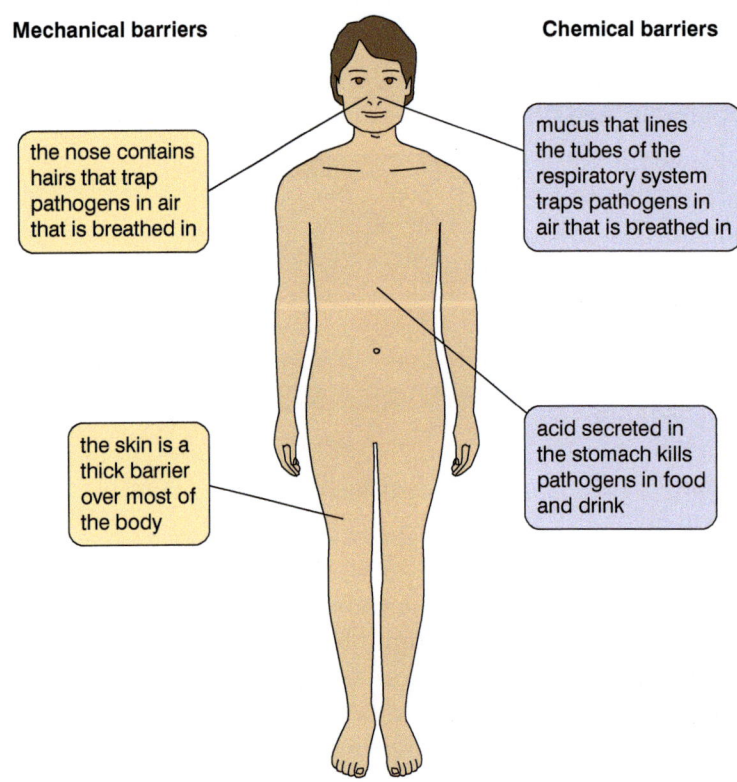

Mechanical barriers

the nose contains hairs that trap pathogens in air that is breathed in

the skin is a thick barrier over most of the body

Chemical barriers

mucus that lines the tubes of the respiratory system traps pathogens in air that is breathed in

acid secreted in the stomach kills pathogens in food and drink

△ Fig. 10.6 Barriers to infection.

If pathogens get past these defences, the body has further ways of protection. White blood cells in the **immune system** will attack pathogens in one of two ways:

- **phagocytosis** – where the white blood cells engulf and surround the pathogens (see Section 9) and then break them down
- they produce chemicals called **antibodies** that attack and destroy the pathogens (also see Section 9).

1. Describe one example of a *mechanical barrier* of the body to infection.

2. Describe one example of a *chemical barrier* in the body to infection.

3. Explain how white blood cells protect us from infection.

The immune system and active immunity

The immune system coordinates the response of the body to infection. This includes the production of **phagocyte** cells that directly attack the invading pathogen, and **lymphocyte** cells that produce antibodies.

Antibodies are a group of proteins that work by shape. Molecules called **antigens** on the surface of a pathogen have a specific shape, which differs depending on the pathogen. An antibody for a particular pathogen has a specific **complementary** shape that matches this, so that it can bind to the antigen and either mark it for phagocytes to engulf the pathogen or directly destroy the pathogen by causing it to break open and die.

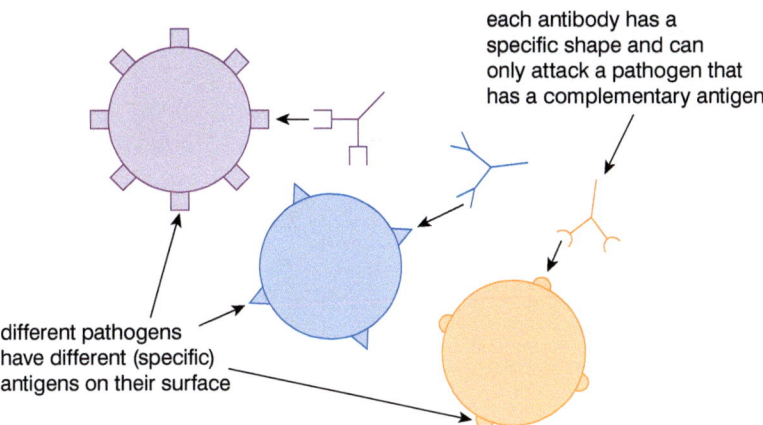

each antibody has a specific shape and can only attack a pathogen that has a complementary antigen

different pathogens have different (specific) antigens on their surface

Δ Fig. 10.7 The shape of an antibody must match the shape of the antigen so that it can attach.

REMEMBER

In Section 5 you learned how enzymes are *specific* because their active sites have a *complementary* shape to that of their particular substrate. In a similar way, antibodies are specific to the particular antigen they have a complementary shape to.

Lymphocytes also produce **memory cells**, which remain in the blood after the pathogen has been destroyed. If you are attacked by the same pathogen another time, the memory cells rapidly respond, causing a rapid increase in the production of antibodies. These antibodies attack and destroy the pathogen, often before you are even aware of another infection. For most infections, it is usually not possible for you to become ill with the infection a second time. We say that you have become **immune** to that kind of infection.

This kind of immunity is called **active immunity** because infection has actively produced the immunity. Active immunity means the body is being defended by antibody production in the body.

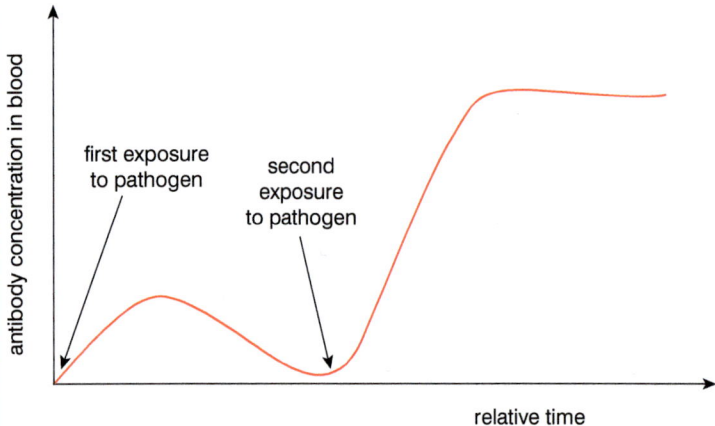

△ Fig. 10.8 Antibody production is much greater the second time you are infected with a pathogen.

SCIENCE IN CONTEXT — MUTATING PATHOGENS

Pathogens can evade the immune system as they mutate and change shape or change the molecules they make. Mutations are usually relatively rare events. However, in some pathogens, such as the viruses that cause the common cold or flu, mutations occur more frequently.

So, even if you are affected by the same kind of pathogen again, your immune system won't recognise it and you will suffer from the disease again. This is because the antibodies that your white blood cells make can no longer fight the infection.

This is why the Coronavirus vaccine needed boosters, as each new variant made it more difficult for our bodies, existing antibodies to fight off the infection.

Challenge Question: There are some transmissible diseases you can get over and over again, like the common cold.

1. Explain what this tells you about the viruses causing these colds.

2. You have a cold and transmit the virus to a friend. When you have recovered, can you catch the same cold back again from your friend? Explain your answer.

Vaccination

A **vaccine** can be given to people to help control the spread of diseases. After being vaccinated, people are less likely to get ill and transmit the pathogen they have been vaccinated against. This means that **vaccinations** slow down and can prevent the spread of some pathogens in the population. However, vaccinations are only available for some pathogens, as they are specific for each type.

Vaccination

Vaccination induces active immunity by making the body respond as if it has been infected. Active immunity can also be induced by making the body respond as if it has been infected. A vaccine is prepared from small amounts of material from the pathogen, which is either dead or weakened so that it is harmless and cannot cause infection. However, it still has the antigens on its surface.

The vaccine is put into the body, either through the mouth (for polio vaccination) or by injection. This is known as vaccination. The antigens in the vaccine stimulate an immune response by lymphocytes which make antibodies to the antigens on the pathogen, and also by producing memory cells. These memory cells remain in the blood after the pathogen has been destroyed, giving long-term immunity. Some memory cells may last for the rest of your life; others may need a booster vaccination after a period of time.

If you are ever infected by the live pathogen, the memory cells recognise it very rapidly and stimulate lymphocytes to produce huge quantities of antibodies very quickly. This response kills off the pathogen rapidly, often before you develop symptoms and realise you have been infected. This rapid response makes you immune to that pathogen.

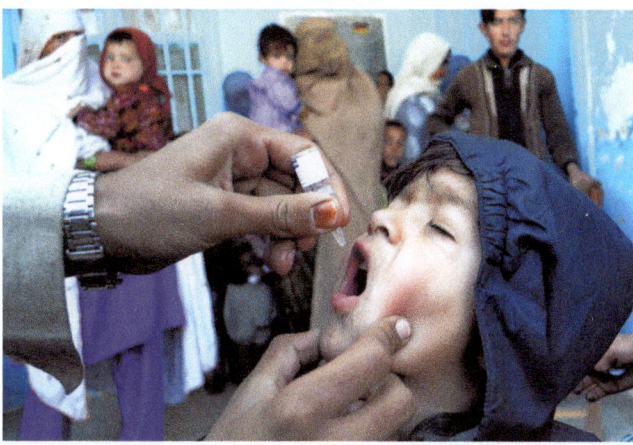

△ Fig. 10.9 A health worker is administering an oral polio vaccine in Pakistan as part of an anti-polio campaign.

Most countries have immunisation programmes in which particular vaccinations are offered to all individuals of a specific age. Young children may receive vaccinations against a wide range of infections, including measles, mumps, rubella, whooping cough, diphtheria, and polio. Teenage girls may receive vaccinations for human papillomavirus (HPV). Many of these infectious diseases are so rarely seen now that few people really appreciate how dangerous they were and how badly people were affected by them.

Parents are usually given the choice whether or not to vaccinate a child. This can cause concern because very rarely a vaccination may cause a reaction and harm the child. However, if someone who has not been vaccinated catches the disease, the risk of permanent damage or even death is far greater.

A very few people, particularly those with damaged immune systems, are at risk of harm from vaccination. However, if most people in the

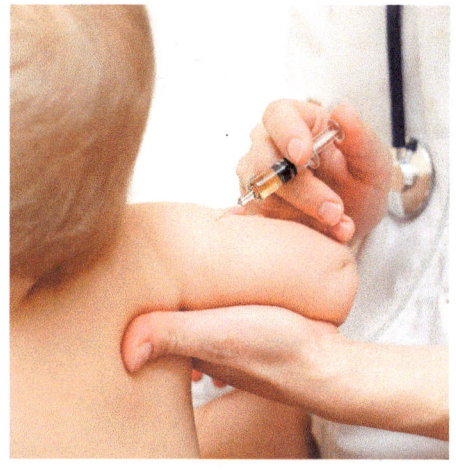

△ Fig. 10.10 An injection of vaccine can be a painful experience, but childhood vaccinations can give life-long protection from dangerous infections.

population have been vaccinated, the risk of transmission is reduced, as vaccinated people are less likely to transmit the pathogens. This means that the risk that anyone will come into contact with the pathogen is so small that everyone is protected. This is known as herd immunity.

SMALLPOX ERADICATION

Smallpox was a devastating disease that killed up to 60% of adults and over 80% of children who became infected with the smallpox virus. It produced large, fluid-filled blisters all over the body. The virus was transmitted between people either through the air or through touching the blisters.

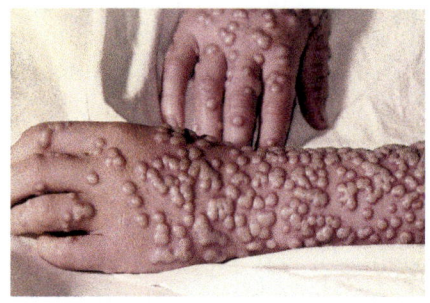

△ Fig. 10.11 Smallpox covered the body in fluid-filled blisters.

In 1796, Edward Jenner produced the first vaccine against the disease, using a similar, but much weaker, virus that caused cowpox. Over the next 100 years, in areas such as Europe and the US, there were great efforts to immunise every person using smallpox vaccine. In 1897 in the US, and around 1900 in Europe, the disease was almost completely eradicated.

In 1958, the World Health Organization (WHO) started a campaign of worldwide immunisation against smallpox. Where an outbreak of the disease occurred, everyone in the area was vaccinated and people with smallpox were isolated until they were no longer infectious. In 1980, WHO confirmed that the disease had been completely eradicated.

Challenge Question: Herd immunity can help protect people from a transmissible disease even if they have not been vaccinated against it.

Suggest an explanation why WHO did not rely on herd immunity to protect unvaccinated people against smallpox.

QUESTIONS

1. Give your own definitions of:

a) antibody

b) pathogen.

2. SUPPLEMENT Explain why you become immune after being given a vaccine.

End of topic checklist

Key terms

antibody, host, hygiene, immune system, mucus, pathogen, phagocytosis, transmissible disease, vaccination, vaccine

SUPPLEMENT active immunity, antigen, complementary, immune, lymphocyte, memory cell, phagocyte

During your study of this topic you should have learned:

○ That a pathogen is a disease-causing organism.

○ That a transmissible disease is a disease in which the pathogen can be passed from one host to another.

○ That a pathogen is transmitted: by direct contact, such as through blood and other body fluids; indirectly, such as from contaminated surfaces, food, animals, and air.

○ That the body's defences include: skin, hairs in the nose, mucus, stomach acid, and white blood cells.

○ How to explain the importance of the following in controlling the spread of disease: a clean water supply; hygienic food preparation; good personal hygiene; waste disposal; sewage treatment.

○ **SUPPLEMENT** That active immunity is defence against a pathogen by antibody production in the body.

○ **SUPPLEMENT** That each pathogen has its own antigens, which have specific shapes.

○ **SUPPLEMENT** That antibodies are proteins that bind to antigens, leading to direct destruction of pathogens or marking of pathogens for destruction by phagocytes.

○ **SUPPLEMENT** That specific antibodies have complementary shapes which fit specific antigens.

○ **SUPPLEMENT** That active immunity is gained after an infection by a pathogen or by vaccination.

○ That vaccinations are available for some pathogens to help control the spread of diseases.

○ **SUPPLEMENT** That the process of vaccination includes the following steps: weakened pathogens or their antigens are put into the body; the antigens stimulate an immune response by lymphocytes which produce antibodies; memory cells are produced that give long-term immunity.

○ **SUPPLEMENT** How to explain the role of vaccination in controlling the spread of diseases.

End of topic questions

1. Which is an example of transmission of a pathogen through direct contact?

   A breathing in droplets from a sneeze

   B by an insect bite

   C eating infected food

   D through sexual contact

2. **SUPPLEMENT** The kitchen of a restaurant displays a list of hygiene rules that everyone must follow.

   a) State the meaning of *hygiene*.

   b) Explain why the restaurant has these rules.

   c) For each of the rules, explain how it helps to maintain good hygiene.

   > *Hygiene rules*
   >
   > 1. *Wash hands before any food preparation.*
   >
   > 2. *Only use clean equipment for food preparation.*
   >
   > 3. *Keep cooked and uncooked food separate – use separate preparation surfaces and utensils.*
   >
   > 4. *Clean any cuts immediately and cover with a waterproof plaster before continuing work.*

3. Many pathogens are microscopic.

   a) Explain the meaning of the words *pathogen* and *microscopic*.

   b) Explain the importance of the development of the microscope in helping us to understand the causes of transmissible diseases.

4. **SUPPLEMENT** Explain why the immune response to an infection of the pathogen that causes measles:

   a) will protect you from a future infection by the measles pathogen

   b) will not protect you against infection by the pathogen that causes chickenpox.

5. **SUPPLEMENT** One of the diseases that health organisations are now trying to eradicate from the world is polio. Polio is a highly infectious disease caused by a virus that is transmitted usually in contaminated food or water. Severe cases may result in paralysis or even death. Many parts of the world are free from polio as a result of vaccination, but around 1000 cases occur each year in a few countries.

   a) Vaccination for polio is with drops placed in the mouth. Explain how this leads to immunity.

b) Polio vaccination and boosters give life-long protection against the disease. Give the name of this type of immunity.

c) Suggest how eradication of polio could be carried out, and explain why your suggestion would work.

6. SUPPLEMENT In many countries parents of young children may choose whether or not to have their child vaccinated against diseases that used to be common. In some countries, vaccination is compulsory for particular diseases.

a) Suggest why some parents might not want their child vaccinated.

b) Describe an argument that a doctor would give to parents to explain why vaccination is in the best interests of the child. Explain your answer.

c) Give arguments for and against compulsory vaccination, and use them to decide whether compulsory vaccination should be carried out.

Our lungs are the organs that allow the body to take in oxygen from the air and expel carbon dioxide that is produced in cells. We breathe in and out about 500 ml of air during every breath. Oxygen from this air passes into tiny air sacs in the lungs, which are called alveoli, and diffuses into the capillaries that surround them. From here, the oxygen-rich blood is passed to the heart, where it is pumped around the rest of the body, before being passed back to the lungs to offload the carbon dioxide it has collected and pick up a fresh supply of oxygen.

STARTING POINTS

1. What is gas exchange and where does it happen in humans?

2. What happens when we breathe?

3. How is breathing affected by physical activity, and why?

SYLLABUS SECTIONS COVERED

11.1 Gas exchange in humans

11
Gas exchange in humans

△ Gas exchange takes place in the alveoli, located in the lungs.

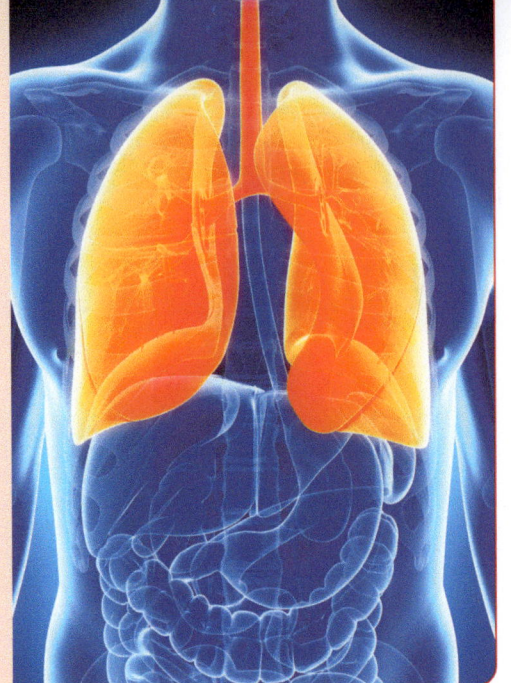

△ Fig. 11.1 The lungs are the site of gas exchange in humans.

Gas exchange in humans

INTRODUCTION

Respiration uses oxygen from the air and produces carbon dioxide that is returned to the environment. These gases must get into and out of the body fast enough to support the rate at which respiration needs to work. For single-celled organisms this isn't a problem. They have a large surface area to volume ratio, and diffusion across the cell membrane can supply and remove the gases at a fast enough rate. Larger organisms cannot do this. Not only do they have a much smaller external surface area to volume ratio, which slows the rate of diffusion, but many of them also live on land, where the delicate surface required for gas exchange would dry out if it was directly exposed to the external environment. Different groups of organisms have different solutions to these problems but all involve structures with a large surface area. Plants exchange gases inside the leaf; insects have internal tubes (a tracheal system) inside the body where they exchange gases; fish have gills; and many vertebrates, including humans, have lungs.

KNOWLEDGE CHECK

✓ Animals use oxygen from the air inspired and give out the carbon dioxide they produce in the air they expire.
✓ Humans use lungs for breathing.

LEARNING OBJECTIVES

✓ Identify in diagrams and images the following parts of the breathing system: lungs, diaphragm, ribs, intercostal muscles, larynx, trachea, bronchi, bronchioles, alveoli and associated capillaries.
✓ **SUPPLEMENT** Describe the features of gas exchange surfaces in humans, limited to: large surface area, thin surface, good blood supply, and good ventilation with air.
✓ Investigate the differences in composition between inspired and expired air using limewater as a test for carbon dioxide.
✓ Describe the differences in composition between inspired and expired air, limited to: oxygen, carbon dioxide, and water vapour.
✓ **SUPPLEMENT** Explain the differences in composition between inspired and expired air.
✓ Investigate and describe the effects of physical activity on the rate and depth of breathing.
✓ **SUPPLEMENT** Explain the link between physical activity and the rate and depth of breathing in terms of: an increased carbon dioxide concentration in the blood, which is detected by the brain, leading to an increased rate and greater depth of breathing.

The human breathing system

Breathing is the way that oxygen is taken into our bodies and carbon dioxide is removed. When we breathe, air is moved into and out of our lungs. This involves different parts of the breathing system within the chest.

When we breathe in, air enters though the nose and mouth. In the nose the air is moistened and warmed. The air passes over the **larynx**, where it may be used to make sounds, for example when we talk. The air travels down the **trachea** to the **lungs**. The air enters the lungs through the **bronchi** (singular: **bronchus**), which branch and divide to form a network of **bronchioles**.

At the end of the bronchioles are air sacs. The bulges on an air sac are called **alveoli** (singular: **alveolus**). The alveoli are covered in tiny blood capillaries. This is where oxygen and carbon dioxide are exchanged between the blood and the air in the lungs.

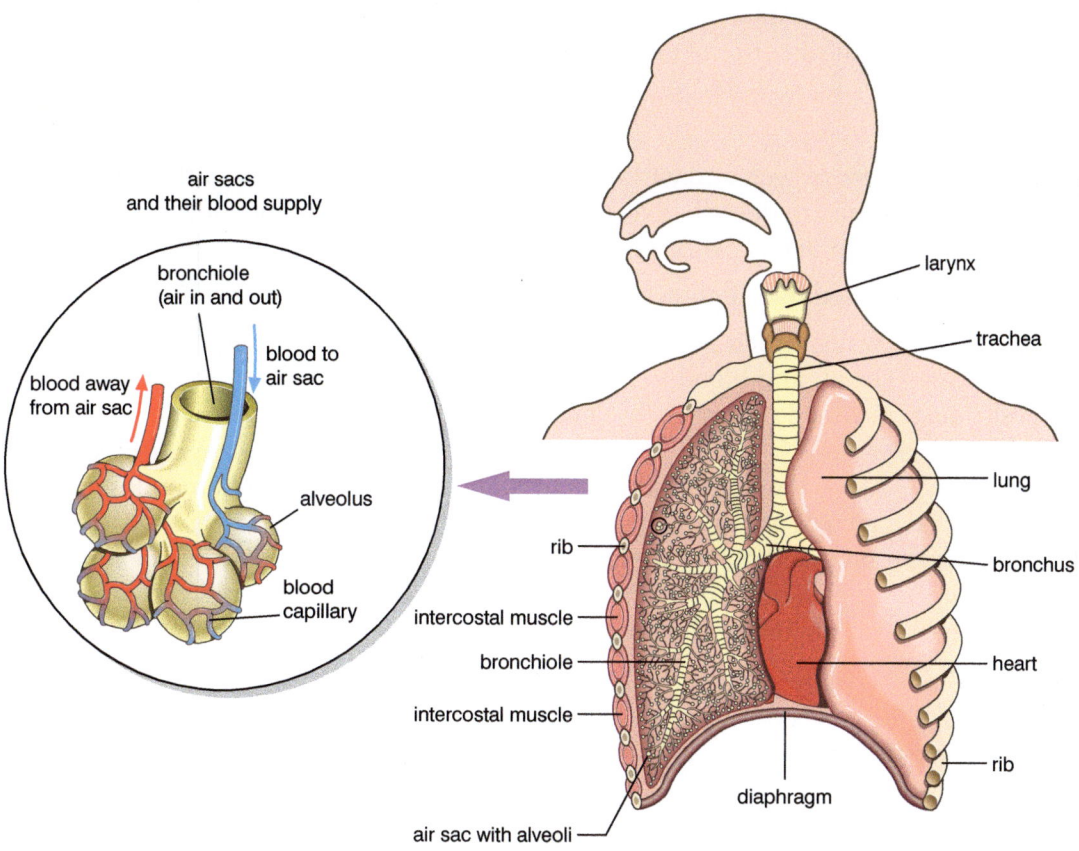

△ Fig. 11.2 The human breathing system.

SUPPLEMENT

Gas exchange

Animals need to exchange gases with the environment, to supply oxygen for respiration in cells, and to remove the waste product of respiration – carbon dioxide. These gases are exchanged at surfaces by diffusion. So **gas exchange surfaces**, such as in the human lungs, need adaptations to maximise the rate at which diffusion occurs.

An effective gas exchange surface has:

• a large surface area

- a short distance over which substances have to diffuse, so cells across which diffusion occurs are usually thin
- a good blood supply and good ventilation to deliver oxygen and remove carbon dioxide from the body rapidly – this maintains high concentration gradients for both gases.

Breathing in and out

Breathing in is known as *inhalation* or **inspiration**, and breathing out as *exhalation* or **expiration**. Breathing in and out involves movements of:

- the **ribs**, which are joined together by **intercostal muscles**
- the **diaphragm**, which is a domed sheet of muscle below the lungs.

Features of alveoli

The alveoli are where oxygen and carbon dioxide diffuse into and out of the blood. For this reason the alveoli are described as the *gas exchange surface*. The movement of air across the alveolar surface is called **ventilation**.

The alveoli are adapted for efficiency in exchanging gases by diffusion. They have:

- thin permeable walls, which keep the distance over which diffusion of gases takes place between the air and blood to a minimum
- a moist lining, in which the gases dissolve before they diffuse across the cell membranes
- a large surface area – there are hundreds of millions of alveoli in a human lung, giving a surface area of around 70 m^2 for diffusion
- high concentration gradients for the gases, because the blood is continually flowing past the air sacs, delivering excess carbon dioxide and taking on additional oxygen, and because of ventilation of the lungs, which refreshes the air in the air sacs.

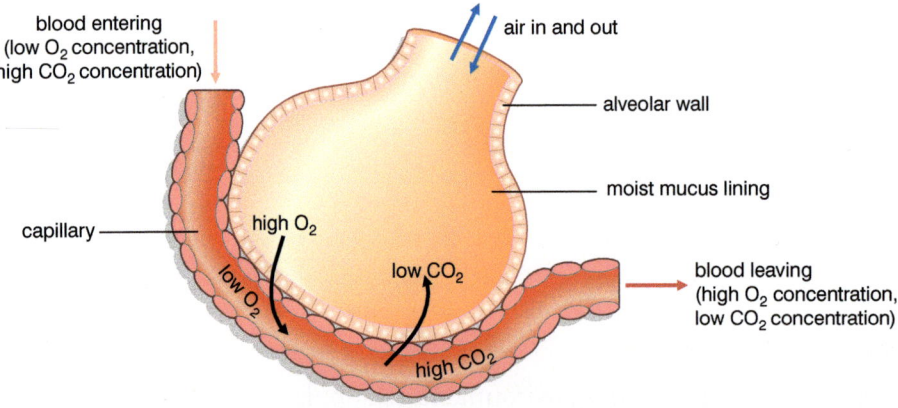

△ Fig. 11.3 Gas exchange in an air-filled alveolus.

1. Give the structures of the human breathing system in the order that an oxygen molecule would travel through.

2. **SUPPLEMENT** Explain as fully as you can how the lungs are adapted for a rapid rate of diffusion for gas exchange.

3. **SUPPLEMENT** Sketch a diagram of an alveolus and annotate it to show how it is adapted for efficient gas exchange. (Hint: remember to refer to diffusion.)

SCIENCE IN CONTEXT

VENTILATION BY MACHINE

Sometimes accident or illness can damage a person's ability to breathe. As exchange of gases is essential for respiration, and so for life, it is crucial that this process is continued artificially until the patient is sufficiently recovered to be able to do it independently again.

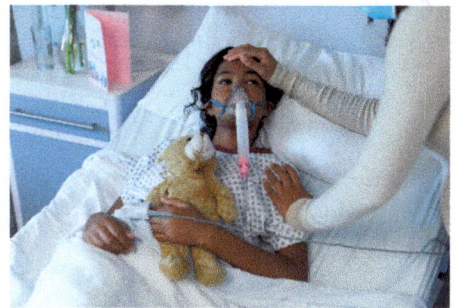

In the past, the patient was sometimes placed inside a large machine, called an *iron lung*. The machine was sealed from the air, and changes in pressure inside the machine caused volume and pressure changes inside the chest cavity caused by movements of the ribs and diaphragm. These movements then resulted in air being drawn in or forced out of the patient's lungs. Today, a sealed mask is placed over the patient's mouth and nose, and air is forced into the lungs by increasing the air pressure. The air is naturally breathed out again as the stretched muscles relax.

△ Fig. 11.4 A ventilator mask forces air into the patient's lungs by increasing air pressure, and allows air out of the patient's lungs by decreasing the air pressure in the mask.

Challenge Question: Mountain climbers at very high altitudes may also need to use ventilator masks. Suggest why it can be difficult to breathe normally at very high altitudes.

Inspired air and expired air

The air we breathe in and out contains many gases. Oxygen is taken into the blood from the air we breathe in. Carbon dioxide and water vapour are added to the air we breathe out. The other gases in the air we breathe in are breathed out almost unchanged, except for being warmer.

	In inspired air	In expired air
oxygen	21%	16%
carbon dioxide	0.04%	4.5%
water	variable	high

△ Table 11.1 Differences in composition of inspired and expired air.

We can compare the carbon dioxide in inspired air and expired air using the apparatus shown in Fig. 11.5. Limewater reacts with carbon dioxide and turns cloudy, so this is a test for carbon dioxide.

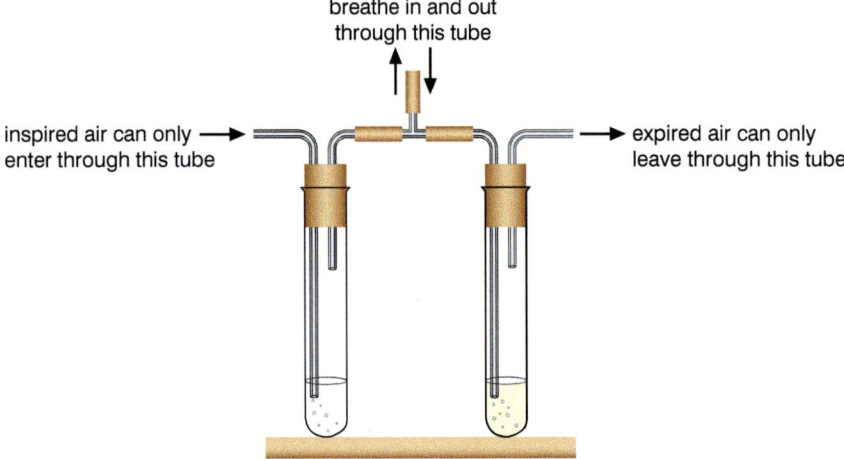

△ Fig. 11.5 Limewater in the tubes show that expired air contains much more carbon dioxide than inspired air.

The composition of inspired and expired air changes because:

- oxygen is removed from blood by respiring cells and used for respiration (see Section 12), so blood returning to the lungs has a lower concentration of oxygen than blood leaving the lungs
- carbon dioxide is produced by respiration (see Section 12) and diffuses into the blood from respiring cells; the blood transports the carbon dioxide to the lungs, where it diffuses into the alveoli
- water vapour concentration increases because water evaporates from the moist linings of the alveoli into the expired air as a result of the warmth of the body.

Other gases in the air we breathe in and out remain unaffected because they are not used or produced by the body.

Investigating the effects of exercise on breathing

There are two aspects of breathing that can change during exercise – the rate of breathing and the depth of breathing (the volume of a breath).

- Rate of breathing is usually counted as number of breaths per minute.
- The volume of a breath can be measured in dm^3 using a spirometer. A simple spirometer can be made using a 2-litre plastic bottle that has been marked down the side with volumes of water. (This can be done by adding 500 cm^3 of water at a time, and marking the volume on the side of the bottle with a waterproof marker.) When the bottle is full of water, turn it upside-down into a water trough without allowing any air into the bottle. Insert a flexible plastic tube into the neck of the bottle and secure the bottle and tube in position. Clean the other end of the tubing with antiseptic solution. (Alternatively, add a mouthpiece to the end of the tubing that can easily be removed and sterilised after each test.) To measure the volume of a breath, ask the person to wear a noseclip and then to breathe out a normal breath

into the tube. The scale on the bottle can be used to measure the volume of air breathed out.

The bottle must be set up again before measuring another breath. This is because carbon dioxide build-up in the air in the bottle over several breaths can be dangerous.

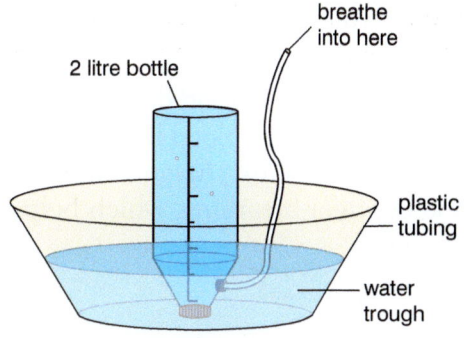

breathe into here

2 litre bottle

plastic tubing

water trough

△ Fig. 11.6 Simple spirometer.

Developing practical skills

Devise and plan investigations

1. Design an investigation into the effect of exercise on breathing. (Hint: think carefully about how many people to test, and how to test them, in order to get valid results.)

Demonstrate and describe techniques

2. This investigation could involve vigorous exercise. What risks will you need to prepare for, and how should they be minimised?

Analyse and interpret data

The data in the table are the results from an investigation into the effect of exercise on breathing in four people. They were first tested at rest and then after 2 minutes of running on a treadmill set at the same speed.

Person	A		B		C		D	
	Rate/ breaths/ minute	Breath volume /dm³	Rate/ breaths/ minute	Breath volume /dm³	Rate/ breaths/ minute	Breath volume /dm³	Rate/ breaths/ minute	Breath volume /dm³
At rest	13	0.5	15	0.4	12	1.2	18	0.6
After exercise	19	1.3	23	0.9	18	1.3	26	1.5

3. Explain how these data should be adjusted before they can give a valid answer to the question 'How does exercise affect breathing?'.

The results of an investigation like the one in the Developing practical skills box should show that both the rate of breathing and depth of breathing increase with the level of activity. However, a trained athlete will change in rate and depth of breathing less than an untrained person.

SUPPLEMENT

The rate and depth of breathing increase with the level of physical activity because as muscles contract faster they respire faster and so make carbon dioxide more quickly. Carbon dioxide is an acidic gas that dissolves easily in water-based solutions, such as the cytoplasm of a cell and blood plasma. The more carbon dioxide there is in solution, the more acidic the solution. A change in pH can affect the activity of many cell enzymes (see Section 5), so it is important that carbon dioxide is removed from the cells and the body as quickly as possible.

The increase in carbon dioxide concentration as a result of increased physical activity is detected as the blood flows past receptors in part of the brain. The receptors send impulses to the diaphragm and intercostal muscles, causing an increase in the rate and depth of breathing, which helps to remove the extra carbon dioxide as quickly as possible.

SCIENCE IN CONTEXT

HOW BREATHING RATE IS CONTROLLED

Rate of breathing is controlled by part of the brain that measures not the oxygen concentration of the blood but the carbon dioxide concentration. This is because a small increase in carbon dioxide concentration in body fluids could have a much more damaging effect on the body than a small decrease in oxygen concentration.

Challenge Question: Explain why an increase in carbon dioxide concentration can affect the activity of cell enzymes.

QUESTIONS

1. Describe the differences in composition between inspired air and expired air.

2. Describe the effects of exercise on the rate and depth of breathing.

3. **SUPPLEMENT** Explain the differences in composition between inspired air and expired air.

4. **SUPPLEMENT** Explain what could happen to cells if rate and depth of breathing did not change during exercise.

End of topic checklist

Key terms

alveoli (singular: alveolus), bronchi (singular: bronchus), bronchioles, diaphragm, expiration, gas exchange, gas exchange surface, inspiration, intercostal muscles, larynx, lungs, ribs, trachea, ventilation

During your study of this topic you should have learned:

○ How to describe the features of gas exchange surfaces in humans: large surface area, thin surface, good blood supply, and good ventilation with air.

○ How to identify in diagrams and images the parts of the breathing system: lungs, diaphragm, ribs, intercostal muscles, larynx, trachea, bronchi, bronchioles, alveoli and associated capillaries.

○ How to investigate the differences in composition between inspired and expired air, using limewater as a test for carbon dioxide.

○ How to describe the differences in composition between inspired and expired air, in terms of: oxygen, carbon dioxide, and water vapour.

○ **SUPPLEMENT** How to explain the differences in composition between inspired and expired air.

○ How to investigate and describe the effects of physical activity on the rate and depth of breathing.

○ **SUPPLEMENT** How to explain the link between physical activity and the rate and depth of breathing in terms of an increased carbon dioxide concentration in the blood, which is detected by the brain, leading to an increased rate and greater depth of breathing.

End of topic questions

1. Which shows the parts of the breathing system in the order that air passes through them when you breathe in?

A larynx trachea bronchi bronchioles alveoli

B larynx trachea bronchioles bronchi alveoli

C trachea larynx bronchi alveoli bronchioles

D trachea larynx bronchi bronchioles alveoli

2. a) Predict what would happen to breathing rate during exercise.

   b) **SUPPLEMENT** Explain why these changes would occur.

3. a) Define the terms *diffusion* and *gas exchange*.

   b) Describe the role of diffusion in gas exchange in humans.

   c) **SUPPLEMENT** Explain how the lungs are adapted to maximise the rate of gas exchange.

4. A sample of expired air is collected in a gas jar. The other gas jar contains normal atmospheric air.
A lighted candle is placed in each jar as shown in the diagram, and the time taken for the candle to go out is measured. Burning is a chemical reaction that requires oxygen so when the level of oxygen drops too low the candle in the jar will go out.

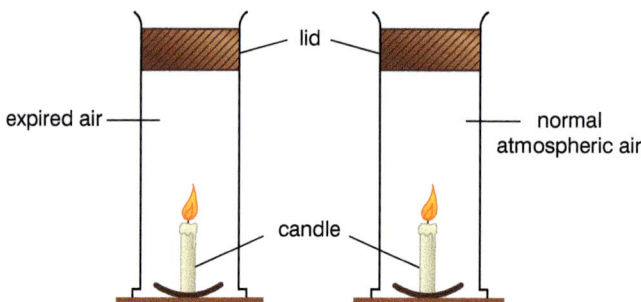

A student repeated this experiment three times. Their results are shown in the table below.

	Time taken for candle to go out (s)		
	Jar 1	Jar 2	Jar 3
Expired air	15	16	17
Atmospheric air	9	7	8

a) Calculate the mean times for the flame to go out in both expired air and atmospheric air.

b) Draw a suitable graph to compare the mean values for expired air and atmospheric air.

c) Describe and explain the difference between the results for the two sets of data.

5. A student investigated the effect of physical activity on the rate and depth of breathing for four volunteers. They measured each person's breathing rate and the volume of one breath at rest and immediately after jumping on the spot for two minutes. The data is shown in the table below:

	Breathing rate (breaths/minute)			
	1	2	3	4
at rest	12	16	13	17
after exercise	20	25	17	27
	Volume of one breath (dm³)			
	1	2	3	4
at rest	0.4	0.5	0.4	0.6
after exercise	1.2	1.0	1.3	1.4

a) Calculate the mean breathing rate and breath volume before and after exercise. Give your answers to one decimal place.

b) Using the mean values, calculate the percentage increase in breathing rate and breath volume before and after exercise.

Blood delivers oxygen from the lungs, and nutrients from the digestive system, to all cells so that they can respire and carry out all the processes needed for life. Respiration is a series of chemical reactions that take place in cells to release the energy that cells need to carry out their specific functions.

STARTING POINTS

1. What is respiration?

2. Why is respiration so important for all living things?

SYLLABUS SECTIONS COVERED

12.1 Respiration

12
Respiration

Δ During vigorous exercise, anaerobic respiration takes place in the muscle cells.

Δ Fig. 12.1 Sperm whales have been recorded around 3 km below the surface of the water and may remain submerged for about 90 minutes, although a normal dive is about 35 minutes long.

Respiration

INTRODUCTION

The current record for a human holding their breath under water is over 11 minutes, although even this can be extended by starting with breaths of pure oxygen. For most people, this length of time would be impossible, and the urge to breathe would become overwhelming after 2 or 3 minutes. Even people who are well trained for this risk damage to cells – particularly brain cells, which need a constant supply of oxygen to function properly. These records are small in comparison with those for diving whales, some of which may be underwater for over an hour.

KNOWLEDGE CHECK

✓ Organisms need energy for all the life processes that keep them alive.
✓ Plants get this energy from the sugar they make in photosynthesis.
✓ Animals get this energy from their food.

LEARNING OBJECTIVES

✓ State the uses of energy in living organisms, including: muscle contraction, protein synthesis, cell division, active transport, growth, the passage of nerve impulses, and the maintenance of a constant body temperature.
✓ Describe aerobic respiration as the chemical reactions in cells that use oxygen to break down nutrient molecules to release energy.
✓ State the word equation for aerobic respiration as:
glucose + oxygen → carbon dioxide + water.
✓ **SUPPLEMENT** State the balanced symbol equation for aerobic respiration as:
$C_6H_{12}O_6 + 6O_2 → 6CO_2 + 6H_2O$.
✓ **SUPPLEMENT** Describe anaerobic respiration as the chemical reactions in cells that break down nutrient molecules to release energy without using oxygen.
✓ **SUPPLEMENT** State that anaerobic respiration releases much less energy per glucose molecule than aerobic respiration.
✓ **SUPPLEMENT** State the word equation for anaerobic respiration in muscles during vigorous exercise as: glucose → lactic acid.
✓ **SUPPLEMENT** State that lactic acid builds up in muscles and blood during vigorous exercise. causing an oxygen debt.
✓ **SUPPLEMENT** Outline how the oxygen debt is removed after exercise, limited to: continuation of fast heart rate to transport lactic acid in the blood from the muscles to the liver; continuation of deeper and faster breathing to supply oxygen for aerobic respiration of lactic acid; aerobic respiration of lactic acid in the liver.

RESPIRATION

Respiration is the release of energy from the chemical bonds in food molecules such as glucose. This only takes place inside cells, and every living cell carries out respiration.

Every cell in a living organism requires energy, and this energy comes from respiration, which is the breakdown of chemical bonds in food molecules such as glucose to release energy in a form that can be used in cells.

In human cells, this energy is used:

- for the contraction of muscle cells
- to produce new chemical bonds during the **synthesis** (formation) of new molecules, for example in **protein synthesis**
- to produce new chemicals needed for **cell division** and for the growth of cells
- SUPPLEMENT for the active transport of molecules across cell membranes
- for the passage of nerve impulses along nerve cells
- for the maintenance of a constant core body temperature.

Note that we usually refer to glucose as the *nutrient molecule* or *food molecule* that is broken down in respiration. This is because it is the molecule most commonly used in this reaction in the body. If glucose is in short supply, then other molecules may be used instead from the breakdown of fats or proteins.

Respiration is a series of reactions and, like other reactions in cells, these are controlled by enzymes. So any change in a cell that affects enzymes (such as a change in temperature or pH) will affect the rate of respiration (see Section 5).

AEROBIC RESPIRATION

Most plant and animal cells use oxygen during respiration. Respiration that uses oxygen to release energy from nutrient molecules such as glucose is called **aerobic respiration**. Water and carbon dioxide are produced as waste products. This is very similar to burning fuel, except that in our bodies, enzymes control the process.

Aerobic respiration can be summarised by a word equation:

$$\text{glucose} + \text{oxygen} \xrightarrow{\text{energy released}} \text{carbon dioxide} + \text{water}$$

SUPPLEMENT

It can also be written as a balanced symbol equation:

$$C_6H_{12}O_6 + 6O_2 \xrightarrow{\text{energy released}} 6CO_2 + 6H_2O$$

The oxygen needed for respiration comes from the air (except for a small proportion in photosynthesising plants, which comes from photosynthesis). The carbon dioxide from respiration is released to the air, and the water is either used in the body or excreted through the kidneys.

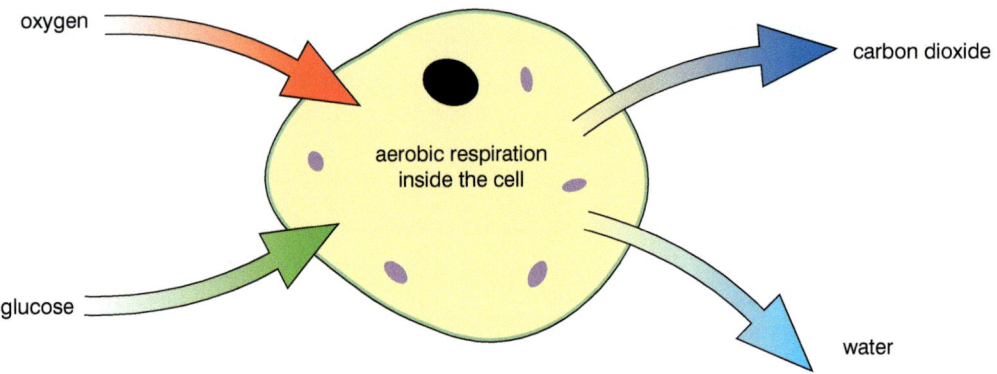

△ Fig. 12.2 Aerobic respiration in a cell.

During aerobic respiration, many of the chemical bonds in the glucose molecule are broken down. This releases a lot of energy: around 2900 kJ of energy are released for each mole of glucose molecules used in aerobic respiration.

SCIENCE IN CONTEXT

WATER FROM RESPIRATION

△ Fig. 12.3 Wild camels live in dry areas and so must go many days without drinking water.

A camel can survive for many days without drinking liquid water, which means it survives well in desert conditions. The camel's hump is not a store of water, but a store of fat. Over a long period without food, the fat is broken down to release substances for aerobic respiration. As water is one of the products of aerobic respiration, this also helps the camel to survive longer without drinking water.

A complete lack of food and drinking water are conditions that would kill a human in a few days, because we don't metabolise fat or retain water as well as the camel does. So, before setting out into the desert for a long trip, make sure your camel has a large hump (and you have plenty of food and water).

Challenge Question: Some animals, such as kangaroo rats, never need to drink water in their whole lives as long as they get enough dry seeds to eat. Explain why they do not need to drink.

QUESTIONS

1. **a)** Write the word equation for aerobic respiration.
   b) Add notes to your equation to show where the reactants come from.
   c) Add notes to your equation to show what happens to the products of the reaction in a human.
   d) Describe how your answer to part **c)** might differ for a camel on a long journey without water, and explain your answer.

2. State where respiration takes place in the body.

3. Give three examples of the use of energy from respiration in the human body.

4. **SUPPLEMENT** Write the balanced symbol equation for aerobic respiration.

SUPPLEMENT

ANAEROBIC RESPIRATION

Aerobic respiration supplies most of the energy that plant cells and animal cells need most of the time. However, there are times when not enough oxygen is available for aerobic respiration to be carried out fast enough to deliver all the energy that is needed – for example:

- in diving animals, such as whales and seals
- in muscle cells, when vigorous exercise requires more energy than can be provided by an increased supply of oxygen from deeper, faster breathing and a faster heart rate
- in organisms that live in conditions that are naturally low in oxygen.

In these cases, the additional energy needed is supplied by **anaerobic respiration**. This kind of respiration also releases energy from nutrient molecules such as glucose, but without the need for oxygen.

In anaerobic respiration, the glucose molecule is only partly broken down, so much less energy is released from each glucose molecule in anaerobic respiration compared with aerobic respiration. Only about 150 kJ is produced from every mole of glucose molecules respired anaerobically in a muscle cell.

Anaerobic respiration in muscle cells

When animal cells respire anaerobically, such as muscle cells during vigorous exercise, the glucose is broken down to **lactic acid**:

$$\text{glucose} \xrightarrow{\text{energy released}} \text{lactic acid}$$

Note that, even when a muscle cell is respiring anaerobically, aerobic respiration is also taking place and using all the oxygen that is available. Where aerobic respiration cannot supply all the energy needed, only the additional energy needed comes from anaerobic respiration.

SCIENCE IN CONTEXT

RESPIRATION IN ATHLETICS

If you watch carefully, you will not see a sprint athlete breathe during a race. At the start of the race there will be some oxygen in their muscle cells, but this is rapidly used up as they start running. Anaerobic respiration provides virtually all of the energy used in a 100 m sprint by a well-trained athlete.

Sprinting cannot be maintained for long, because the muscle cells also need a rapid

△ Fig. 12.4 A fit athlete in the middle of a sprint is using almost entirely anaerobic respiration.

supply of glucose for respiration, and the build-up of lactic acid causes muscle fatigue and pain. So longer distance races are managed using a combination of aerobic and anaerobic respiration. In marathons, most of the race is run aerobically, with only the last stretch being managed as a sprint using anaerobic respiration.

Challenge Question: Explain why a marathon runner may manage a sprint at the end of the race but would not try to sprint earlier in the race.

During anaerobic respiration, the concentration of lactic acid builds up in muscle cells and in the blood. When exercise has finished and sufficient oxygen is available again, a fast heart rate continues for a while to help transport lactic acid in the blood from the muscles to the liver. Deeper and faster breathing also continue to supply the additional oxygen for breaking down the lactic acid by aerobic respiration in both muscles and the liver.

The additional oxygen needed after exercise to get rid of the build up of lactic acid is called the **oxygen debt**.

QUESTIONS

1. **SUPPLEMENT** Explain why muscle cells sometimes need to respire anaerobically.

End of topic checklist

Key terms

aerobic respiration, cell division, protein synthesis, respiration, synthesis

SUPPLEMENT anaerobic respiration, lactic acid, oxygen debt

During your study of this topic you should have learned:

○ That the uses of energy in living organisms include: muscle contraction, protein synthesis, cell division, active transport, growth, the passage of nerve impulses, and the maintenance of a constant body temperature.

○ That aerobic respiration is the chemical reactions in cells that use oxygen to break down nutrient molecules to release energy.

○ That the word equation for aerobic respiration is:

glucose + oxygen → carbon dioxide + water

○ **SUPPLEMENT** That the balanced chemical equation for aerobic respiration is:

$C_6H_{12}O_6 + 6O_2 \rightarrow 6CO_2 + 6H_2O$

○ **SUPPLEMENT** That anaerobic respiration is the chemical reactions in cells that break down nutrient molecules to release energy without using oxygen.

○ **SUPPLEMENT** That the word equation for anaerobic respiration in muscles during vigorous exercise is: glucose → lactic acid

○ **SUPPLEMENT** That lactic acid builds up in muscles and blood during vigorous exercise, causing an oxygen debt.

○ **SUPPLEMENT** That oxygen debt is removed after exercise by:
- continuation of fast heart rate to transport lactic acid in the blood from the muscles to the liver
- continuation of deeper and faster breathing to supply oxygen for aerobic respiration of lactic acid
- aerobic respiration of lactic acid in the liver.

End of topic questions

1. What are the products of aerobic respiration in human muscle cells?

    A carbon dioxide and oxygen

    B glucose and carbon dioxide

    C glucose and water

    D carbon dioxide and water

2. **a)** State the body systems in a human that are involved in supplying the reactants of respiration.

    b) State the body systems in a human that are involved in removing the products of respiration.

3. **SUPPLEMENT** Draw up a table to summarise the similarities and differences between aerobic and anaerobic respiration.

4. A whale takes a deep breath of air and then dives for half an hour. Suggest how energy would be generated in the whale's muscles over the period of the dive. Explain your answer.

To stay alive and keep healthy, any organism needs to be able to respond appropriately to any changes, both inside its body, as well as in its surroundings. In mammals, detecting and responding to such changes involves both the nervous and hormone systems. Different systems, or parts of systems, working together is known as coordination. Coordination is important in keeping conditions inside the body within healthy limits – this is also known as homeostasis.

STARTING POINTS

1. Why do we need a nervous system?

2. What do hormones do?

3. What is homeostasis?

SYLLABUS SECTIONS COVERED

13.1 Coordination and response

13.2 Hormones

13.3 Homeostasis

13

Coordination and response

△ A micrograph showing a motor nerve.

Δ Fig. 13.1 Professional tennis players serve so fast that a radar gun is used to measure the speed of their serves.

Coordination and response

INTRODUCTION

Some professional tennis players can hit a tennis ball so hard that it travels at over 250 km per hour – only a little less than the top speed of a Ferrari. In order to return the ball successfully, their opponents have only a fraction of a second to work out where to stand and how best to return the ball. Their responses are built on years of training, so that they can respond without thinking.

KNOWLEDGE CHECK

✓ Animals detect changes in their surroundings using sense organs, and respond to those changes, for example by movement.
✓ Nerve cells are specialised cells.

LEARNING OBJECTIVES

✓ State that electrical impulses travel along neurones.
✓ Describe the mammalian nervous system in terms of: the central nervous system (CNS) consisting of the brain and the spinal cord; the peripheral nervous system (PNS) consisting of the nerves outside of the brain and spinal cord.
✓ Describe the role of the nervous system as coordination and regulation of body functions.
✓ Identify in diagrams and images sensory, relay, and motor neurones.
✓ Describe a simple reflex arc in terms of: receptor, sensory neurone, relay neurone, motor neurone, and effector.
✓ Describe a reflex action as a means of automatically and rapidly integrating and coordinating stimuli with the responses of effectors (muscles and glands).
✓ Describe sense organs as groups of receptor cells responding to specific stimuli: light, sound, touch, temperature, and chemicals.
✓ Describe a hormone as a chemical substance, produced by a gland and carried by the blood, which alters the activity of one or more specific target organs.
✓ Identify in diagrams and images specific endocrine glands and state the hormones they secrete, limited to: adrenal glands and adrenaline; pancreas and insulin; testes and testosterone; ovaries and oestrogen.
✓ **SUPPLEMENT** State that glucagon is secreted by the pancreas.
✓ Describe adrenaline as the hormone secreted in 'fight or flight' situations and its effects, limited to: increased breathing rate; increased heart rate; increased pupil diameter.
✓ **SUPPLEMENT** Describe homeostasis as the maintenance of a constant internal environment.
✓ **SUPPLEMENT** Explain the concept of homeostatic control by negative feedback with reference to a set point.

✓ **SUPPLEMENT** Describe the control of blood glucose concentration by the liver and the roles of insulin and glucagon.

✓ **SUPPLEMENT** Identify in diagrams and images of the skin: hairs, hair erector muscles, sweat glands, receptors, sensory neurones, blood vessels, and fatty tissue.

✓ **SUPPLEMENT** Describe the maintenance of a constant internal body temperature in mammals in terms of: insulation, sweating, shivering, the role of the brain; vasodilation and vasoconstriction of arterioles supplying skin surface capillaries.

Sensitivity

Sensitivity is the ability to recognise and respond to changes in external and internal conditions, and is one of the characteristics of living organisms.

A change in conditions is called a **stimulus**. For a coordinated response to occur to that stimulus there must be a **receptor** organ, which recognises the stimulus, and an **effector**, which is a mechanism to carry out the **response**.

There are two systems involved in **coordination** and response in humans.

- One is the *nervous system*, which includes the brain, the spinal cord, the peripheral **nerves**, and specialist sense organs such as the eye and the ear.
- The other is the *hormonal* (or endocrine) *system*, which uses chemical communication by means of **hormones**.

COORDINATION AND RESPONSE

The human **nervous system** consists of:

- the **central nervous system** (**CNS**; the brain and **spinal cord**), which processes nervous impulses from the body and coordinates any response
- nerves (large bundles of many neurones) of the **peripheral nervous system** (**PNS**) that connect the central nervous system to other parts of the body.

To carry out its functions, the nervous system is connected to:

- specialised **receptor organs** (**sense organs** such as the eye and ear) that contain receptor cells, and which sense stimuli (changes in conditions)
- specialised effectors (muscles or glands), which produce the response to the stimulus, such as the contraction of muscles and the secretion of hormones from glands.

The nervous system coordinates and regulates many functions in the body.

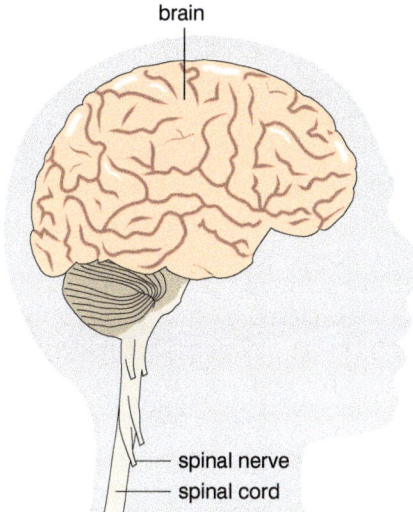

Δ Fig. 13.2 The human central nervous system (CNS).

Neurones

Nerves connect the sense organs to the central nervous system. Nerves, and the brain and spinal cord are made of specialised nerve cells called **neurones** (Fig. 13.4). These cells are specially adapted for their function because they have many endings that connect with other neurones, for passing signals called electrical **impulses**.

- Neurones that link sense organs to the central nervous system are called **sensory neurones**.
- Neurones within the central nervous system may be very short, and are called **relay neurones**.
- Neurones that connect the central nervous system to an effector, such as a muscle, are called **motor neurones**.

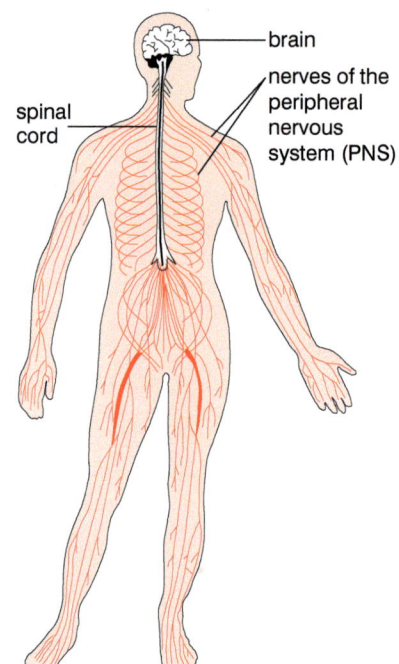

Δ Fig. 13.3 The human nervous system.

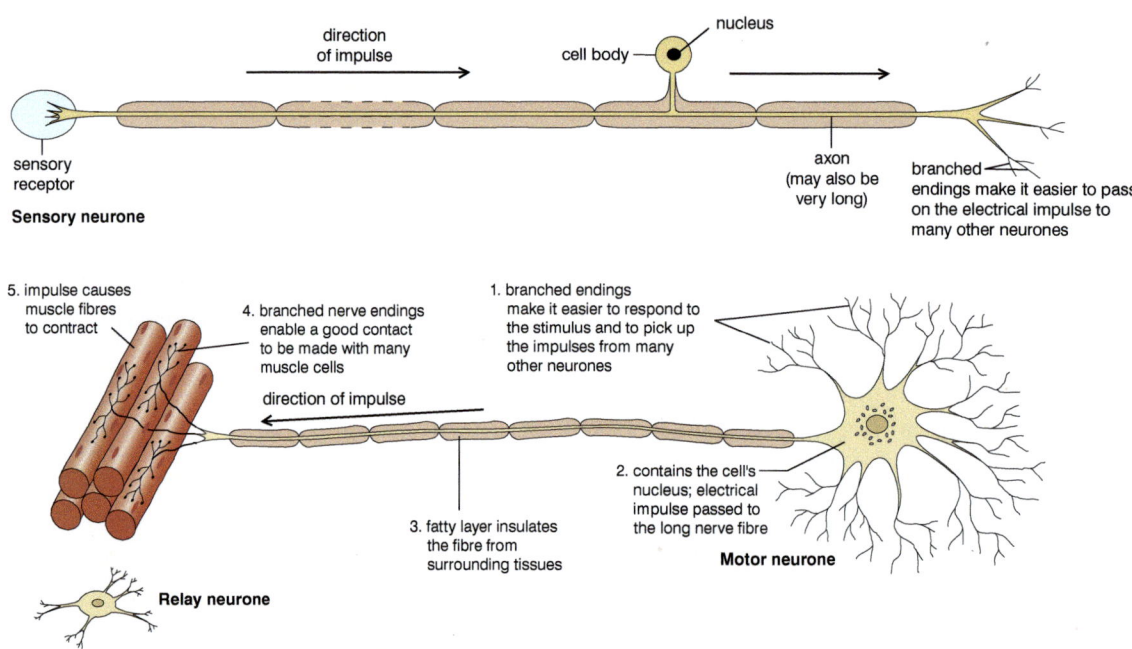

Δ Fig. 13.4 Different types of neurone.

QUESTIONS

1. Explain what is meant by the term *sensitivity*.

2. Use the following words in one or two sentences to explain how the body responds to change: *effector, receptor, response, stimulus*.

3. Describe two different kinds of effector organ in the human body.

4. Give three different types of neurone and describe the function of each in the nervous system.

Reflex response

The simplest type of response to a stimulus is a **reflex** action. Reflexes are rapid, automatic responses to a specific stimulus that often act to protect you in some way – for example, blinking if something gets in your eye or sneezing if you breathe in dust.

The pathway that impulses travel along during a reflex is called a **reflex arc**:

stimulus → receptor → sensory neurone → relay neurone in CNS → motor neurone → effector → response

Simple reflexes are usually spinal reflexes, which means that the impulses are processed by the spinal cord, not the brain (Fig. 13.5). The spinal cord sends an impulse back to the effector. Effectors are the parts of the body that respond, either muscles or glands. Examples of spinal reflexes include standing on a pin or touching a hot object.

stand on pin → nerve endings → sensory neurone → spinal cord → motor neurone → leg muscles → leg moves

When the spinal cord sends an impulse to an effector, other impulses are sent on to the brain so that it is aware of what is happening. It also allows the brain to over-ride the reflex response. For example, if you were holding a large bowl of hot food that you were looking forward to eating, you might look around quickly for somewhere to put it down rather than drop it immediately and risk breaking the bowl.

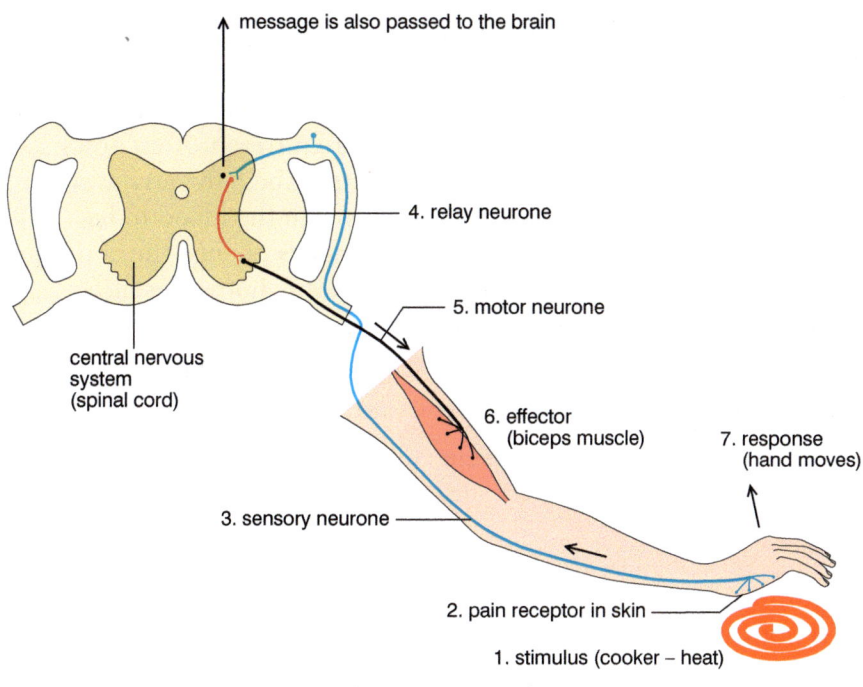

△ Fig. 13.5 A diagrammatic spinal reflex.

REMEMBER

You should now be able to understand the structure and functioning of any particular reflex arc, and be able to interpret diagrams and describe what happens at each step.

1. Explain what is meant by the term *reflex action.*

2. Explain why reflex actions are important for survival.

SENSE ORGANS

Different sense organs contain different specialised receptor cells that respond to different specific stimuli. Table 13.1 shows the different sense organs in humans.

Sense organ	Sense	Stimulus
Skin	touch	pressure, pain, hot/cold temperatures
Tongue	taste (chemicals)	chemicals in food and drink
Nose	smell (chemicals)	chemicals in the air
Eyes	sight	light
Ears	hearing	sound
	balance	movement/position of head

Δ Table 13.1 Human sense organs and their responses.

HORMONES

Hormones are chemical messengers used in the hormonal (or endocrine) system. They are produced in **endocrine glands**. Endocrine glands are different from other glands (such as salivary glands) in that they do not have ducts (tubes) to carry away the hormones they make: the hormones are secreted directly into the blood to be carried around the body dissolved in the blood plasma. Hormones change the activity of other specific parts of the body, called the **target organs**. Many hormones affect several target organs; some may only affect one target organ.

Different hormones produce different responses by the body.

- **Insulin**, produced by the **pancreas**, causes muscle cells and liver cells to take glucose out of the blood (such as after a meal), and so reduce blood glucose concentration.
- **SUPPLEMENT** Glucagon, produced by the pancreas, increases blood glucose concentration (see later in this section).
- **Oestrogen**, produced in the **ovaries**, helps to produce the female secondary sexual characteristics and control the menstrual cycle in women (see Section 15).
- **Testosterone**, produced in the **testes**, controls the development of the male secondary sexual characteristics (see Section 15).
- **Adrenaline**, produced by the **adrenal glands**, prepares the body in times of excitement, anger, fright, or stress (see below).

Adrenaline

Adrenaline is a hormone that is produced in the adrenal glands just above the kidneys. This hormone is secreted in the crucial moments when an animal must instantly decide whether to attack or run for its life – in what are known as 'fight or flight' situations.

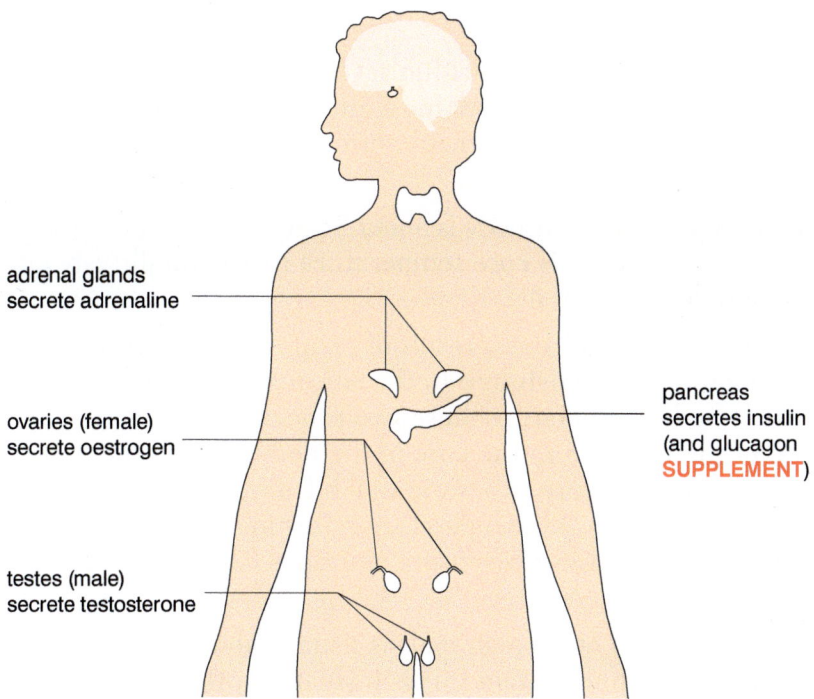

△ Fig. 13.6 The position of some endocrine glands in the human body and the hormones they secrete.

Some of the effects of adrenaline are:

- increased heart rate to circulate blood more rapidly around the body
- increased rate (and depth) of breathing to take more oxygen into the body and remove carbon dioxide more rapidly from the body
- dilated pupils (wider diameter) for better vision.

All these changes prepare the body for action.

QUESTIONS

1. Explain the meaning of the following terms:

   a) hormone

   b) endocrine gland

   c) target organ.

2. Describe in what conditions adrenaline might be released in the body.

3. Explain the advantages of adrenaline in preparing the body for action.

SUPPLEMENT

HOMEOSTASIS

For our cells to carry out all the life processes properly they need the conditions in and around them, such as the temperature and amount of water and other substances, to stay relatively constant. Keeping the internal environment constant is called **homeostasis**.

Temperature control

Homeostasis is the control of internal conditions within set limits. Staying within these limits helps the processes in the body to work most effectively. For example, as you saw in Section 5, enzymes and the processes they control work best at an optimum temperature.

The temperature in the core (middle) of your body is about 37 °C, regardless of how hot or cold you may feel on the outside. This **core temperature** may naturally vary a little, but it never varies a lot unless you are ill.

Energy is constantly being released in cells as a result of respiration and other chemical reactions in the body, and is transferred to the surroundings outside the body. To maintain a constant body temperature these two processes have to balance. The temperature of the blood from the core of the body is monitored as it flows through the brain. If the temperature varies too much from 37 °C, the brain causes changes to happen that result in the temperature returning to about 37 °C. The brain also receives electrical impulses via sensory neurones from temperature receptor cells in the skin surface.

The body temperature of humans and other mammals is usually above that of their surroundings. One factor that helps to maintain this is insulation provided by fatty tissue under the skin. Some mammals may build up extra fat reserves to help them survive cold winters, but all mammals have other quicker ways of maintaining their core temperature if it starts to rise or fall.

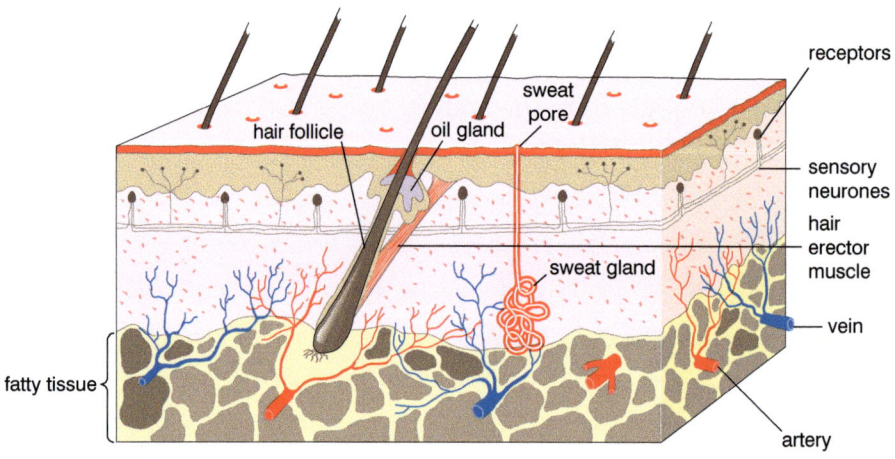

Δ Fig. 13.7 The structure of skin.

If core temperature rises too far, the following events occur.

- Sweat is released on to the surface of the skin from sweat glands. Sweat is mostly water, and this water evaporates. Evaporation needs energy, so energy is removed from the skin surface as the sweat evaporates, cooling the skin.
- **Arterioles** (small arteries) carrying blood near the surface of the skin dilate (get wider) so more blood flows through the capillaries at the skin surface. This is known as **vasodilation**, and is what can make some people's skin look redder when they are hot. This makes it easier for thermal energy to be transferred to the skin surface and from there to the environment by radiation and conduction.

If core temperature falls too far, the following events occur.

- Body hair may be raised by hair erector muscles in the skin. This has little effect in humans (often called 'goose bumps') but is more effective in mammals with fur and in birds, because the fur or feathers trap air next to the skin. Air is not a good thermal conductor, so this still layer of air acts as insulation.
- You may start to shiver. This means that your muscles produce rapid, small contractions. Respiration is used to produce these contractions, releasing energy at the same time, which warms the blood flowing through the muscles.
- Arterioles carrying blood near the surface of the skin constrict (get narrower), which reduces the amount of blood flowing through the surface capillaries. This is known as **vasoconstriction**. As the warm blood is kept deeper in the skin, this reduces the rate of thermal energy transfer by conduction to the skin surface and from there to the environment.

A cold day

A hot day

Δ Fig. 13.8 The skin responds to temperature changes.

Developing practical skills

We can use a test tube of warm water wrapped in wet paper towel as a model to investigate whether sweating really does cool the body, measuring how the temperature of the water changes over time.

Devise and plan investigations

1. Explain how the tube models sweating in a human.

2. Suggest how you would set up the control for this investigation. Explain your answer.

The table shows the results of an investigation like the one described.

	Time in min	0	2	4	6	8	10	12	14	16
Temperature of water in tube/°C	wet towel	56	50	46	42	39	36	34	32	31
	dry towel	56	52	49	46	44	44	41	40	39

3. Use the results to draw a suitable graph.

Analyse and interpret data

4. Make a conclusion from the graph, and explain the conclusion using your scientific knowledge.

Negative feedback

The control of core body temperature is an example of **negative feedback**. This is where a change in a condition from the ideal value or **set point** causes a response that produces the opposite change. It depends on a monitoring control centre that detects changes and initiates responses, such as the brain. We can summarise a negative feedback response in a diagram like that shown in Fig. 13.9.

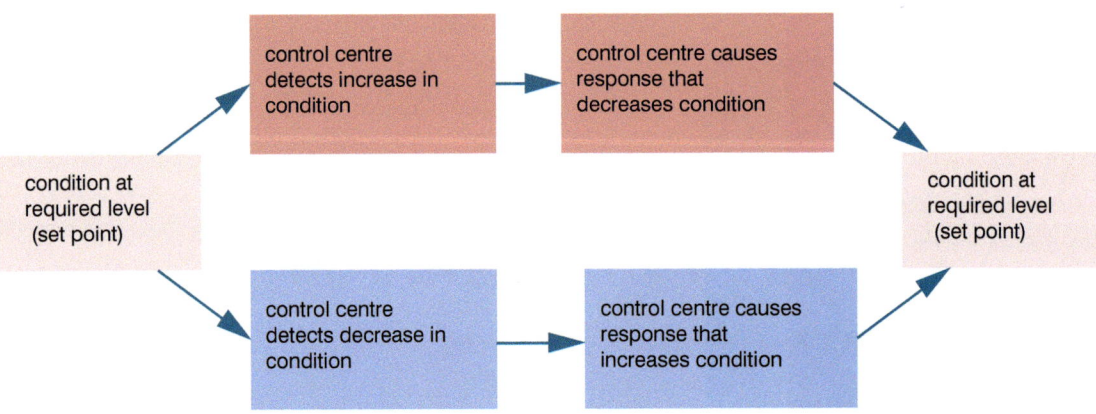

Δ Fig. 13.9 Negative feedback response.

The effect of a negative feedback response is to prevent large increases or decreases from the best conditions for the body. This is how the temperature and pH of the tissues are controlled so that enzymes work most effectively.

Control of blood glucose concentration

It is important that the blood glucose concentration remains within a small range. If it rises or falls too much, you can become very ill very quickly.

After a meal, the blood glucose concentration rises rapidly as glucose from digested food is absorbed from the small intestine. Cells in the pancreas detect this increase and respond by releasing the hormone insulin, which travels in the blood to the liver. Here it causes any excess glucose to be converted to another carbohydrate, glycogen, which is insoluble and is stored in the liver.

Between meals, glucose in the blood is constantly diffusing into cells for use in respiration. So the blood glucose concentration falls. When a low level of glucose is detected by the pancreas, the insulin-secreting cells stop secreting insulin and other cells start to secrete the hormone glucagon instead. Glucagon converts some of the stored glycogen back into glucose, which is released into the blood to raise the blood glucose concentration again.

QUESTIONS

1. Define the term *homeostasis*.
2. Give one example of homeostasis in the human body.
3. **SUPPLEMENT** Explain the role of skin blood vessels in maintaining core body temperature.
4. **SUPPLEMENT** Define the term *negative feedback*.
5. **SUPPLEMENT** Use a negative feedback diagram like the one in Fig. 13.9 to show how blood glucose concentration is kept within a safe range.

End of topic checklist

Key terms

adrenal gland, adrenaline, central nervous system (CNS), coordination, core temperature, effector, endocrine gland, hormone, impulse, insulin, motor neurone, nerve, nervous system, neurone, oestrogen, ovaries, pancreas, peripheral nervous system (PNS), receptor, reflex, receptor organ, reflex arc, relay neurone, response, sense organ, sensory neurone, spinal cord, stimulus, target organ, testes, testosterone

SUPPLEMENT arteriole, glucagon, homeostasis, negative feedback, set point, vasoconstriction, vasodilation

During your study of this topic you should have learned:

○ That electrical impulses travel along neurones.

○ How to describe the mammalian nervous system in terms of:
   ● the central nervous system (CNS) consisting of the brain and the spinal cord
   ● the peripheral nervous system (PNS) consisting of the nerves outside of the brain and spinal cord.

○ That the role of the nervous system is the coordination and regulation of body functions.

○ How to identify in diagrams and images sensory, relay, and motor neurones.

○ How to describe a simple reflex arc in terms of: receptor, sensory neurone, relay neurone, motor neurone, and effector.

○ That a reflex action is a means of automatically and rapidly integrating and coordinating stimuli with the responses of effectors (muscles and glands).

○ That sense organs are groups of receptor cells responding to specific stimuli: light, sound, touch, temperature, and chemicals.

○ That a hormone is a chemical substance, produced by a gland and carried by the blood, which alters the activity of one or more specific target organs.

○ How to identify in diagrams and images specific endocrine glands and state the hormones they secrete: adrenal glands (adrenaline); pancreas (insulin); testes (testosterone); ovaries (oestrogen).

○ **SUPPLEMENT** That glucagon is secreted by the pancreas.

○ That adrenaline is the hormone secreted in 'fight or flight' situations and its effects include: increased breathing rate; increased heart rate; increased pupil diameter.

○ **SUPPLEMENT** How to describe the role of adrenaline in the control of metabolic activity: increasing the blood glucose concentration; increasing heart rate.

○ That homeostasis is the maintenance of a constant internal environment.

○ **SUPPLEMENT** How to explain the concept of homeostatic control by negative feedback with reference to a set point.

○ **SUPPLEMENT** How to describe the control of blood glucose concentration by the liver and the roles of insulin and glucagon.

○ **SUPPLEMENT** How to identify in diagrams and images of the skin: hairs, hair erector muscles, sweat glands, receptors, sensory neurones, blood vessels, and fatty tissue.

○ **SUPPLEMENT** How to describe the maintenance of a constant internal body temperature in mammals in terms of: insulation, sweating, shivering, and the role of the brain.

○ **SUPPLEMENT** How to describe the maintenance of a constant internal body temperature in mammals in terms of vasodilation and vasoconstriction of arterioles supplying skin surface capillaries.

End of topic questions

1. Where is the hormone insulin made?

    A adrenal glands

    B brain

    C liver

    D pancreas

2. Use the terms *stimulus*, *neurone*, *reflex arc*, *effector* and *response*, to describe the response of someone who touches something very hot.

3. a) Describe the sequence of sensing and response in the nervous system of a tennis player who returns a serve successfully.

    b) State if this is a reflex action. Explain your answer.

4. SUPPLEMENT One example of homeostasis in humans is the control of core body temperature.

    a) Identify the receptors, monitoring area, and effectors in the response to a change in external temperature.

    b) Describe the changes in the effectors if core body temperature is too high.

    c) Describe the changes in the effectors if core body temperature is too low.

    d) Explain why the changes that occur in the body when it is too hot help to return the core body temperature to normal.

5. SUPPLEMENT Using the example of blood glucose concentration, explain why negative feedback control is an essential feature of homeostasis.

Many of the medicinal drugs that we can take to help treat, prevent, or cure illnesses are chemical compounds that scientists can make in the laboratory. A number of these were originally discovered in plants, animals, and microorganisms, and have been found to help treat human diseases. The anticancer drug Taxol, for example, was originally discovered back in the 1960s in the bark of a particular species of yew tree. Some very poisonous chemicals have also led to the development of life-saving drugs. An example is a drug known as Lisinopril, which is used to treat high blood pressure and heart failure. This drug was originally developed from the venom of a poisonous snake in Brazil.

STARTING POINTS

1. What do we mean by a drug?

2. What are antibiotics?

3. Can we prevent bacteria becoming resistant to antibiotics?

SYLLABUS SECTIONS COVERED

14.1 Drugs

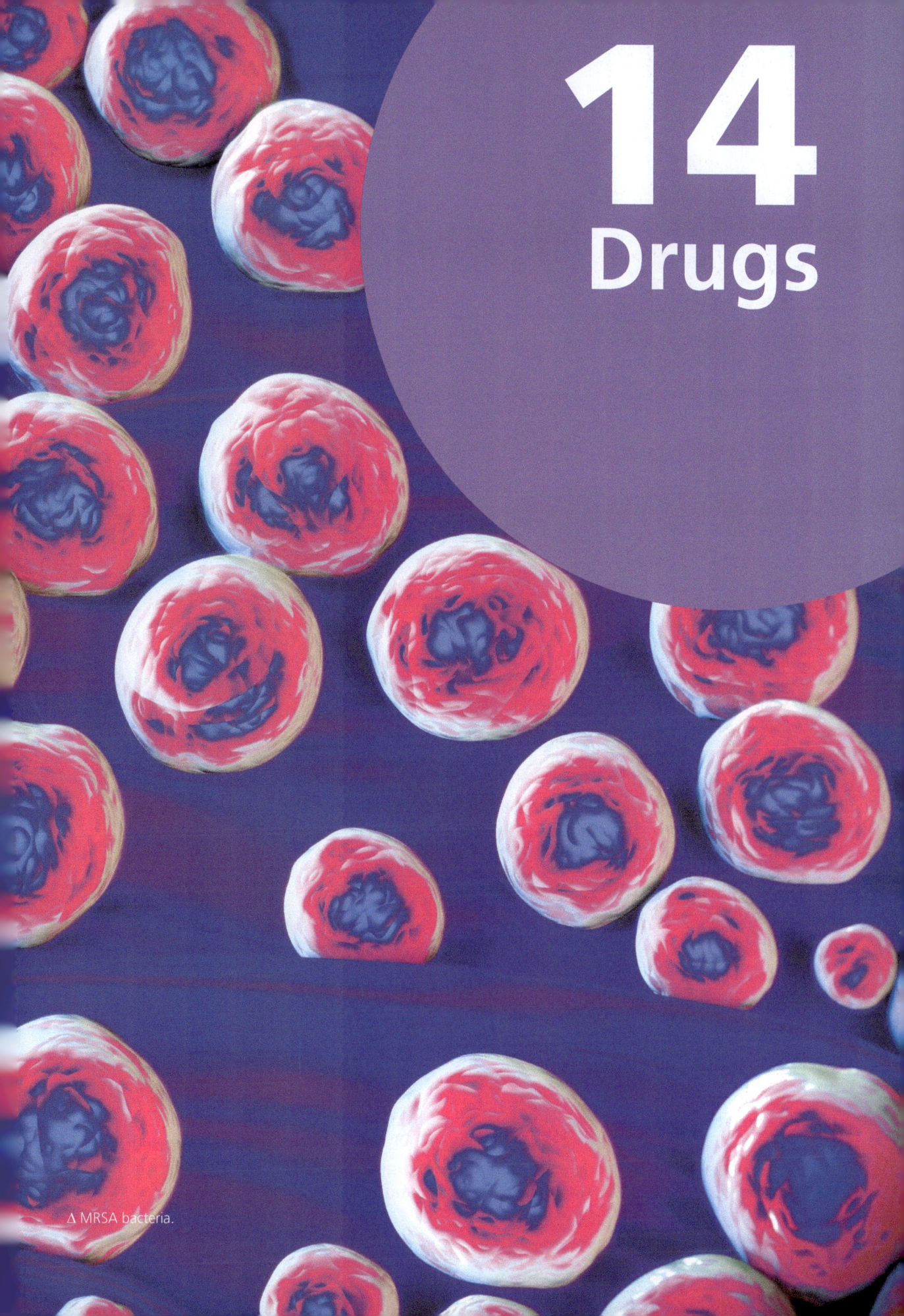

14
Drugs

△ MRSA bacteria.

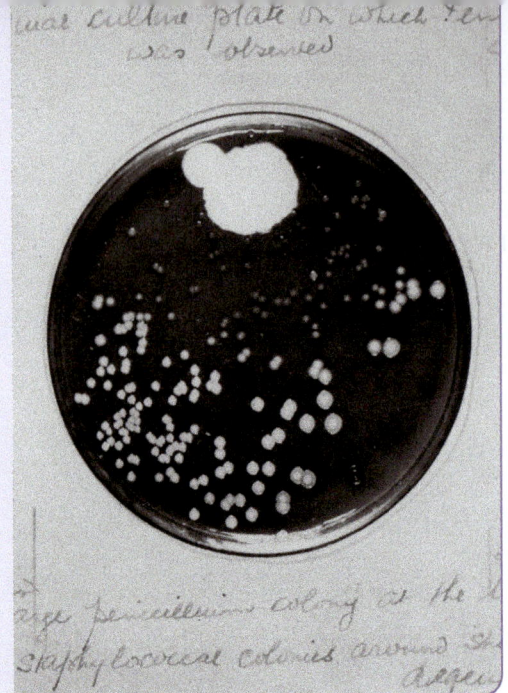

△ Fig. 14.1 Alexander Fleming's record of his discovery of penicillin. The large blob at the top of the plate is the mould, and the smaller blobs on the rest of the plate are bacterial colonies. This shows how the bacteria nearest to the mould have been killed.

Drugs

INTRODUCTION

The first antibiotic was developed from an accidental discovery. In 1928, Alexander Fleming was studying the properties of some kinds of bacteria. After returning from a holiday he noticed that one of the bacterial plates he had made earlier was contaminated with a common mould fungus. What was strange was that bacteria near the mould had been killed, whereas those further away were still growing. Fleming grew the mould separately and extracted a chemical that killed many kinds of bacteria. This chemical was then developed as the first antibiotic, penicillin.

KNOWLEDGE CHECK

✓ Bacteria are single-celled organisms with a simple cell structure.
✓ Viruses infect the cells of organisms and reproduce within those cells.
✓ Some microorganisms are pathogenic and can cause disease.
✓ Some drugs can be used to treat disease.

LEARNING OBJECTIVES

✓ Describe a drug as any substance taken into the body that modifies or affects chemical reactions in the body.
✓ Describe the use of antibiotics for the treatment of bacterial infections.
✓ State that some bacteria are resistant to antibiotics, which reduces the effectiveness of antibiotics.
✓ SUPPLEMENT Explain how using antibiotics only when essential can limit the development of resistant bacteria such as MRSA.
✓ State that antibiotics kill bacteria but do not affect viruses.

DRUGS

A **drug** is any substance that, when taken into the body, influences the way the body works by modifying or affecting chemical reactions in the body.

Medicinal drugs

Some drugs are medicinal drugs that are used to treat the symptoms or causes of disease. Many medicinal drugs can only be prescribed by a doctor because they are potentially harmful, or because their use should be restricted. For example, **antibiotics** are drugs that kill bacteria in the body or stop them from reproducing. Antibiotics can treat harmful bacterial diseases such as tetanus, pneumonia, and some kinds of meningitis.

Antibiotics

Antibiotics are drugs that are used to control many kinds of bacterial infection. Different antibiotics are used to kill different kinds of bacteria. Examples of antibiotics include penicillin, methicillin, and erythromycin.

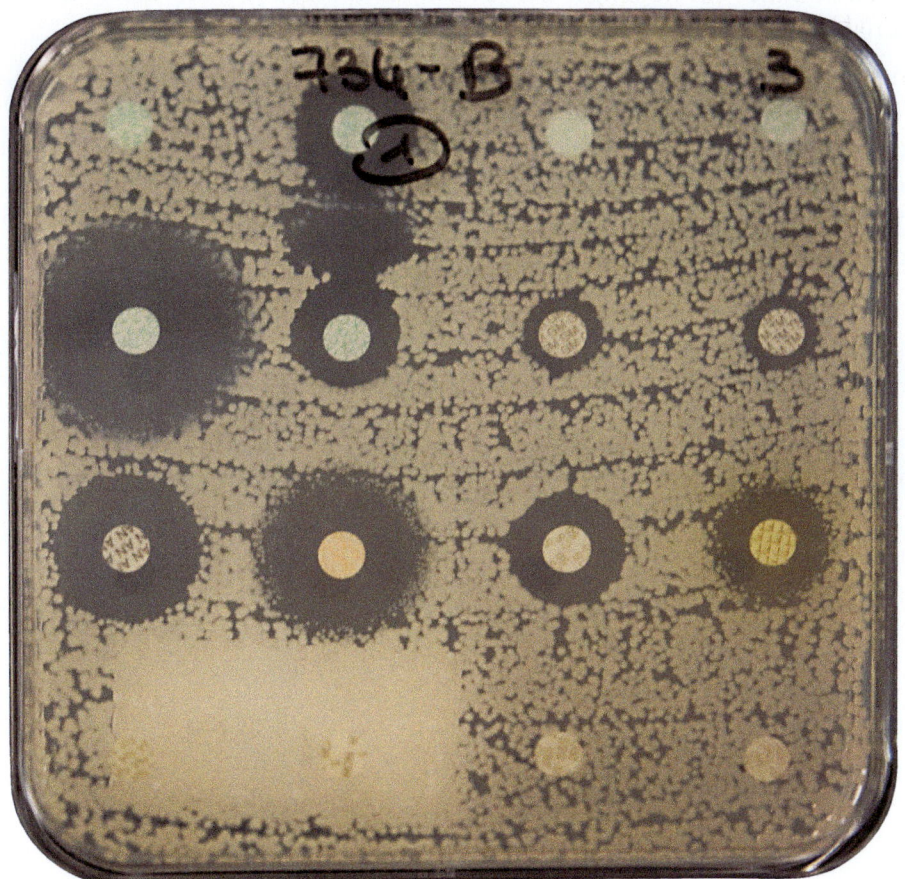

Δ Fig. 14.2 There are many different kinds of antibiotic, which are effective against different types of bacteria.

Antibiotics *do not* affect viruses and so cannot be used to treat viral infections such as a cold or flu. The way antibiotics work against bacteria is not effective against viruses.

Antibiotic resistance

Some bacteria are not affected by antibiotics. We describe them as having **antibiotic resistance**. As the use of antibiotics has increased, so there has also been an increase in the development of antibiotic-resistant bacteria.

SUPPLEMENT

This means that antibiotics are no longer as effective as they once were. **MRSA** is one kind of antibiotic resistant bacterium. The initials 'MRSA' mean 'methicillin-resistant *Staphylococcus aureus*', which refers to a strain of a common bacterium found in and on the human body that has become resistant not just to methicillin but also to other kinds of antibiotic. Normally *Staphylococcus aureus* grows on human skin and in the nose and throat without causing problems. In people who are not healthy for other reasons, it may cause skin infections, nose and throat infections, even pneumonia (infection of the lungs) and death.

The development of antibiotic resistance in bacteria such as MRSA is an example of evolution through natural selection (see Section 17).

As bacteria develop resistance to more and more antibiotics, doctors have fewer ways to treat a person with a bacterial infection. Many people are becoming worried that we may get to a point when antibiotics are no longer useful. This could lead to many more deaths from simple bacterial infections, as there were before antibiotics were developed.

It is important that we try to minimise the development of antibiotic-resistant bacteria, such as MRSA, so that the drugs are effective when they are most needed. One way of limiting the development of resistant bacteria is by only using antibiotics when essential. If bacteria are not exposed to an antibiotic, there will be no development of resistance. So, only using the most suitable antibiotic to treat an infection, and only using antibiotics when really needed, will help to minimise the rate of development of resistance.

QUESTIONS

1. Define the following terms:

    a) *drug*

    b) *antibiotic*

    c) *antibiotic resistance.*

2. Explain why some infections can be treated with antibiotics and others cannot.

3. **SUPPLEMENT** Describe one way of minimising the development of antibiotic resistance and explain why it works.

End of topic checklist

Key terms

antibiotic, antibiotic resistance, drug

SUPPLEMENT MRSA

During your study of this topic you should have learned:

- ◯ That a drug is any substance taken into the body that modifies or affects chemical reactions in the body.
- ◯ That antibiotics are used for the treatment of bacterial infections.
- ◯ That some bacteria are resistant to antibiotics, which reduces the effectiveness of antibiotics.
- ◯ That antibiotics kill bacteria but do not affect viruses.
- ◯ **SUPPLEMENT** That using antibiotics only when essential can limit the development of resistant bacteria such as MRSA.

End of topic questions

1. Which is the best definition of a drug?

   A a substance that affects chemical reactions in the body

   B a substance that affects your behaviour

   C a substance that cures disease

   D a substance that is addictive

2. The following shows advice from a medical website about the treatment of some conditions.

Influenza Cause: influenza virus Treatment: bed rest for 24–48 hours, drink plenty of fluids; contact a doctor after this time if symptoms have not improved.	Strep throat Cause: *Streptococcus* bacterium Treatment: may require antibiotics, so see a doctor in case a prescription is needed.

Explain the difference in advice for treatment of these two diseases.

3. a) Explain what is meant if some bacteria are *antibiotic resistant*.

   b) SUPPLEMENT Explain why doctors do not prescribe antibiotics as frequently as they did in the past.

Scientists believe that there has been life on Earth for over 3500 million years. Nobody knows yet what triggered non-living molecules to become organised into living things that can reproduce themselves, but scientists have found traces of bacteria-like structures that may represent the earliest forms of life, in very ancient rocks.

Reproduction allows information from one generation to pass on to the next and despite the varied ways it can happen, it occurs in every living thing, and is a key characteristic of life.

STARTING POINTS

1. What is meant by reproduction?

2. What are the differences between asexual reproduction and sexual reproduction?

3. What happens when plants reproduce sexually?

4. What are sexually transmitted infections?

SYLLABUS SECTIONS COVERED

15.1 Asexual reproduction

15.2 Sexual reproduction

15.3 Sexual reproduction in plants

15.4 Sexual reproduction in humans

15.5 Sexually transmitted infections

15
Reproduction

△ Each poppy plant can produce up to 60 000 seeds, as a result of sexual reproduction.

Reproduction

INTRODUCTION

Most multicellular organisms are able to reproduce sexually. This requires the joining of gametes from the male and the female in fertilisation. Many plants and animals can also reproduce *a*sexually. This is where gametes and fertilisation are not needed to produce new individuals. Until recently, it was thought that asexual reproduction in animals was something that only happened as an addition to sexual reproduction. However, we now know that some species of rotifers (microscopic aquatic animals) have not reproduced sexually for at least 40 million years. Males of these species simply do not exist.

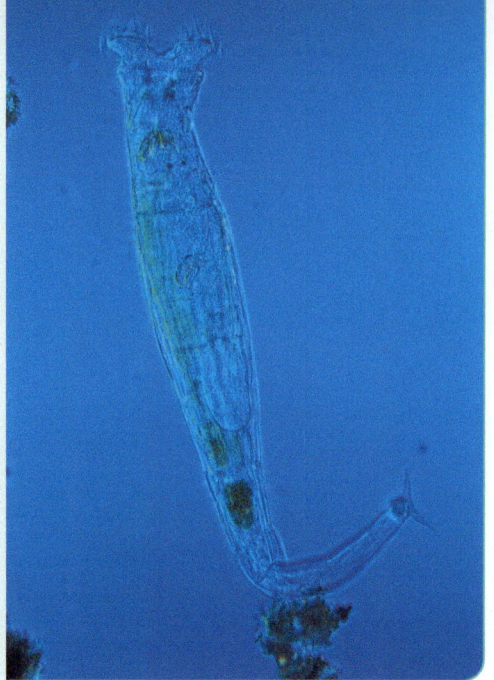

Δ Fig. 15.1 Scientists think that this species of rotifer has not reproduced sexually for over 40 million years.

KNOWLEDGE CHECK

✓ Flowering plants have flowers, which contain organs used for reproduction.
✓ Reproduction is the life process that makes more of the same kind of organism.
✓ In human reproduction, egg and sperm cells are involved.
✓ The male and female human reproductive organs are different.
✓ In humans, reproduction is the production of new individuals as a result of fertilisation.

LEARNING OBJECTIVES

✓ Describe asexual reproduction as a process resulting in the production of genetically identical offspring from one parent.
✓ Identify examples of asexual reproduction in diagrams, images, and information provided.
✓ **SUPPLEMENT** Discuss the advantages and disadvantages of asexual reproduction to a population of a species in the wild.
✓ Describe sexual reproduction as a process involving the fusion of the nuclei of two gametes to form a zygote and the production of offspring that are genetically different from each other.
✓ **SUPPLEMENT** State that nuclei of gametes are haploid and that the nucleus of a zygote is diploid.
✓ **SUPPLEMENT** Discuss the advantages and disadvantages of sexual reproduction to a population of a species in the wild.
✓ Identify in diagrams and images, and draw the following parts of an insect-pollinated flower: sepals, petals, stamens, filaments, anthers, carpels, style, stigma, ovary and ovules.
✓ State the functions of the structures listed above.
✓ **SUPPLEMENT** Identify in diagrams and images, and describe the anthers and stigmas of a wind-pollinated flower.
✓ Describe pollination as the transfer of pollen grains from an anther to a stigma.
✓ State that fertilisation occurs when a pollen nucleus fuses with a nucleus in an ovule.

- ✓ Describe the structural adaptations of insect-pollinated and wind-pollinated flowers.
- ✓ Investigate and describe the environmental conditions that affect germination of seeds, limited to the requirement for: water, oxygen, and a suitable temperature.
- ✓ Identify on diagrams and state the functions of the following parts of the male reproductive system: testes, scrotum, sperm ducts, prostate gland, urethra, and penis.
- ✓ Identify on diagrams and state the functions of the following parts of the female reproductive system: ovaries, oviducts, uterus, cervix, and vagina.
- ✓ Describe fertilisation as the fusion of the nuclei from a male gamete (sperm) and a female gamete (egg cell).
- ✓ **SUPPLEMENT** Explain the adaptive features of sperm, limited to: flagellum, mitochondria, and the presence of enzymes in the acrosome.
- ✓ **SUPPLEMENT** Explain the adaptive features of egg cells, limited to: energy stores and the jelly coat that changes at fertilisation.
- ✓ **SUPPLEMENT** Compare male and female gametes in terms of: size, structure, motility, and numbers.
- ✓ Describe the roles of testosterone and oestrogen in the development and regulation of secondary sexual characteristics during puberty.
- ✓ Describe the menstrual cycle in terms of changes in the ovaries and in the lining of the uterus (knowledge of sex hormones is **not** required).
- ✓ Describe a sexually transmitted infection (STI) as an infection that is transmitted through sexual contact.
- ✓ State that human immunodeficiency virus (HIV) is a pathogen that causes an STI.
- ✓ State that HIV infection may lead to AIDS.
- ✓ Describe the methods of transmission of HIV.
- ✓ Explain how the spread of STIs is controlled.

ASEXUAL REPRODUCTION

Some organisms increase in number by **asexual reproduction**. For this type of reproduction it is not necessary to have two parents. During asexual reproduction, cells from an adult organism divide to produce the offspring. This means that offspring produced by asexual reproduction are genetically identical to their parent and to each other.

Asexual reproduction is used by many different organisms. Bacteria reproduce asexually

Δ Fig. 15.2 The toadstools we see growing are specialised spore-producing bodies of fungi.

using binary fission. When they are large enough, their genetic material copies itself exactly and then the cell splits in half. The process then begins all over again. This can occur very rapidly to produce large numbers of identical bacteria.

Almost all fungi can reproduce asexually. Different types of fungi use different means of asexual reproduction but by far the most important type is that of spore formation (Fig. 15.2). This can be seen in *Mucor*, the common pin mould, which often grows on bread. When this fungus has a plentiful supply of nutrients, a hypha grows up vertically and the tip swells with cytoplasm containing many nuclei. This tip releases many spores into the atmosphere. If they find the right conditions for growth, each spore can develop into a new mycelium.

Another form of asexual reproduction is seen in plants such as potatoes that produce tubers (Fig. 15.3). Tubers form from the end of stems that grow underneath the soil surface. The stems swell into storage organs filled with starch. When the leaves and stems of the plant die back at the end of the growing season, the tubers stay dormant until the next season. Each tuber then produces several potato plants from the buds on the side of the tuber. Each potato plant gives rise to several tubers and each tuber produces a number of plants, so several new plants are formed from one parent.

Δ Fig. 15.3 Each of the potato tubers formed by this plant could produce new plants in the next growing season.

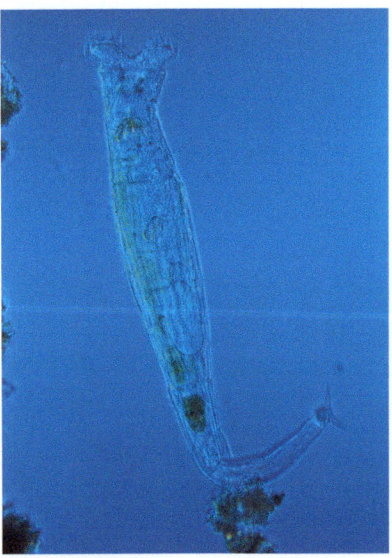

Δ Fig. 15.4 Rotifers are microscopic animals that live in water all over the world. Some types of rotifer are not believed to have reproduced sexually for at least 40 million years.

Δ Fig. 15.5 Some plant cuttings will even grow roots when placed in water. When the roots are large enough, the new plant can be planted in compost.

◁ Fig. 15.6 Aphids are insects that reproduce asexually during the summer.

△ Fig. 15.7 New strawberry plants can be produced asexually from runners.

Advantages and disadvantages

Asexual reproduction has advantages and disadvantages.

Advantages:

- Only one parent is required; there is no need for a parent animal to find a mate or for pollination in plants.
- Often large numbers of organisms can be produced in a relatively short time.
- All the offspring produced are identical so should survive well in the conditions in which the parent lives.

Disadvantages:

- The lack of variation in the offspring means that any adverse change in conditions will affect all equally.
- Because the offspring do not vary they are not suited to moving away and exploiting environments with different conditions.

There is more about how genetic variation between individuals provides an advantage in Section 17.

There are some crop plant species that we use for food that do not reproduce sexually very easily. For example, new banana plants are produced from offshoots of older plants. When the offshoot has well-developed roots, it is removed from the parent plant and planted separately. Like the production of tubers by potato plants, this is an example of asexual reproduction.

parent plant

offshoots produced by asexual reproduction of parent plant

Δ Fig. 15.8 Banana plants reproduce asexually by growing offshoots from the base of the plant.

REMEMBER

Offspring produced by asexual reproduction are genetically identical to their parent and to each other (if they share the same parent). This does not mean, however, that there will be no differences between them. Different environmental conditions may cause quite noticeable physical differences.

QUESTIONS

1. Define the term *asexual reproduction*.
2. Explain why binary fission of bacteria is an example of asexual reproduction.
3. **SUPPLEMENT** Explain the advantages and disadvantages to the banana species of reproducing asexually.

SEXUAL REPRODUCTION

Sexual reproduction is the most common method of reproduction for the majority of larger organisms, including almost all animals and plants. It occurs when there is **fertilisation**, which is when the nucleus of a male **gamete** (sex cell) fuses (joins) with the nucleus of a female gamete to form a **zygote**. The zygote will contain some of the genetic information of each of its parents. So it will be genetically different from each of the parents. It will also be genetically different from all other offspring produced by those parents (unless it has an identical twin).

There is more about how genetic variation between individuals provides an advantage in Section 17.

Organisms in a species share many features. We define a species as a group of organisms that can reproduce to produce fertile offspring. 'Fertile' means that their offspring are also able to reproduce when they are adult. Some species, like the horse and donkey, share many features and can sexually reproduce, but they produce offspring called mules that are not fertile. Therefore, horses and donkeys are classified as two different species.

Sexual reproduction involves **haploid** cells, which have just one set of genetic information, and **diploid** cells that have two sets of genetic information (see Section 16). The gametes have haploid nuclei, and when they fuse they produce a zygote with a diploid nucleus. The cell of the zygote will divide repeatedly to produce all the cells of the new organism, and all these cells will also be diploid.

Advantages and disadvantages

There are advantages and disadvantages to sexual reproduction.

Advantages:

- Fusion of gametes brings genetic information from two parents, which results in variety in the offspring. This produces individuals that may be better adapted to different conditions than the parents and each other, which makes the chance of survival of the species in changing conditions more likely.

Disadvantages:

- Sexual reproduction usually requires a second parent for fertilisation. Finding a mate can take time and energy for the individual, and failure to mate means that the individual produces no offspring.
- The need to find a mate also means that sexual reproduction takes longer to produce offspring than asexual reproduction.

QUESTIONS

1. Define the following terms:

   a) *fertilisation*

   b) *sexual reproduction.*

2. **SUPPLEMENT** Give an example of:

   a) a haploid cell

   b) a diploid cell.

3. **SUPPLEMENT** Describe an advantage and a disadvantage of sexual reproduction to a population of species in the wild.

SEXUAL REPRODUCTION IN PLANTS

The most successful group of plants is the flowering plants. These are the only plants to have true flowers and produce **seeds** with a tough protective coat. During sexual reproduction, flowering plants:

- produce male and female gametes – some species may produce male and female gametes in the same flowers; other species may have male-only flowers and female-only flowers on the same plant; and in other species male flowers and female flowers are produced on different plants
- male pollen is transferred to the female part of the flower so that **pollination** can take place

- the male gamete and female gamete fuse during fertilisation to form a zygote
- the zygote develops to form an **embryo** within a seed, which protects the embryo and provides food during germination of the seed
- seeds are dispersed, so that they germinate and grow away from the parent.

Structure of flowers

All flowers have a similar basic arrangement. They have structures stacked one on top of each other along a short stem, arranged either in a spiral or in separate rings.

The centre of the flower contains the female parts, which consist of the carpels, containing the ovules, inside which are the female gametes.

The stamens, which contain anthers, are the male parts of the flower. The anthers contain cells that produce the male gametes (inside the pollen grains) as the flowers mature.

The **sepals** are the green leaf-like structures that protect the flower in the bud. They are the lowest ring of structures.

The **petals** are modified leaves. Large, colourful petals attract insects for pollination. Some petals are shaped to guide insects to particular parts of the flower.

Δ Fig. 15.9 Structure of an insect-pollinated flower.

The male part of a flower is the ring of **stamens**. There may be up to 100 stamens, or fewer than a dozen. Each stamen consists of two parts – the **anther** at the top and a stalk called the **filament**. **Pollen grains** develop inside the anthers. Inside each pollen grain is a male gamete. As a grain matures, it develops a thick outer wall to protect the delicate male gamete inside. When all the pollen grains in the anther are mature, the anther splits open to release them.

The female part of the flower is the **carpel**. A flower can contain more than one carpel, each with its own **style** and **stigma**. The style supports the stigma, which is where the pollen lands during pollination. The **ovary** at the base of the carpel protects the female gamete or gametes from the dry air outside. The ovary contains one or more **ovules**, and each ovule contains an egg sac that surrounds the egg cell (female gamete).

Δ Fig. 15.10 Alder catkins contain flowers that shed pollen into the air to be transported to other flowers.

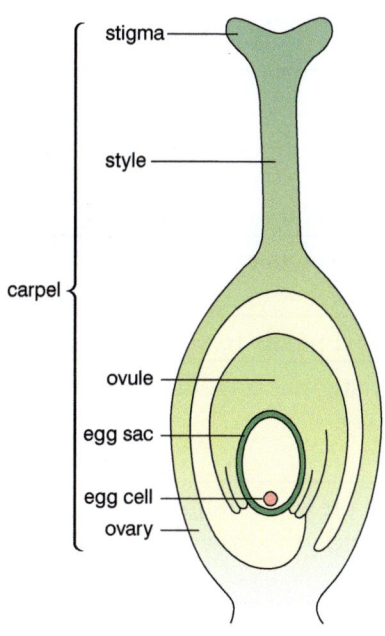

△ Fig. 15.12 Even complex flowers like daisies, which contain thousands of male and female parts, have carpels surrounded by stamens.

△ Fig. 15.11 The carpel.

QUESTIONS

1. Give the names of the female parts of a flower, and state the function of each part.
2. Give the names of the male parts of a flower, and state the function of each part.

Pollination

Before fertilisation can take place, the male gametes have to reach the female gametes. This involves transferring the pollen grains from an anther to a stigma, in a process known as pollination. In many plants this means transferring the pollen from one flower to another. Some plants use the wind to transfer their pollen between flowers; others use animals, especially insects, to carry the pollen. Flowers have different features depending on whether they are pollinated by wind or by insects.

△ Fig. 15.13 In insect-pollinated plants, nectaries secrete a sugary liquid to attract insects.

Wind-pollinated flowers	Insect-pollinated flowers
no scent	often scented to attract insects
no nectaries	nectaries present at the base of the flower produce a sugary liquid to attract insects, e.g. bees and butterflies
small or no petals, so pollen dispersal is not obstructed	large petals for insects to land on
green, inconspicuous or no petals	brightly coloured petals to attract insects
SUPPLEMENT stigmas are large and feathery; often hanging outside the flower to trap pollen	**SUPPLEMENT** stigmas are small and held inside the flower
SUPPLEMENT many anthers, which are often large and hang outside the flower so that pollen is easily dispersed	**SUPPLEMENT** a few small anthers, usually held inside the flower

Δ Table 15.1 Comparison of wind-pollinated and insect-pollinated flowers.

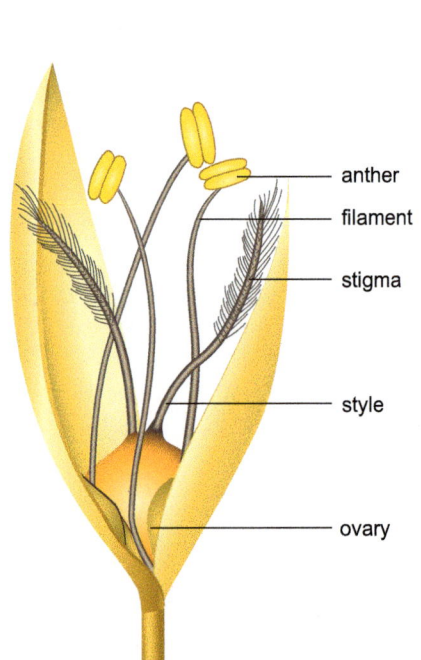

anther
filament
stigma
style
ovary

Δ Fig. 15.14 **SUPPLEMENT** These grass plants have anthers that hang outside the flowers and release large amounts of pollen to the wind. The stigmas also hang outside the flower to collect pollen from other grass plants.

FLOWERS AND POLLINATORS

Different features of animal-pollinated flowers attract different pollinators. Tube-shaped flowers attract insects with a long tongue, such as butterflies, or birds with a long bill, such as hummingbirds. Blue and violet flowers are more attractive to bees, whereas butterflies often prefer red. Plants pollinated by moths or bats tend to open at night and may not be brightly coloured but instead produce a strong sweet scent. Plants that rely on flies to pollinate them often smell like rotting flesh.

One of the most bizarre partnerships between flower and insects occurs between a particular species of orchid and a wasp. Male wasps are attracted to the flowers to mate with what they think are female wasps. During the 'mating' the flowers deposit pollen on the insect, which then carries it to the next flower that it is attracted to.

Δ Fig. 15.15 A male wasp receiving pollen while 'mating' with an orchid flower.

Challenge Question: Suggest why it benefits a plant to only attract particular animal species, that are less likely to visit flowers of other species.

REMEMBER

Be very careful not to confuse *pollination* with *fertilisation*.

THE PROBLEM WITH BEES

About a third of all plants that we use for food or other uses depends on bees for pollination. This includes plants such as oilseed rape, cotton, coffee, apples, and pears. If there are few bees, the crop harvest can be reduced by up to 75%. During the flowering season of crop plants, farmers and growers may place bee hives close to the crop to encourage successful pollination of most flowers. This helps to ensure a good harvest.

Δ Fig. 15.16 By encouraging bees to build their hives in portable boxes, the farmer can move the hives to where the flowers of a crop are ready for pollination.

Recently people have become concerned about a large decrease in bee populations. There are many possible reasons for this. In some places, it has been suggested that the lack of a range of food plants, including weeds, has been the cause. An increase in the use of pesticides that also kill bees may be another cause of the fall in their numbers.

Without bees, food production will be greatly affected. So there are many studies being done to identify why bee numbers are decreasing and to work out how to improve the environment for bees.

Challenge Question: Suggest why placing hives close to crop plants also benefits bees.

Fertilisation

After pollination, fertilisation occurs when the nucleus from a pollen grain fuses with an egg cell nucleus in an ovule.

QUESTIONS

1. Explain the difference between *pollination* and *fertilisation* in a plant.

2. **SUPPLEMENT** Describe three differences in structure between wind-pollinated and insect-pollinated flowers.

3. **SUPPLEMENT a)** Suggest one advantage to a plant of having adaptations for attracting insects rather than relying on wind for pollination.

   b) Suggest one disadvantage for an insect-pollinated plant that relies on one or just a small number of insect species for pollination.

Germination

Germination is when the seed coat breaks open and the embryo starts to grow and develop into a new plant.

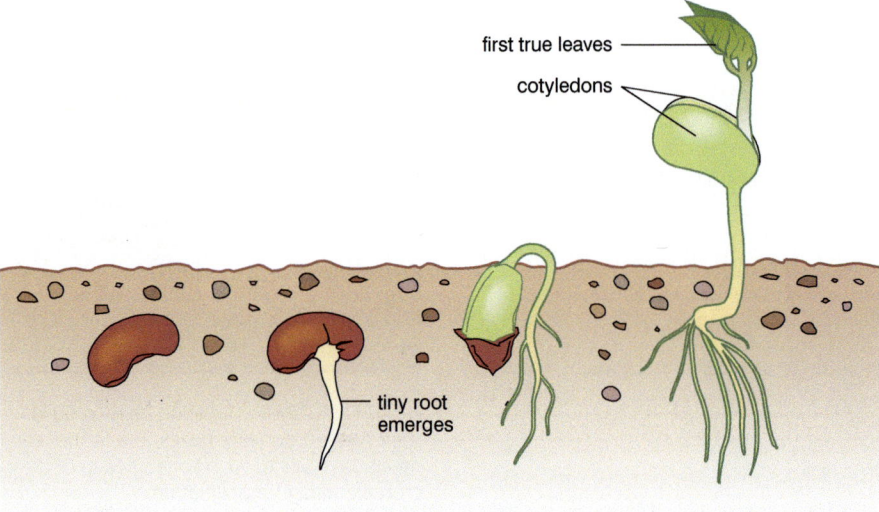

first true leaves

cotyledons

tiny root emerges

△ Fig. 15.17 Germination of a bean seed.

There are three environmental conditions that need to be right for seeds to germinate:

- temperature
- water
- oxygen.

The presence of light is not usually needed for germination. This is because most seeds germinate below ground, so they cannot get their food from photosynthesis.

Temperature

A seed will not start to germinate until the conditions around it reach a suitable temperature. Many seeds lie dormant for a long time during cold periods, such as winter, and start to grow as the earth warms. However, if the temperature becomes too hot the seed may be killed. This is why it is very important to store seeds in the correct conditions and to control the temperature in glasshouses carefully, e.g. through the use of ventilation and shading.

SCIENCE IN CONTEXT **CONDITIONS FOR GERMINATION**

Different plants are suited to different climates. Those that are adapted to colder climates will germinate at lower temperatures. They may also need a very cold period followed by an increase in temperature before they will germinate.

Other seeds will not germinate until they have been exposed to very high temperatures, such as the heat from a forest fire. The extreme heat weakens the seed coat so that water can enter the seed and germination can begin.

Δ Fig. 15.18 Fire clears the ground of competing plants, and stimulates some seeds to germinate in ideal conditions.

Germinating after a fire means that there is likely to be less competition with other species that usually cover the ground. Also, the ash left from the burning acts as a natural fertiliser for the new plants.

Challenge Question: If a winter is warmer than usual, some plant seeds may germinate and start to flower earlier than normal. Suggest why this could be a disadvantage for the plants.

Water

Water is required to swell the seed and burst the seed coat. All seeds contain some water but during germination metabolic reactions are being carried out rapidly. More water is needed for:

- activation of hormones and enzymes
- breakdown of storage compounds, e.g. conversion of starch to glucose
- transport of materials to be used for respiration and growth
- metabolic reactions and enzyme actions that occur in solution.

Oxygen

Active living cells respire and the most useful form of respiration, aerobic respiration, requires oxygen. Seeds can use anaerobic respiration for a short while, but the rate at which energy is released is very slow (not useful in an actively growing organism) and the by-products are toxic. That is why most seeds will only germinate successfully if there is plenty of oxygen in the soil.

△ Fig. 15.19 Waterlogged soil excludes oxygen, making it difficult for seeds to germinate and grow.

Developing practical skills

We can investigate the particular conditions for germination.

Devise and plan investigations

1. Using the apparatus shown, write a plan to investigate the effect of

   a) light

   b) water

   c) temperature

   on the germination of seeds. Think carefully about what controls to use in each case.

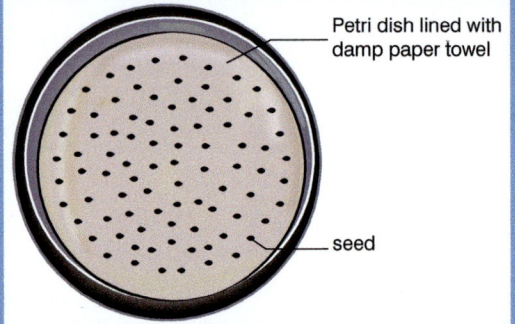

△ Fig. 15.20 A simple set-up for investigating seed germination.

Make observations and measurements

An investigation was carried out using two Petri dishes containing 20 seeds of the same species. Both dishes received the same amount of light and water, but they were kept at different temperatures. The table shows the number of seeds that germinated over a period of 8 days.

Day	Total number germinated seedlings	
	Cool/10 °C	Warm/20 °C
1	0	0
2	0	0
3	0	5
4	1	11
5	6	15
6	16	17
7	18	17
8	18	17

2. Display the results of this investigation in a suitable way.

QUESTIONS

1. What is *germination*?

2. Explain the effects that the following conditions have on the germination of seeds:

   a) oxygen

   b) water

   c) warmth.

SEXUAL REPRODUCTION IN HUMANS

Male reproductive system

A human male has two **testes** (singular: testis) in which **sperm** are produced. The testes are supported outside the body in the **scrotum** to keep them cooler, because at higher temperatures fewer sperm are produced.

Sperm ducts carry the sperm from the testes to the **penis**. The **prostate gland** and seminal vesicles together produce the liquid in which the sperm are able to swim. Semen is the mixture of sperm cells and fluids.

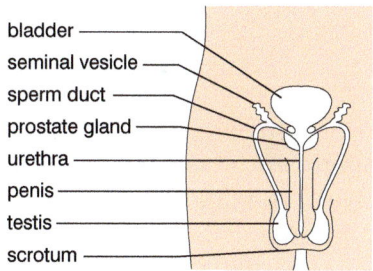

Semen passes along the sperm ducts to the **urethra** to outside the body. The urethra also carries urine from the bladder to outside the body. When the man is sexually excited, large spaces in the penis fill with blood. This causes the penis to become larger and stiffer causing an erection. At the same time a muscle ring (sphincter) at the top of the urethra contracts, preventing urine entering the urethra from the bladder.

△ Fig. 15.21 The male reproductive system. (Note that the bladder is not part of the reproductive system.)

The erection makes it possible for the man to insert his penis into the vagina of the woman for sexual intercourse. Rapid contractions of muscles in the penis during ejaculation send the sperm shooting out into the vagina.

Female reproductive system

The two **ovaries** are the organs in humans that produce the **egg cells**. They are positioned within the abdominal cavity, either side of the **uterus** and joined to it by the **oviducts**.

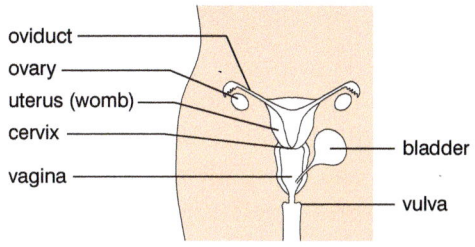

Every month from puberty until menopause, when a woman is around 50 years old, one ovary usually releases one egg, which travels down the oviduct to the uterus (womb). If it is not

△ Fig. 15.22 The female reproductive system. (Note that the bladder is not part of the reproductive system. Note also that in the diagram the bladder is shown misplaced to one side.)

fertilised, the egg will be flushed from the uterus during the monthly period (bleed). At the lower end of the uterus is the **cervix**. This canal produces mucus which changes during the menstrual cycle, allowing sperm to pass through at some times and not at others. It also keeps the developing baby secure in the uterus until birth. The cervix leads into the vagina. The **vagina** is an elastic muscular tube where sperm is received from the penis during sexual intercourse.

SUPPLEMENT

Human gametes

The human gametes are specialised for their roles in reproduction. Fig. 15.23 shows the main adaptive features of the human sperm cell and egg cell.

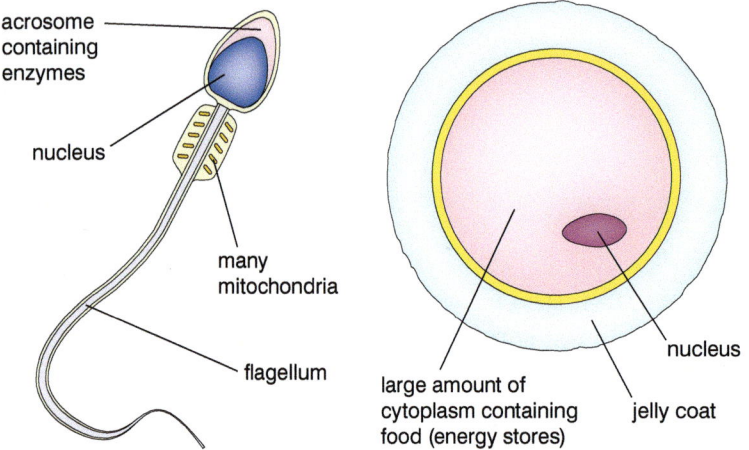

△ Fig. 15.23 Left: human sperm. Right: human egg. (Not to scale: the egg cell is around five times wider than the length of a sperm cell.)

Sperm are among the smallest cells in the human body, at about 45 micrometres long. Over 100 million sperm cells are produced each day. They have very little cytoplasm surrounding the nucleus, because they carry out few functions other than travelling to the egg cell for fertilisation. The mid-piece of the sperm cell contains many mitochondria. The mitochondria provide energy to move the **flagellum** (tail), which moves the sperm through the female uterus towards the egg cell. Since they can move of their own accord, sperm cells are described as being **motile**.

At the front tip of the sperm is a small sac of enzymes called the **acrosome**. When a sperm reaches an egg cell, the acrosome bursts open to release the enzymes. The enzymes digest through the jelly coat and cell membrane of the egg cell, allowing the male nucleus to enter the egg cell.

The egg cell is one of the largest human cells, at about 0.2 mm in diameter. It cannot move on its own, but is wafted along the oviduct by cilia (hair-like structures) on the inside of the tube. Since they can move, but not of their own accord, egg cells are described as being mobile (not motile). An ovary may contain thousands of egg cells, but only one is usually released from one ovary at **ovulation** each month. Within the egg cell is the nucleus and a large amount of cytoplasm. The cytoplasm contains energy stores for the dividing zygote after fertilisation. Surrounding the cell membrane is a jelly coat that protects the cell. Immediately after fertilisation by one sperm, the jelly coat changes to an impenetrable barrier. This prevents other sperm nuclei entering the egg cell.

1. **a)** Sketch a diagram of the human male reproductive system.

   b) Add labels to your sketch to name the main parts of the system.

   c) Describe the role of each of the main parts of the system in human reproduction.

2. **a)** Sketch a diagram of the human female reproductive system.

   b) Add labels to your sketch to name the main parts of the system.

   c) Describe the role of each of the main parts of the system in human reproduction.

3. **SUPPLEMENT** Draw up a table to compare the size, numbers and mobility of human egg and sperm cells.

Fertilisation

During sexual intercourse, sperm deposited near the cervix swim up into the uterus, and then along the oviduct to the egg. Many sperm fail to make the journey, but some will reach the oviducts at the top end of the uterus.

The egg will have been travelling along the oviduct while the sperm have been swimming up from the uterus. Fertilisation takes place in the oviduct. The nucleus of one sperm cell fuses with the nucleus of the egg cell, forming a fertilised egg, or zygote.

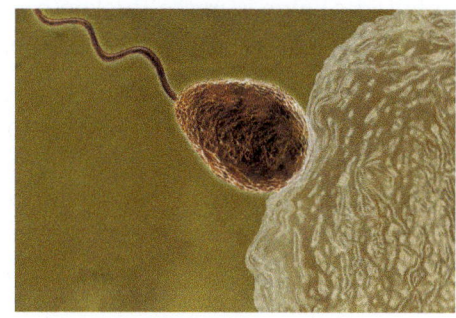

△ Fig. 15.24 The moment just before fertilisation: sperm approaching an egg.

QUESTIONS

1. **a)** Where in the human body does fertilisation of the egg cell occur?

   b) Give the names of the male and female gametes.

2. Describe the process of fertilisation.

SEXUAL HORMONES IN HUMANS

Testosterone is the male sex hormone.

Testosterone is secreted from the testes.

At puberty, the increased secretion of testosterone causes the development of the following **secondary sexual characteristics** in boys:

- an increase in rate of growth until adult size
- hair growth on face and body, including pubic hair
- penis, testes, and scrotum growth and development
- deepening of voice
- increased muscle development
- sperm production.

The female sex hormones include **oestrogen** which is secreted from the ovaries.

At puberty, increased secretion of oestrogen causes the development of the following secondary sexual characteristics in girls:

- an increase in rate of growth until adult size
- breast development
- vagina, oviducts, and uterus development
- start of menstrual cycle (periods)
- hips widening
- pubic hair and under-arm hair growth.

These hormones also control the changes that occur during the **menstrual cycle**.

The menstrual cycle

The menstrual cycle is a sequence of changes that occur in a woman's body every month. The average cycle is 28 days long, but it is normal for it to vary in different women.

The cycle begins with the monthly period, or bleeding (also known as *menstruation*), which is produced from the breakdown of the thickened lining of the uterus. After this, the uterus lining starts to thicken again. Ovulation occurs about halfway through the cycle, when an egg is released from one of the ovaries. The egg travels along the oviduct to the uterus.

If the egg is fertilised during this time, the egg will implant in the uterus lining and the lining will continue to develop for pregnancy. If the egg is not fertilised, the cell and the uterus lining are shed during the monthly period at the start of the next cycle.

QUESTIONS

1. State the name of the male sex hormone.

2. Name the female sex hormone produced in the ovary.

3. Explain why the development of secondary sexual characteristics are important.

4. Define the following terms:
   a) *ovulation*
   b) *menstrual cycle*.

5. In the menstrual cycle, describe the changes in the:
   a) ovaries
   b) lining of the uterus.

SEXUALLY TRANSMITTED INFECTIONS

Unfortunately, sexual contact, such as sexual intercourse, is a method by which infection can spread, because of the exchange of body fluids, which may contain pathogens (see Section 10). There are many such **sexually transmitted infections** (**STIs**), including **human immunodeficiency virus** (**HIV**), which may lead to acquired immunodeficiency disease (**AIDS**).

AIDS

AIDS is a disease of the immune system that means it cannot protect the body against other infections. The pathogen that causes AIDS is a virus called HIV. The virus in an infected person is present in sexual fluids such as semen and vaginal fluids, and so can be transmitted during sexual contact. It may also be passed to another person in blood, either through a scratch, or through the sharing of needles for intravenous injection of drugs such as heroin. Infection can also pass from a mother to her fetus, through the placenta, or to her baby through breast-feeding after birth.

There is no cure for AIDS, so prevention of infection is essential. This is most easily done by abstinence from sex, or by limiting sexual partners to those who do not carry the virus. As a person may have no obvious symptoms early in infection, barrier methods such as the **condom** or **femidom** are most effective in reducing the risk of infection during intercourse. These methods are also effective against other STIs. There is no cure for AIDS, and antibiotics have no effect on the disease because antibiotics do not affect viruses. However, other STIs that are caused by bacterial pathogens, such as gonorrhoea, can be successfully treated using antibiotics.

QUESTIONS

1. Explain what is meant by a *sexually transmitted infection*.

2. Describe the methods of transmission of HIV.

3. Describe the effect of HIV on the body.

4. a) Describe two ways to prevent the transmission of infections that are transmitted during sexual intercourse.

   b) Explain how these actions help control the spread of this infection in the population.

End of topic checklist

Key terms

AIDS, anther, asexual reproduction, carpel, cervix, condom, egg cell, embryo, femidom, fertilisation, filament, gamete, germination, human immunodeficiency virus (HIV), menstrual cycle, oestrogen, ovary, oviduct, ovulation, ovule, penis, petal, pollen grain, pollination, prostate gland, scrotum, secondary sexual characteristics, seed, sepal, sexual reproduction, sexually transmitted infection (STI), sperm, sperm ducts, stamen, stigma, style, testes, testosterone, urethra, uterus, vagina, zygote

SUPPLEMENT acrosome, diploid, flagellum, haploid, motile

During your study of this topic you should have learned:

○ That asexual reproduction is a process resulting in the production of genetically identical offspring from one parent.

○ How to identify examples of asexual reproduction in diagrams, images, and information provided.

○ **SUPPLEMENT** How to discuss the advantages and disadvantages of asexual reproduction to a population of a species in the wild

○ That sexual reproduction is a process involving the fusion of the nuclei of two gametes to form a zygote and the production of offspring that are genetically different from each other.

○ That fertilisation is the fusion of the nuclei of gametes.

○ **SUPPLEMENT** That nuclei of gametes are haploid and that the nucleus of a zygote is diploid.

○ **SUPPLEMENT** How to discuss the advantages and disadvantages of sexual reproduction to a population of a species in the wild

○ How to identify in diagrams and images, and draw the following parts of an insect-pollinated flower and state the functions of: sepals, petals, stamens, filaments, anthers, carpels, style, stigma, ovary, and ovules.

○ **SUPPLEMENT** How to identify in diagrams and images, and describe the anthers and stigmas of a wind-pollinated flower.

○ That pollination is the transfer of pollen grains from an anther to a stigma.

○ That fertilisation occurs when a pollen nucleus fuses with a nucleus in an ovule.

○ How to describe the structural adaptations of insect-pollinated and wind-pollinated flowers.

- How to investigate and describe the environmental conditions that affect germination of seeds: water, oxygen, and a suitable temperature.

- How to identify on diagrams and state the functions of the following parts of the male reproductive system: testes, scrotum, sperm ducts, prostate gland, urethra, and penis.

- How to identify on diagrams and state the functions of the following parts of the female reproductive system: ovaries, oviducts, uterus, cervix, and vagina.

- That fertilisation is the fusion of the nuclei from a male gamete (sperm) and a female gamete (egg cell).

- **SUPPLEMENT** How to explain the adaptive features of sperm: flagellum, mitochondria, and enzymes in the acrosome.

- **SUPPLEMENT** How to explain the adaptive features of egg cells: energy stores and the jelly coat that changes at fertilisation.

- **SUPPLEMENT** How to compare male and female gametes in terms of: size, structure, motility, and numbers.

- How to describe the roles of testosterone and oestrogen in the development and regulation of secondary sexual characteristics during puberty.

- How to describe the menstrual cycle in terms of changes in the ovaries and in the lining of the uterus.

- That a sexually transmitted infection (STI) is an infection that is transmitted through sexual contact.

- That human immunodeficiency virus (HIV) is a pathogen that causes an STI.

- That HIV infection may lead to AIDS.

- How to describe the methods of transmission of HIV.

- How to explain how the spread of STIs is controlled.

End of topic questions

1. What is formed by the fusion of the nuclei of an egg cell and a sperm cell?

A gamete

B pollen grain

C seed

D zygote

2. The photograph in Fig. 15.25 shows a catkin on a goat willow tree. A catkin is formed from a group of flowers.

a) State the purpose of the flowers on a goat willow tree.

b) State the name of the yellow parts of the flowers shown in this photograph.

c) Describe their purpose in a flower.

d) State whether goat willow flowers are pollinated by the wind or by insects. Explain your answer using clues from the photograph.

△ Fig. 15.25 A goat willow catkin.

3. a) Suggest one advantage to a plant of having adaptations for attracting insects rather than relying on wind for pollination.

b) Suggest one disadvantage for an insect-pollinated plant of relying on one or just a small number of insect species for pollination.

4. A gardener has some packets of seeds for planting. The packets explain how to plant the seeds to get the best chances of germination.

a) Define *germination*.

b) The packets say that the seeds should be planted in moist compost and kept indoors. Explain why the seeds need these conditions.

c) Explain why the seeds will not germinate successfully in waterlogged soil.

d) The larger seeds need to be planted deeper in the compost, and the tiniest seeds need to be scattered on the surface of the compost. Suggest why different seeds should be planted at different depths. (Hint: think about food reserves.)

e) Some seeds that come from plants in high-latitude regions (such as Canada or Russia) need to be placed in the freezer for a few weeks before they will germinate. This makes them respond as if they had been through a cold winter. Suggest the survival advantage of this adaptation if they were growing naturally.

5. a) State where sperm cells are made in the human body.

 b) State where egg cells are made in the human body.

 c) State where an egg cell is fertilised by a sperm cell.

 d) Starting from the point of their formation, describe how a sperm cell reaches the egg cell for fertilisation.

6. Draw the menstrual cycle as a circle of 28 days. On your diagram label:

 a) ovulation

 b) menstruation

 c) when the uterine layer would be thickest

 d) when the uterine layer would be thinnest.

7. a) Explain why HIV is classified as a sexually transmitted infection.

 b) Describe two ways that HIV can be transmitted from a mother to her baby.

 c) Describe two ways to prevent the transmission of HIV during sexual intercourse.

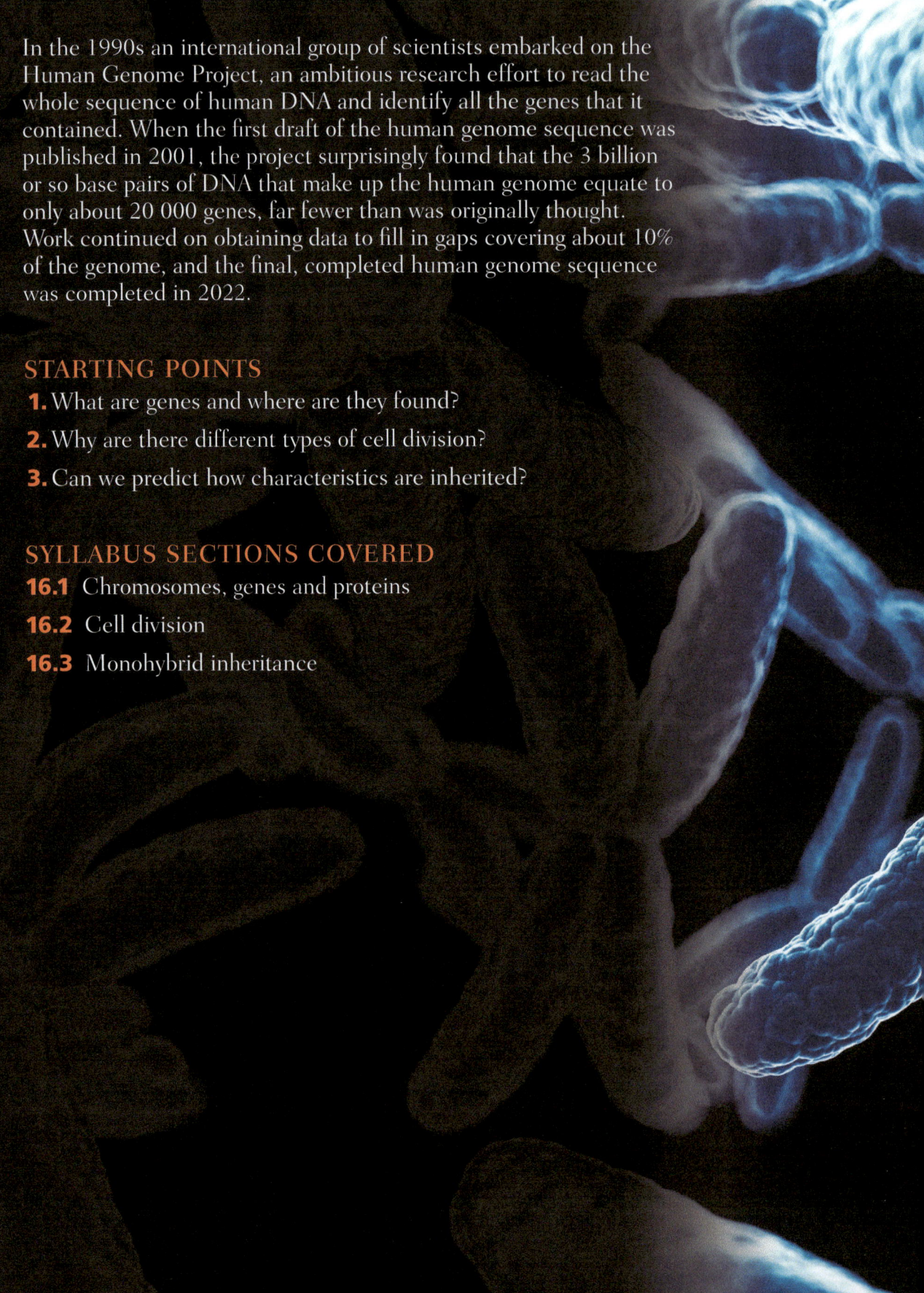

In the 1990s an international group of scientists embarked on the Human Genome Project, an ambitious research effort to read the whole sequence of human DNA and identify all the genes that it contained. When the first draft of the human genome sequence was published in 2001, the project surprisingly found that the 3 billion or so base pairs of DNA that make up the human genome equate to only about 20 000 genes, far fewer than was originally thought. Work continued on obtaining data to fill in gaps covering about 10% of the genome, and the final, completed human genome sequence was completed in 2022.

STARTING POINTS

1. What are genes and where are they found?

2. Why are there different types of cell division?

3. Can we predict how characteristics are inherited?

SYLLABUS SECTIONS COVERED

16.1 Chromosomes, genes and proteins

16.2 Cell division

16.3 Monohybrid inheritance

16
Inheritance

△ Just before nuclear division, each chromosome is copied exactly. The two copies then separate as new cells are made.

△ Fig. 16.1 This baby will have inherited some characteristics from her mother and some from her father.

Inheritance

INTRODUCTION

Unless a zygote (fertilised egg) divides completely into two, and develops as two identical twins, the baby that develops from that zygote is genetically unique. Each cell in a zygote contains genetic information, half of which came from the father and half from the mother. And during gamete formation, some changes may have occurred in some of that genetic information. So the baby may have some variations in its genes that neither of its parents has.

Although the baby is genetically unique, virtually all of the cells in its body are genetically identical.

KNOWLEDGE CHECK

✓ Some characteristics, like skin tone and eye colour, are passed from one generation to the next generation.
✓ The nucleus of a cell contains genetic information (DNA).

LEARNING OBJECTIVES

✓ State that chromosomes are made of DNA, which contains genetic information in the form of genes.
✓ Define a gene as a length of DNA that codes for a protein.
✓ Define an allele as an alternative form of a gene.
✓ Describe the inheritance of sex in humans with reference to XX and XY chromosomes.
✓ **SUPPLEMENT** Describe a haploid nucleus as a nucleus containing a single set of chromosomes.
✓ **SUPPLEMENT** Describe a diploid nucleus as a nucleus containing two sets of chromosomes.
✓ **SUPPLEMENT** State that in a diploid cell, there is a pair of each type of chromosome and in a human diploid cell there are 23 pairs.
✓ **SUPPLEMENT** Describe mitosis as nuclear division giving rise to genetically identical cells (details of the stages of mitosis are **not** required).
✓ **SUPPLEMENT** State the role of mitosis in growth, repair of damaged tissues, replacement of cells and asexual reproduction.
✓ **SUPPLEMENT** State that the exact replication of chromosomes occurs before mitosis.
✓ **SUPPLEMENT** State that during mitosis, the copies of chromosomes separate, maintaining the chromosome number in each daughter cell.
✓ **SUPPLEMENT** State that meiosis is involved in the production of gametes.
✓ **SUPPLEMENT** Describe meiosis as a reduction division in which the chromosome number is halved from diploid to haploid resulting in genetically different cells (details of the stages of meiosis are **not** required).
✓ Describe inheritance as the transmission of genetic information from generation to generation.
✓ Describe genotype as the genetic make-up of an organism and in terms of the alleles present.
✓ Describe phenotype as the observable features of an organism.
✓ Describe homozygous as having two identical alleles of a particular gene.

✓ State that two identical homozygous individuals that breed together will be pure-breeding.
✓ Describe heterozygous as having two different alleles of a particular gene.
✓ State that a heterozygous individual will not be pure-breeding.
✓ Describe a dominant allele as an allele that is expressed if it is present in the genotype.
✓ Describe a recessive allele as an allele that is only expressed when there is no dominant allele of the gene present in the genotype.
✓ Interpret pedigree diagrams for the inheritance of a given characteristic.
✓ Use genetic diagrams to predict the results of monohybrid crosses and calculate phenotypic ratios, limited to 1 : 1 and 3 : 1 ratios.
✓ Use Punnett squares in crosses which result in more than one genotype to work out and show the possible different genotypes.

Inheritance

Inheritance is the passing on (transmission) of characteristics from one generation to the next, from parents to offspring. As characteristics are coded for by genetic information in the cell nucleus, inheritance is also the transmission of **genes** from the parents' gametes to the offspring.

CHROMOSOMES AND GENES

Inside virtually every cell in the body is a nucleus, which contains long threads called **chromosomes**. These threads are usually stretched out and fill the nucleus but, when the chromosomes condense (gather into bundles) just before cell division, they can be seen through a microscope. Chromosomes usually occur in pairs. The two in each pair look very similar to each other, except for the pair of **sex chromosomes**, X and Y (see Fig. 16.3). (There is more about the X and Y chromosomes and how they are involved in the inheritance of sex later in this section.)

Chromosomes are made of a chemical called deoxyribonucleic acid (**DNA**). DNA contains genetic information in the form of genes. Each gene is a length of the DNA that codes for a particular **protein**. Different kinds of protein are made by different cells in the body. For example, as you saw in Section 5, the enzymes that catalyse metabolic reactions are protein molecules, and haemoglobin found in red blood cells is a protein molecule. There are millions of different protein molecules, each with a different role in the body. During cell division the string of genes on each chromosome is copied and passed on to the new cells.

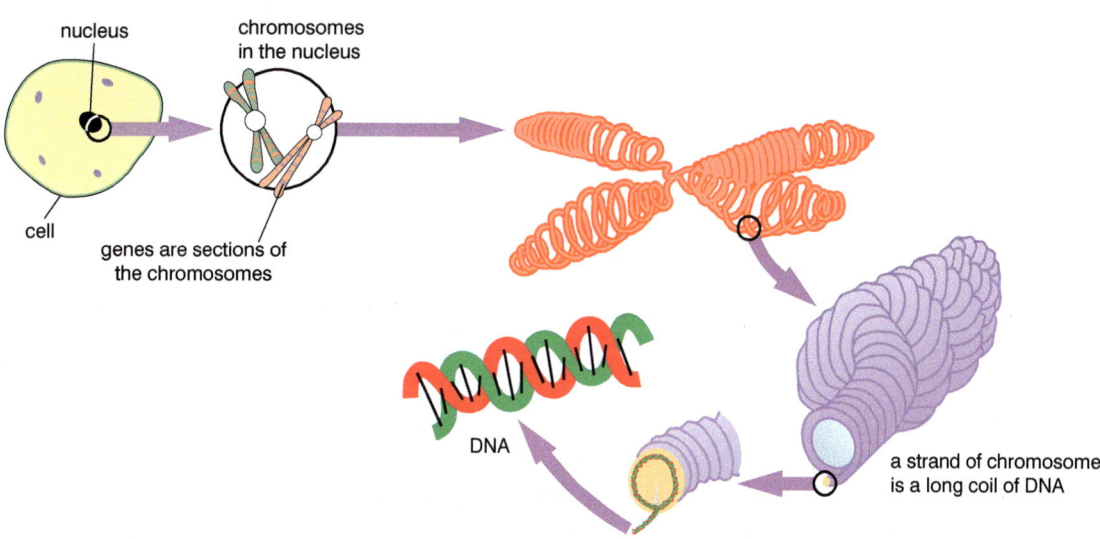

△ Fig. 16.2 The relationship between cell, nucleus, chromosome, and DNA.

Genes code for particular characteristics, such as eye colour. However, variations in a characteristic are caused by different alternative forms of a gene, called **alleles**. For example, some people have alleles that code for brown eye colour; others have alleles that code for hazel eyes, or blue eyes, or grey eyes.

In each pair of chromosomes, both chromosomes have the same genes in the same order, but the genes on each chromosome may be in the form of the same allele or different alleles. This can cause differences in inheritance (as you will see later).

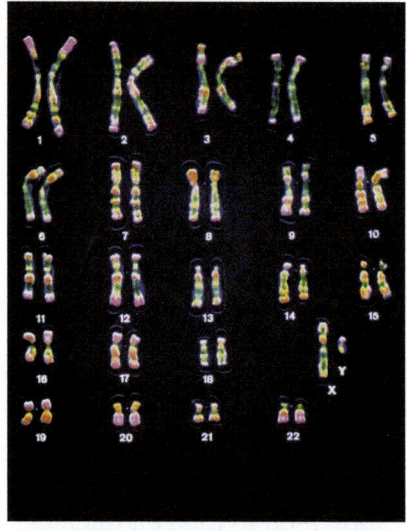

△ Fig. 16.3 The chromosomes from a man's body cell arranged in their pairs. The chromosomes in each pair look very similar except in the sex chromosome pair (marked X and Y).

QUESTIONS

1. Give the following in order of size, starting with the smallest: cell, chromosome, gene, nucleus.

2. Using an example, explain the difference between *gene* and *allele*.

3. Define the term *inheritance* as fully as you can.

SUPPLEMENT

Chromosome sets

Most of the cells in the body contain two sets of chromosomes. This is because each individual that is produced by sexual reproduction receives one set of chromosomes from their mother and one set from their father. This means you will inherit some characteristics from each of your parents.

You only inherit one set of chromosomes from each parent because each gamete only contains one set.

If you remember from Section 15, body cells are **diploid** – meaning that their nucleus contains two sets of chromosomes. Each chromosome is paired.

The gametes, however, are **haploid** – the nucleus of these cells (sperm, egg cell, male gamete in pollen grain) contains only one set of chromosomes. This is the result of the way that they are produced during meiosis (see later).

In humans, the nucleus of diploid cells contains 46 chromosomes (23 pairs), and the nucleus of the haploid gamete cells contains 23 chromosomes.

QUESTION

1. Chimpanzees have 48 chromosomes in each body cell. State how many chromosomes are in a chimpanzee egg cell.

MITOSIS

Organisms grow by the division of cells, when the body cells split in two. Before the cell can split, its nucleus must first divide in two. This kind of division is used to produce new body cells in growth, the repair of damaged tissues, and replacement of cells in the body. It is also the way that single-celled organisms reproduce and is the only type of cell division involved during asexual reproduction (reproduction that does not involve sex cells), as you saw in Section 15. This type of division, giving rise to new cells which are genetically identical to the original cell and to each other, is known as **mitosis**.

Before mitosis occurs, the cell's chromosomes are replicated (copied). During mitosis, the copies of the chromosomes separate, so that one copy of each chromosome ends up in each of the new (daughter) cells. This is why the **daughter cells** have the same chromosome number and are genetically identical to the original cell.

MEIOSIS

During sexual reproduction, a male gamete fuses with a female gamete. If each gamete had the same number of chromosomes as a normal body cell, the zygote would end up with twice as many chromosomes as normal. Instead, gametes are produced by a different form of cell division called **meiosis**. This is a **reduction division** because the cells produced by meiosis have a reduced number of chromosomes, just one chromosome of each pair – half the normal diploid number of chromosomes. These cells are called haploid cells. When the gametes fuse during fertilisation, they restore the normal number of chromosomes, creating a diploid cell with pairs of chromosomes again.

Cells produced by meiosis are genetically different from each other. This means that, during fertilisation where there is a random chance which male gamete will fuse with a female gamete, the offspring produced will be genetically different from each other. We say they show variation.

QUESTIONS

1. a) State which form of cell division produces new body cells.

   b) Explain why this is important to the organism.

2. The diagram shows the life cycle of a human.

Copy the diagram and add notes to show:

- when meiosis and mitosis occur
- which cells are diploid and which are haploid.

3. State the form of cell division found when organisms reproduce by:
   a) sexual reproduction, **b)** asexual reproduction.

4. Explain the importance of meiosis in producing variation in offspring.

MONOHYBRID INHERITANCE

Some characteristics, such as the colour of your eyes, are passed down (inherited) from your parents, but other characteristics may not be passed down. Sometimes characteristics appear to miss a generation: for instance, you and your grandmother might both have dry earwax but both of your parents may have wet earwax.

Leopards occasionally have a cub that has completely black fur instead of the usual spotted pattern. It is known as a black panther but is still the same species as the ordinary leopard. Just as in humans, leopard chromosomes occur in pairs. One pair carries a gene for fur colour.

There are two copies of the gene in a normal body cell (one on each chromosome). The two versions (alleles) of the gene may be identical but sometimes they are different, one being for a spotted coat and the other for a black coat.

Leopard cubs receive half their genetic information from each parent. Eggs and sperm cells contain only half the number of chromosomes of normal body cells. This means that egg and sperm cells contain only one of each pair of alleles. When an egg and sperm join together at fertilisation, forming a zygote that will develop into the new individual, it now has two alleles of each gene.

Different combinations of alleles will produce different fur colour:

spotted coat allele	+	spotted coat allele	=	spotted coat
spotted coat allele	+	black coat allele	=	spotted coat
black coat allele	+	black coat allele	=	black coat

The black coat only appears when *both* of the alleles for the black coat are present. As long as there is at least one allele for a spotted coat, the coat will be spotted because the allele for a spotted coat over-rides the allele for a black coat. It is the **dominant** allele. A dominant allele is an allele that is always expressed if it is present. Alleles like the one for the black coat are described as **recessive**. A recessive allele is an allele that is not expressed in the phenotype unless there are two copies present. It is only expressed when there is no dominant allele of the gene present.

An individual with two identical alleles for a particular gene is said to be **homozygous** (*homo* means 'the same') for that gene.

An individual with two different alleles for a particular gene is said to be **heterozygous** for that gene (*hetero* means 'different').

△ Fig. 16.4 Two spotted leopard parents may produce offspring with a spotted coat or a black coat.

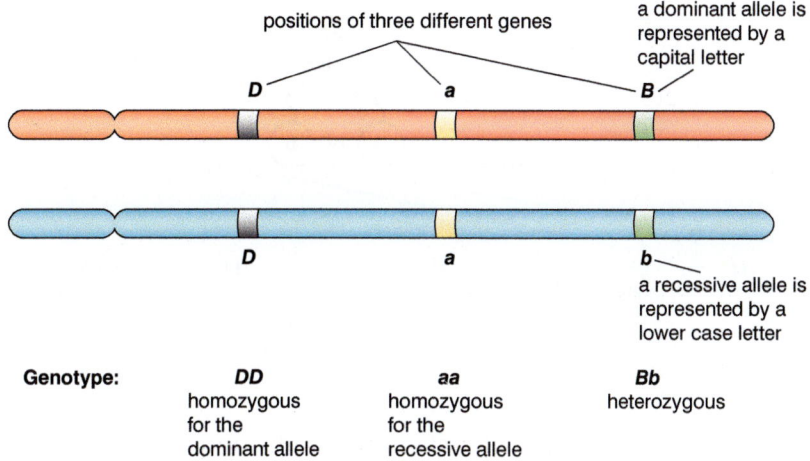

Genotype:	DD	aa	Bb
	homozygous for the dominant allele	homozygous for the recessive allele	heterozygous

△ Fig. 16.5 A pair of chromosomes showing some definitions in genetics.

A leopard with a spotted coat may be homozygous for the spotted allele, or heterozygous. A leopard with a black coat can only be homozygous for the black allele.

QUESTIONS

1. Define the following terms in your own words:

   a) *dominant*   **b)** *recessive*   **c)** *homozygous*   **d)** *heterozygous.*

2. How many alleles for a particular gene would be found in:

   a) a body cell   **b)** a gamete   **c)** a zygote?

Monohybrid crosses

An organism's genetic make-up, or combination of alleles present, is its **genotype**. An organism's combination of observable features is known as its **phenotype**. An organism's phenotype is influenced by both its genotype and its environment. (There is more about the effects of genes and the environment on phenotypes in Section 17.)

We can show the influence of the genotype in a **genetic diagram**. This uses a capital letter for the dominant allele and a lower case letter for the recessive allele.

REMEMBER

To avoid the risk of confusion when drawing genetic diagrams, choose a letter that is easily distinguished in capital and lower case, for example, A/a or B/b. It might seem reasonable to use S or C for the coat gene, but this gives alleles S/s or C/c, which could be confused if written by hand.

Using the example of the leopards, let us choose **B** for the dominant allele for a spotted coat and letter **b** for the recessive allele for the black coat. Two spotted parents who have a black cub must each be carrying a **B** and a **b**. The genetic diagram below shows the possible genotypes and phenotypes of the offspring.

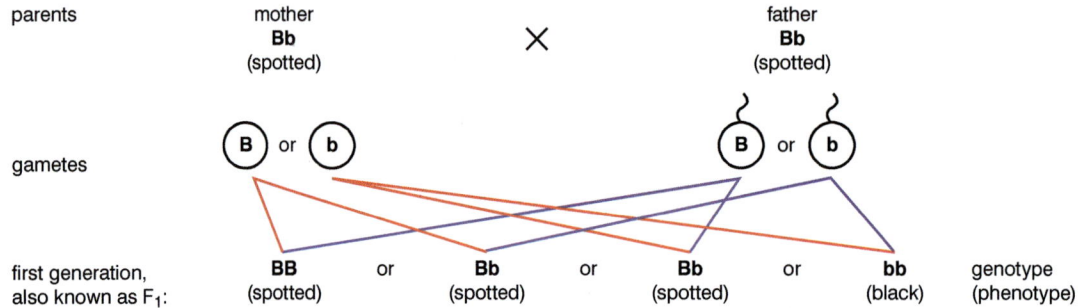

△ Fig. 16.6 Two different phenotypes are possible from the cross in this diagram. The probabilities of the phenotypes are 3 spotted : 1 black.

Because we are looking at a characteristic (fur colour) controlled by one gene, this is an example of a **monohybrid cross**. (*Mono* means 'one', and a *hybrid* is produced when two different types breed or cross.)

Another type of genetic diagram is known as a **Punnett square**. The example above can be shown in a Punnett square. The four boxes at the bottom right show the possible combinations in the offspring.

		male Bb spotted	
		gametes	
		B	b
female Bb spotted	gametes	B	BB spotted	Bb spotted
		b	Bb spotted	bb black

Probabilities and predictions

If you cross two individuals that are homozygous for the same alleles, then you will always get offspring that are homozygous for those alleles. We say that these individuals are **pure-breeding**. If you cross heterozygous individuals, you will get variety in the offspring. So heterozygous individuals are not pure-breeding.

When two heterozygous parents are crossed, the offspring showing the dominant characteristic and the offspring showing the recessive characteristic appear in the phenotypic ratio of 3 : 1. The 3 : 1 ratio refers to the probabilities of particular combinations of alleles, so the likelihood of having an offspring with the phenotype of the dominant characteristic is three times the likelihood of having an offspring with the phenotype of the recessive characteristic.

In the example of the leopards, there is a 1 in 4, 25%, or 0.25 probability of a leopard cub being black. This is because the offspring must inherit two recessive alleles (one from each parent) in order for the offspring to be homozygous and for the phenotype to be expressed (be observable).

If you had a large number of offspring from an *actual* cross of two heterozygous leopard parents, you might expect something near the 3 : 1 ratio of spotted to black cubs. However, because it is a matter of chance which sperm cell fertilises which egg cell, you should not be too surprised if a small litter – for example, of four cubs – contained two black cubs, or none.

Using the example of leopard coat colour and a Punnett square, we can also look at what happens if we cross homozygous recessive and heterozygous individuals:

			male Bb spotted	
			gametes	
			B	b
female bb black	gametes	b	Bb spotted	bb black
		b	Bb spotted	bb black

The predicted outcome from this cross is a 1 : 1 phenotypic ratio of spotted to black colouring. This gives a 1 in 2, 50%, or 0.5 probability of a cub from these two parents being spotted or being black.

REMEMBER

You can predict probabilities of outcomes from any monohybrid genetic cross by using a genetic diagram. There are different types of genetic diagrams you can practise, though many people find it easiest to draw Punnett squares.

Can you predict what will happen if a homozygous spotted coat leopard is crossed with a heterozygous spotted coat leopard?

			male Bb spotted	
			gametes	
			B	b
female BB spotted	gametes	B	BB spotted	Bb spotted
		B	BB spotted	Bb spotted

In this case, although some of the cubs born are likely to be homozygous and some heterozygous, they will all have spotted coats. They will have the same phenotype but not the same genotypes.

QUESTIONS

1. Define *monohybrid inheritance* in your own words.

2. Rabbits have a gene for coat colour – the allele for brown coat is dominant over the allele for black coat colour. Using the letter B for the dominant allele, and b for the recessive allele, write down all the possible genotypes and phenotypes for this gene. Explain your answers.

3. Using your answers from Question 2:

a) Construct a genetic diagram to show the possible offspring from a cross between a male rabbit that is homozygous for the dominant allele and a female rabbit that is homozygous for the recessive allele.

b) Predict the probability of producing a black baby rabbit from this cross.

Developing practical skills

In an investigation into the inheritance of a characteristic, students used red beads to represent dominant alleles and blue beads to represent recessive alleles.

As a homozygous dominant individual produces gametes that only contain the dominant allele, all the red beads were placed into a beaker to represent the gametes for this individual. As a homozygous recessive individual produces gametes that only contain the recessive allele, all the blue beads were placed into another beaker to represent the gametes for this individual.

To model what would happen in a cross between these two individuals, students took one gamete (bead) from one pot and paired it with one gamete (bead) from the other pot and wrote down the genotype and phenotype for that 'offspring'. This showed that all the offspring from these parents would be heterozygous (one red, one blue bead).

Devise and plan investigations

1. Describe and explain how you would adapt this method to represent a cross between two heterozygous individuals. (Hint: Make sure you use enough beads to get a reasonable approximation of the actual result to the expected result.)

Analyse and interpret data

Some students carried out an investigation like this that started with two 'parents', each heterozygous for a characteristic.

2. Each pot started with 40 beads. How many red beads and how many blue beads were in each of the two pots? Explain your answer.

20 selections were made from the two beakers, to produce the 'offspring'. The results are shown in the table.

Number of red/red pairs in 'offspring'	5
Number of red/blue pairs in 'offspring'	12
Number of blue/blue pairs in 'offspring'	3

3. Draw a genetic diagram for this cross, to show the predicted probabilities of genotypes and phenotypes. (Hint: remember to choose letters for the alleles and explain which allele is modelled by the red beads and which by the blue beads.)

4. Describe how the actual results differ from the expected results.

Evaluate data and methods

5. Comment on the difference between the expected and actual results.

6. Explain how you could adjust the method to get results closer to the expected.

MENDEL'S PEAS

Gregor Mendel (1822–1884) was the first person to study genetic inheritance in a thorough and scientific manner. He chose characteristics in peas to study because he could see clear differences in characteristics and patterns in their inheritance. He started by crossing plants with the same characteristics many times, until he was certain that they were pure-breeding.

He then made hundreds of crosses of the same kind. He started by removing the anthers of each flower. Then he brushed pollen from a plant he had chosen as one parent onto the stigma of the other parent and covered the flower to prevent other pollen getting in.

From his results, Mendel was able to show that parental characteristics generally do not blend together in the offspring phenotype (as was commonly believed), but that a dominant allele in a heterozygote prevents the recessive allele being expressed.

Δ Fig. 16.7 The results of one of Mendel's crosses for pea form and colour. (Note that each pea is the result of a separate cross between a pollen grain and an egg cell.)

1. Explain why it was important that the parent plants were pure-breeding.

2. Explain why Mendel needed to carry out hundreds of crosses before drawing a conclusion.

3. Pea flowers are pollinated by insects. Explain why Mendel could be certain that no chance fertilisations took place.

4. At the time Mendel carried out his work, people couldn't understand how a characteristic could be present in one generation, 'disappear' in the next generation and then reappear in the next. Using genetic diagrams, and a characteristic of your choice, show how this happens when starting with pure-breeding parents for the dominant and recessive characteristics.

5. Explain the importance of a thorough and scientific method for drawing valid conclusions.

Sex determination and inheritance

If you take all the chromosomes from a body cell, you can arrange them into pairs. This is because you inherit one chromosome of each pair from your father and one from your mother.

In the nucleus of a human body cell there are 46 chromosomes that form 23 pairs. In all but one of these pairs, the two chromosomes of each pair look almost identical. We call these 22 pairs the autosomal chromosomes. The chromosomes of the other pair are identical in women, but differ in men. These are called the sex chromosomes. The sex chromosomes are called X and Y. Women have the genotype XX and men have the genotype XY.

When gametes are produced, by meiosis, they each receive one of the sex chromosomes. So egg cells all contain an X chromosome, but sperm cells may contain an X chromosome or a Y chromosome. About 50% of sperm cells carry an X chromosome and 50% a Y chromosome.

During fertilisation, one sperm cell fuses with one egg cell. We can use a Punnett square to show the possible combinations of sex chromosomes in the offspring.

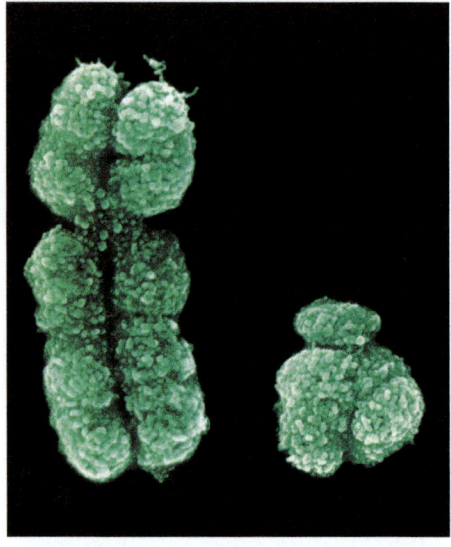

Δ Fig. 16.8 The human X and Y chromosomes are different shapes.

		father's gametes	
		X	Y
mother's gametes	X	XX (female)	XY (male)
	X	XX (female)	XY (male)

This shows that there is a 50%, or 1 in 2, or 0.5 probability of any child being a boy or a girl. The ratio of boys to girls born in a family is often not 1 : 1, but over the whole human population about equal numbers of baby boys and baby girls are born.

QUESTIONS

1. State which sex chromosomes would be found in a body cell of an adult woman.

2. State which sex chromosomes would be found in a body cell of a baby boy.

3. A couple have three boys. Give the probability of their next child being a girl. Explain your answer.

Pedigree diagrams

A genetic diagram predicts the possible outcomes of a cross for alleles of a particular gene. This is useful in species that produce large numbers of offspring, because chance is averaged out by the large number of crosses, so that the predicted and actual outcomes of a cross are similar. But in species like humans, where parents produce only a few offspring, it can be difficult to work out whether an allele is dominant or recessive from actual results.

Instead, we can use a **pedigree diagram** to investigate the inheritance of a particular characteristic. This shows the relationship between individuals in a family and which of them has which version of that characteristic.

One example of a characteristic we can study like this is earwax type. Some people have wet earwax, which is an orange colour and sticky. Other people have dry earwax, which is paler and flaky.

Here is a pedigree diagram that shows the inheritance of earwax type. Note the key, which explains whether each individual in the diagram is male or female and which form of earwax they have.

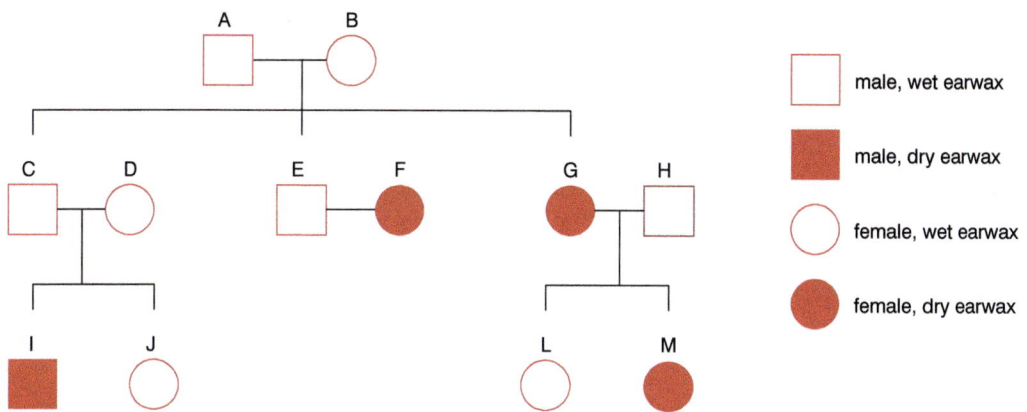

△ Fig. 16.9 A pedigree diagram for earwax.

This pedigree diagram shows the following information.

- There are three generations shown for this family – A and B are the eldest generation; I, J, L, and M are the youngest generation.
- Couple A and B had three children: C, E, and G. Couples C/D and G/H each have two children.
- This family includes four people with dry earwax; one is male and three are female.

This pedigree diagram also shows that, although A and B do have wet earwax, they had a child (G) who has dry earwax. Similarly, couple C/D have a child with dry earwax. Using genetic diagrams we can show the following.

- Since G has dry earwax, at least one of the two parents must have the allele for dry earwax.
- If the dry earwax allele was dominant, then either A or B should have dry earwax.
- Since neither A or B have dry earwax, the allele for dry earwax must be recessive.
- Since G has dry earwax, A and B must both be heterozygous for the dry earwax allele.

If you aren't sure about any of these statements, draw a genetic diagram to help you confirm what they are saying.

Using this kind of argument, you can work out the genotypes of all the individuals in the pedigree diagram except E and J (because they haven't had any children and may be homozygous dominant or heterozygous).

One human gene controls the presence of freckles on skin. Use the pedigree diagram to answer the following questions.

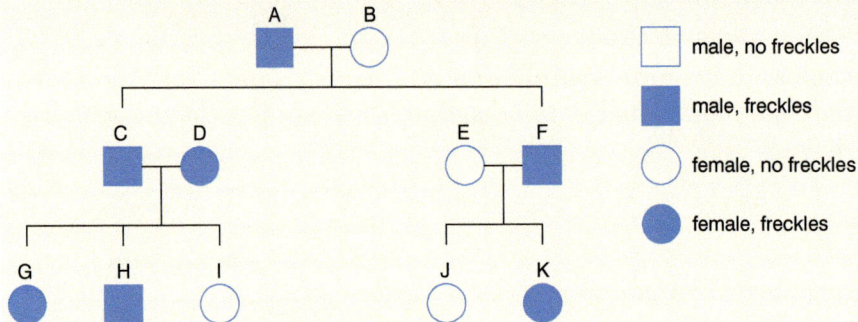

△ Fig. 16.10 A pedigree diagram for freckles.

1. State how many generations are shown.

2. State how many children couple A and B had.

3. State how many daughters couple C and D had.

4. Identify which allele seems to be the dominant one. Explain your answer.

End of topic checklist

Key terms

allele, chromosome, DNA, dominant, gene, genetic diagram, genotype, heterozygous, homozygous, inheritance, monohybrid cross, pedigree diagram, phenotype, Punnett square, pure-breeding, recessive, sex chromosome

SUPPLEMENT daughter cell, diploid, haploid, meiosis, mitosis, reduction division

During your study of this topic you should have learned:

◯ That chromosomes are made of DNA, which contains genetic information in the form of genes.

◯ That a gene is a length of DNA that codes for a protein.

◯ That an allele is an alternative form of a gene.

◯ How to describe the inheritance of sex in humans with reference to X and Y chromosomes.

◯ **SUPPLEMENT** That a haploid nucleus is a nucleus containing a single set of chromosomes.

◯ **SUPPLEMENT** That a diploid nucleus is a nucleus containing two sets of chromosomes.

◯ **SUPPLEMENT** That in a diploid cell, there is a pair of each type of chromosome, and in a human diploid cell there are 23 pairs.

◯ **SUPPLEMENT** That mitosis is nuclear division giving rise to genetically identical cells.

◯ **SUPPLEMENT** How to state the role of mitosis in growth, repair of damaged tissues, replacement of cells, and asexual reproduction.

◯ **SUPPLEMENT** That the exact replication of chromosomes occurs before mitosis.

◯ **SUPPLEMENT** That during mitosis, the copies of chromosomes separate, maintaining the chromosome number in each daughter cell.

◯ **SUPPLEMENT** That meiosis is involved in the production of gametes.

◯ **SUPPLEMENT** That meiosis is a reduction division in which the chromosome number is halved from diploid to haploid, resulting in genetically different cells.

◯ That inheritance is the transmission of genetic information from generation to generation.

End of topic checklist continued

○ How to describe genotype as the genetic make-up of an organism, and in terms of the alleles present.

○ That phenotype is the observable features of an organism.

○ That homozygous is having two identical alleles of a particular gene.

○ That two identical homozygous individuals that breed together will be pure-breeding.

○ How to describe heterozygous as having two different alleles of a particular gene.

○ That a heterozygous individual will not be pure-breeding.

○ That a dominant allele is an allele that is expressed if it is present in the genotype.

○ That a recessive allele is an allele that is only expressed when there is no dominant allele of the gene present in the genotype.

○ How to interpret pedigree diagrams for the inheritance of a given characteristic.

○ How to use genetic diagrams to predict the results of monohybrid crosses and calculate phenotypic ratios, limited to 1 : 1 and 3 : 1 ratios.

○ How to use Punnett squares in crosses which result in more than one genotype to work out and show the possible different genotypes.

End of topic questions

1. What are alleles?

   A all the genes on a chromosome

   B alternative forms of a gene

   C the recessive versions of genes

   D versions of genes passed to offspring

2. Write two sentences that correctly link all the following words to explain how they are related: *characteristic, chromosome, DNA, gene, nucleus, protein.*

3. Explain why there are two sexes in humans, using the inheritance of sex to support your answer.

4. The form of earwax in humans is controlled by one gene. The dominant allele produces wet earwax, and the recessive allele produces dry earwax.

   a) Using appropriate symbols, draw a genetic diagram to show the inheritance of earwax between a man with dry earwax and a woman who is heterozygous for the characteristic.

   b) Describe the predicted probability of genotypes and phenotypes in their children.

   c) Explain why it is possible that their three children all have dry earwax.

5. **SUPPLEMENT** **a) i)** Which type of cells have a haploid nucleus?

   ii) In human body cells, how many chromosomes does a haploid nucleus have?

   b) What is a diploid nucleus?

   c) If you counted the number of chromosomes in a human cell, how could you tell if the nucleus was diploid or haploid?

6. **SUPPLEMENT** Draw up a table to summarise the similarities and differences between mitosis and meiosis in terms of: number of cells produced, whether daughter cells are genetically identical or different, whether daughter cells are haploid or diploid, and what its purpose is.

7. **SUPPLEMENT** Explain why a life cycle needs a stage in which meiosis occurs before fertilisation.

Note: practice questions, sample answers and comments have been written by the authors. The marks awarded for these questions indicate the level of detail required in the answers. In examinations, the way marks are awarded may be different. References to assessment and/ or assessment preparation are the publisher's interpretation of the syllabus requirements and may not fully reflect the approach of Cambridge Assessment International Education.

Example answer

Question 1

People are either able to taste a chemical called PTC, or not. Being able to taste PTC is controlled by a single gene which has two alleles, T and t.

a) Genotype TT can taste PTC.

i) Give the term that describes the allele for being able to taste PTC. Explain your answer. (2)

Dominant ✓ ①

They are capitals. ✗

ii) Give the other possible genotypes related to tasting PTC, along with their phenotypes. (2)

Tt – The phenotype is that they can taste PTC. ✓ ①

tt – The phenotype is that they cannot taste PTC. ✓ ①

b) A couple who can both taste PTC have children.

i) Give the possible genotypes of the man and the woman. (2)

The couple could be TT ✓ ① *or Tt.* ✓ ①

ii) Their first child cannot taste PTC; the second one can.

Explain fully what this tells you about the genotypes of the couple. (4)

COMMENTS

a) i) Correct identification, but the explanation should be more specific, such as 'The two letters making up the genotype are written in upper case.'

ii) This is a good answer.

b) i) The answer is correct, but could be better worded. Rather than saying 'the couple', it is better to say that the man could be TT or Tt, and the woman could be TT or Tt.

ii) Both statements are correct, but the statements need explaining. The answer is correct, that unless both parents were Tt, all children would be able to taste PTC but the answer would benefit from two statements of explanation:

First of all, it should be made clear that as both the man and woman can taste PTC, they must each have at least one T allele.

There should then be a sentence of explanation to link the statements, such as:

'Without the presence of a t allele in both parents, all the children would be able to taste PTC.'

iii) The diagram illustrates the cross correctly, but lacks detail.

The best way of illustrating the cross is to use a Punnett square, showing each stage of the cross:

the genotypes of the parents

the different alleles that could be passed on to the offspring from the mother and father (the alleles in the egg cells and sperm cells)

the possible combinations of alleles in the offspring (genotypes)

the possible phenotypes produced.

		Mother (can taste PTC) possible alleles in eggs	
		T	t
Father (can taste PTC) possible alleles in sperm	T	TT can taste PTC	Tt can taste PTC
	t	Tt can taste PTC	tt cannot taste PTC

A further point is the way in which the answer shows the third possible genotype (tT). Although not incorrect, the convention is to write the dominant allele first, so it should be written Tt.

The genotype of both the man and women must be Tt. ✓

Because otherwise all the children would be able to taste PTC. ✓ ①

iii) Draw a genetic diagram to show the parents and their possible children. Include genotypes and phenotypes. (4)

Tt   x   Tt ✓

↓

TT   Tt   tT   tt ✓ ①

(Total 14 marks)

⑨⁄₁₄

Question 2

In which part of the breathing system does gas exchange occur? (1)

A alveoli

B bronchioles

C larynx

D trachea

(Total 1 mark)

Question 3

The circulatory system has several functions, including the transport of substances, temperature regulation, and defence against disease.

a) The diagram shows the structure of the mammalian heart.

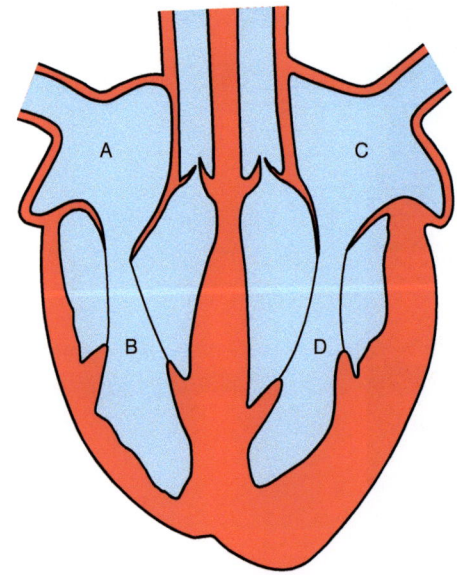

Identify the chambers of the heart, labelled A, B, C, and D. (4)

b) i) Describe the function of red blood cells. (1)

ii) Which component of blood is responsible for defence against infectious disease? (1)

iii) What is the role of platelets? (1)

iv) Name three things that can be dissolved in blood plasma. (3)

(Total 10 marks)

SUPPLEMENT Question 4

a) Name the part of the heart that controls the direction of the blood. (1)

b) Describe how backflow of blood in the heart is prevented. (2)

c) The blood system is involved in the body's immune defences.

Explain how a vaccination can make a person immune to a particular disease. **(5)**

(Total 8 marks)

Question 5

The nervous system is involved in the body's response to stimuli.

a) When a person puts her hand on a hot object, she removes it quickly using a reflex action.

   i) Draw a flow chart to describe the reflex arc involved. **(5)**

   ii) Almost immediately after she removes her hand, she realises that she has touched a hot object. Explain how this occurs. **(2)**

b) The hormone system is also involved in coordinating the body's responses.

Explain what a hormone is. **(4)**

(Total 11 marks)

SUPPLEMENT Question 6

The diagram shows the human female reproductive system.

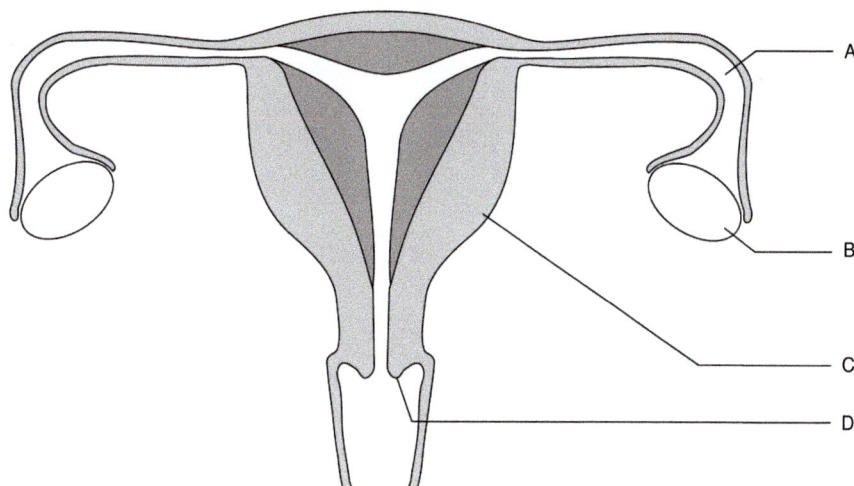

a) Identify the organs labelled A, B, C, and D **(4)**

b) i) Which organ is the site of egg production? **(1)**

  ii) Which organ is where the growing embryo develops into a baby? **(1)**

  iii) Describe the function of organ A. **(1)**

c) Using the key words 'gametes' and 'nuclei', describe the process of fertilisation. **(2)**

(Total 9 marks)

Flowers are adapted to be pollinated by insects or by the wind.

a) **i)** Identify parts A, B, C, and D on the diagram. (4)

　ii) Which structure from parts E–J are colourful modified leaves? (1)

　iii) Which structures from parts E–J are parts of the male reproductive system of the flower? (3)

b) Explain how each of the following structures is different in wind-pollinated flowers:

　i) anthers (2)

　ii) stigmas. (2)

(Total 12 marks)

SUPPLEMENT Question 8

This question is about human reproduction.

The diagram shows a human sperm.

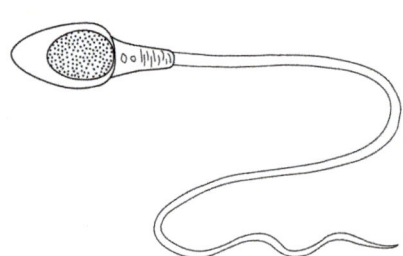

Explain how the sperm is adapted to fertilising a human egg. (4)

(Total 4 marks)

Question 9

A disease called cystic fibrosis is caused by a recessive allele of a gene.

The pedigree diagram shows the occurrence of cystic fibrosis in a family.

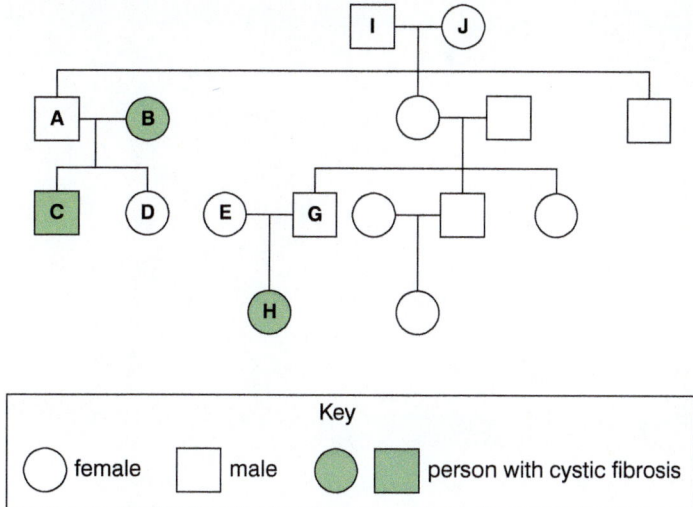

Key

○ female □ male ● ■ person with cystic fibrosis

a) Using the letters f for the allele causing cystic fibrosis and F for the 'healthy' allele, give the genotypes of the following members of the family. Explain how you came to your decisions.

   i) Father A.   (2)

   ii) Son C.   (2)

   iii) Parents E and G.   (3)

b) Explain what conclusions you can make about the genotypes of the couple I and J.   (4)

(Total 11 marks)

SUPPLEMENT Question 10

A horse has a chromosome number of 64.

a) State whether each statement refers to mitosis, meiosis, both or neither.   (5)

   i) The chromosome number in each daughter cell is 64.

   ii) The daughter cells are haploid.

   iii) Two identical cells are produced.

   iv) Some variability occurs in the alleles of parent and daughter chromosomes.

   v) Occurs when new red blood cells are produced in the blood of the horse.

b) Explain why a horse's gametes are produced by meiosis.   (5)

(Total 10 marks)

Even as children we recognise that some organisms are only usually found in particular places: fish in water, thick-furred mammals where it is cold, camels where it is dry. The adaptations of each organism help it to survive and produce young in that particular environment. Within each species, though, individuals are not identical. They show variation, that is little differences between them. Some are better suited to the environment than others. Natural selection by the environment will lead to the better-adapted individuals producing more young.

STARTING POINTS

1. What are some of the ways living things show variation?

2. What does it mean to say that organisms are adapted to their environments?

3. How can selection change the features of living things?

SYLLABUS SECTIONS COVERED

17.1 Variation

17.2 Selection

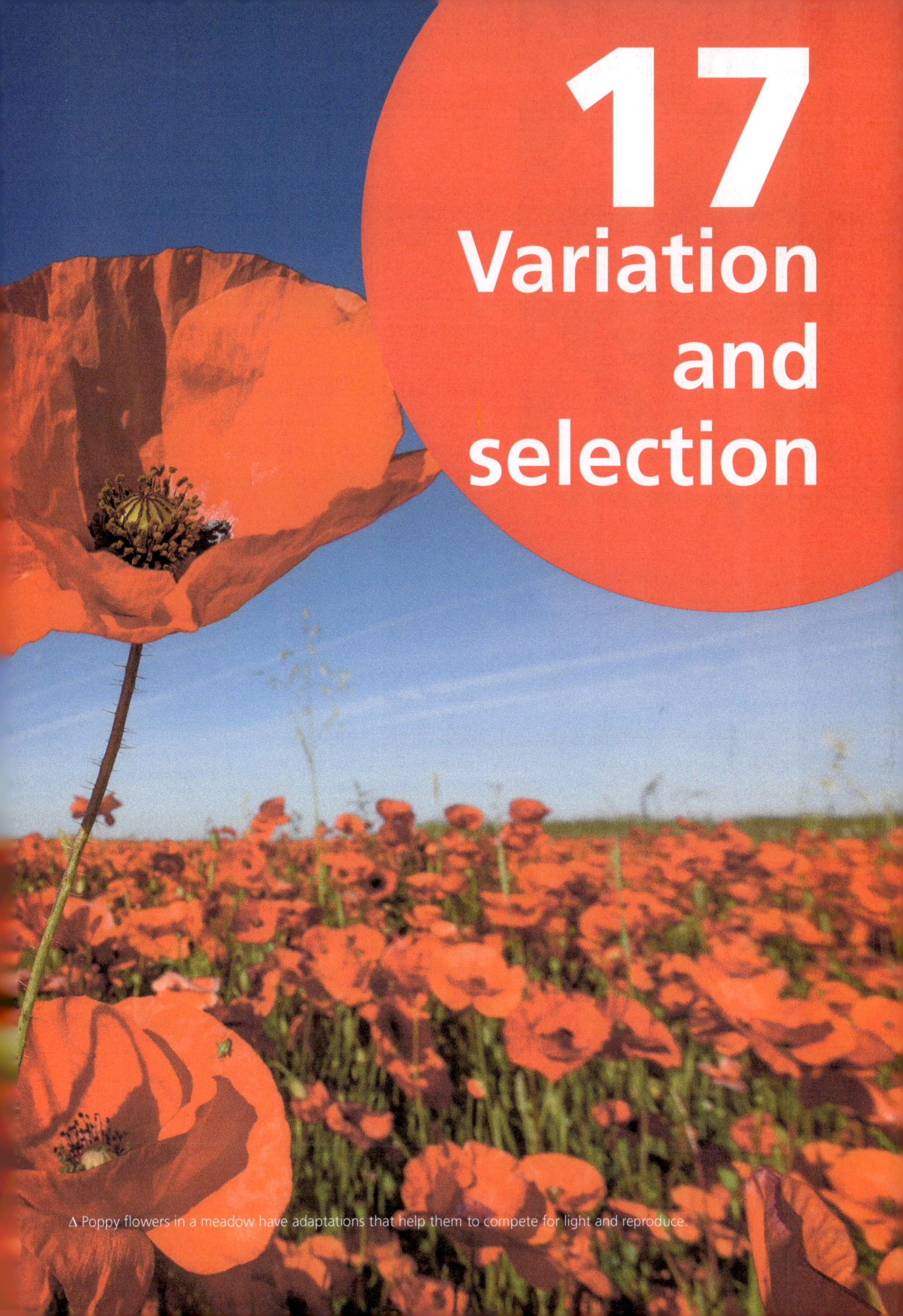

17
Variation and selection

Δ Poppy flowers in a meadow have adaptations that help them to compete for light and reproduce.

Variation and selection

Δ Fig. 17.1 How many variations can you spot between these individuals of the human species?

INTRODUCTION

There is a huge range in variation between all the people on Earth today. We differ in many characteristics, such as size, skin colour, strength, and hair type. Some of these characteristics are caused by genes and may stay the same throughout life. Some are also affected by the environment and so may change as we grow. Even though there is so much variation between people, we all belong to one species, *Homo sapiens*.

KNOWLEDGE CHECK

✓ Sexual reproduction results in offspring that are genetically different from each other.
✓ Individuals of the same species show variation in their features.

LEARNING OBJECTIVES

✓ Describe variation as differences between individuals of the same species.

✓ State that continuous variation results in a range of phenotypes between two extremes; examples include body length.

✓ State that discontinuous variation results in a limited number of phenotypes with no intermediates; examples include ABO blood groups.

✓ Describe mutation as a genetic change.

✓ State that mutation is the way in which new alleles are formed.

✓ Describe natural selection with reference to: genetic variation within populations; production of many offspring; struggle for survival, including competition for resources; a greater chance of reproduction by individuals that are better adapted to the environment than others; these individuals pass on their alleles to the next generation.

✓ **SUPPLEMENT** Describe the development of strains of antibiotic resistant bacteria as an example of natural selection.

✓ Describe selective breeding with reference to: selection by humans of individuals with desirable features; crossing these individuals to produce the next generation; selection of offspring showing the desirable features.

✓ Outline how selective breeding by artificial selection is carried out over many generations to improve crop plants and domesticated animals, and apply this to given contexts.

VARIATION

No two people are the same. Similarly, no two trees (even of the same species) are exactly the same in every way. For example, they have different heights, different trunk widths, and different numbers of leaves. **Variation** is what we call these differences between individuals of the same species.

Variation can be divided into two different types depending on how you are able to group the measurements.

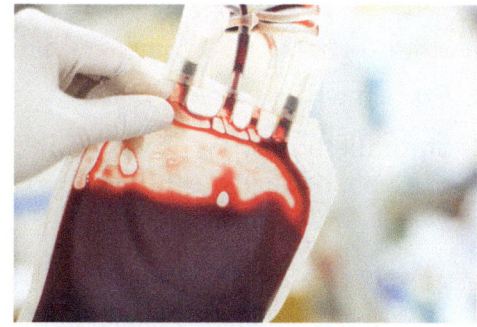

△ Fig. 17.2 ABO blood groups are an example of discontinuous variation.

- **Discontinuous variation** (sometimes called discrete variation) is where a characteristic can have one of a limited number of specific phenotypes with no intermediates. For example, ABO blood groups (you are group A, B, AB, or O).
- **Continuous variation** is where a characteristic can have any value in a range of phenotypes between two extremes. For example, body length (height).

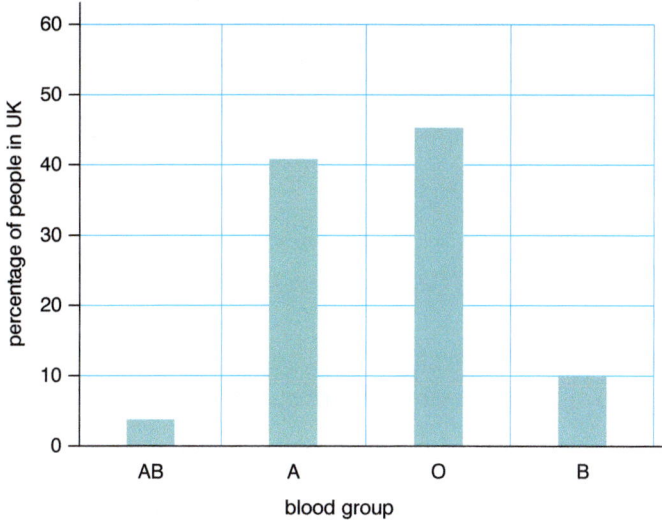

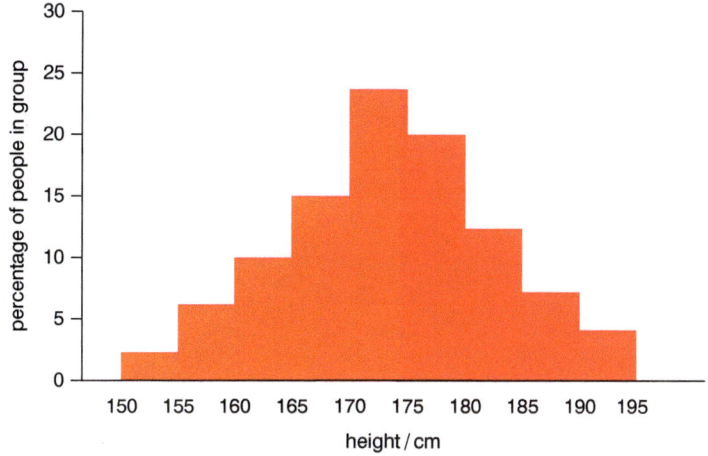

△ Fig. 17.3 Charts showing discontinuous variation in human blood groups (top) and continuous variation in human height (bottom).

1. Explain what is meant by *variation*.

2. State the two types of variation of characteristics in a species, and give an example of each.

Mutation

A **mutation** is a genetic change in the DNA of an individual. This is the way that new alleles (see Section 16) are formed.

1. Define the term *mutation*.

SELECTION

Different environments offer different challenges for survival and reproduction, including:

- effects of temperature
- effects of water
- lack of light.

In any environment the individuals within a population that have the best **adaptations** to the environment are most likely to survive and reproduce. This results in **natural selection** as follows.

- In any population of organisms there is genetic variation between the individuals.
- When organisms reproduce, they usually produce many offspring, more than the environment is able to support.
- This means that there is competition between the offspring for food, or for mates, or for other resources that are limited. This competition results in a 'struggle for survival'.
- Individuals that are better adapted to the environment than others will have a better chance of surviving to adulthood and reproducing.
- So the better-adapted individuals have a better chance of passing on their alleles to the next generation through reproduction. This means that in the next generation there will be a greater proportion of individuals with these alleles.

Example of natural selection

In any population of organisms, such as a species of flowering plant in a field, there will be variation in characteristics as a result of the variation in their alleles. For our example, let's consider height in sunflower plants.

- Some individual plants of this species will grow taller than others, as a result of variation in the alleles of their genes (see Fig. 17.4).

Δ Fig. 17.4 Taller plants are better adapted to receiving more sunlight on their leaves.

- Taller plants grow their leaves higher up the stalk, capturing more sunlight for photosynthesis and shading shorter plants so that those get less sunlight.
- So taller plants are able to make more food, and therefore make more seeds than shorter plants.
- Embryos in the seeds from the tall plants will have inherited the 'tallness' alleles from the parent plants.
- There will be more seeds with 'tallness' alleles, so when they germinate there are likely to be a greater proportion of tall plants in the next generation.
- Over more generations the average height of sunflower plants in this area will increase because taller plants are more likely to survive and reproduce than shorter plants.

It is a factor of the environment (light intensity) that has caused this change in the population. We call this *natural* selection, because a natural factor appears to 'select' individuals with some characteristics more than others, making it possible for them to pass on their alleles to the next generation more successfully.

SUPPLEMENT

Antibiotic resistance in bacteria

Antibiotics are chemicals that are used to kill bacteria when they cause infection (see Section 14). The first antibiotic that was developed was penicillin. Antibiotics were first used widely to treat injured soldiers during the Second World War and since then have saved millions of lives. However, more recently, the development of strains of **antibiotic-resistant bacteria** has become a major problem, causing many human deaths from infections each year (see Section 14 again).

The development of strains of antibiotic-resistant bacteria is a good example of natural selection in action. It happens as follows.

- A patient suffering from a bacterial infection is treated with an antibiotic, for example penicillin.
- The bacterial infection is caused by millions of bacteria of one species, and the individual bacteria within that population will show variation.
- Some bacteria, as a result of random mutation, may have an allele for a new characteristic that means the penicillin doesn't kill them as quickly as the other bacteria – we say these bacteria have developed antibiotic resistance.
- The few bacteria that are more resistant are more likely to survive and reproduce.
- The number of resistant bacteria in the patient may increase enough that the resistant bacteria escape from the body and are passed to another person.
- The newly infected person, if they become ill as a result of the infection, cannot be successfully treated with penicillin because it will not kill all the resistant bacteria; so the doctor will have to use a different antibiotic to control the infection.

Over time, some bacteria have developed resistance to a larger range of antibiotics, and many species show multiple resistance – resistance to many kinds of antibiotics. Now there are very few new antibiotics to use on multiple-resistant types of bacteria and doctors are concerned that there will be an uncontrollable increase in the numbers of deaths from bacterial infections that used to be treatable.

The rate of adaptation of a population is related to how strongly the new characteristic is favoured by natural selection. Antibiotics are an example of a *strong selecting factor* – only those bacteria that are most resistant to them will survive and the rest will die. So spread of antibiotic resistance happens quickly.

1. Define *natural selection*.

2. Explain the importance of the following factors in natural selection:

   a) variation in the population, **b)** competition, **c)** adaptive features.

3. **SUPPLEMENT** Explain how natural selection results in populations becoming better suited to the environment.

4. **SUPPLEMENT** Draw a diagram, such as a cartoon strip, to explain how antibiotic resistance develops in a bacterial species.

Selective breeding

Selective breeding (also known as **artificial selection**) can be used to improve crop plants and domesticated animals. It involves a human breeder choosing which parent animals or plants to breed from.

The steps in selective breeding are as follows.

- The human animal or plant breeder selects (chooses) individuals with desirable features.
- These individuals are crossed to produce the next generation.
- From the offspring, the breeder selects those individuals with the most desirable features.
- These individuals are then themselves used for breeding, and the whole process is repeated over many generations.

The choice is usually for desirable features that have economic importance. Examples include increasing the size of farm animals kept for meat, or increasing egg size in chickens, or increasing yields from crop plants. It may be done for other reasons too, such as to develop different breeds of dogs, or to produce new colours or shapes of flowers in plants grown for horticulture.

As many characteristics are controlled by genes, by breeding together organisms that show the nearest form to a desired characteristic the breeder is more likely to get offspring that also have this characteristic (such as large size). If the breeder continues to select individuals with the largest size, for example, over time the average size of the organisms will increase. Artificial selection may also be done to combine particular multiple characteristics. For example, an apple grower might want to produce apple trees that have fruit that are large, and sweet, and brightly coloured.

Example of selective breeding

One example of the improvement of a crop plant is the breeding of wheat plants to develop a variety that has a large seedhead (that gives lots of grain, which we use for food) but a short stalk (which means the plant wastes less energy in producing parts that we don't use, but also helps to support the large seedhead better in strong winds).

The plant breeders start by crossing a wheat plant with a long stalk and a large seedhead with a wheat plant that has a short stalk and small seedhead; there may be some offspring with the advantageous combination of short stalk and large seedhead.

Selecting from the offspring that have the best combinations of these characteristics over many generations can produce a new variety with the perfect combination of these characteristics.

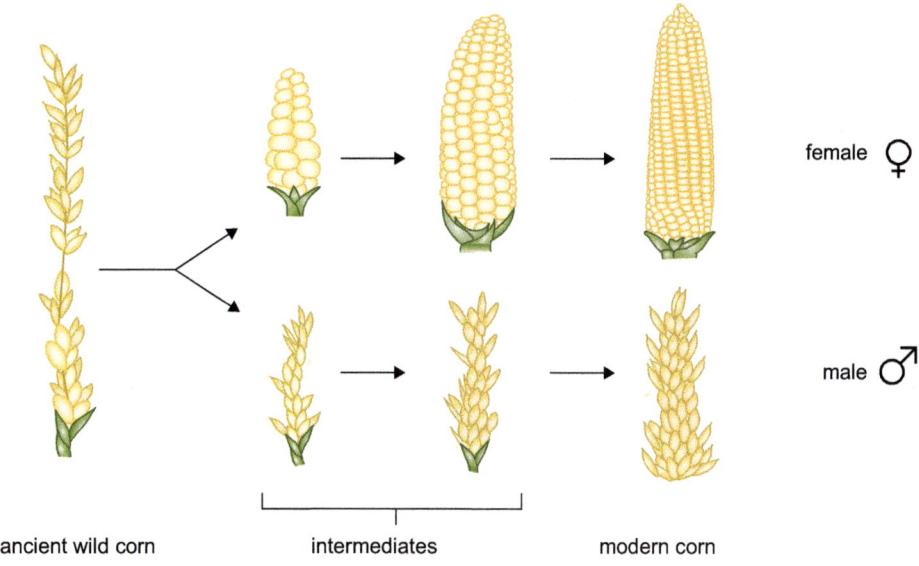

ancient wild corn   intermediates   modern corn

Δ Fig. 17.5 Over centuries of selective breeding, the ancient wild corn (maize) has developed into the modern varieties of corn that have the large seed cobs that we harvest.

The same process can be done with domesticated animals, by breeding together individuals that show the best features, to produce new breeds. For example, a farmer might want to produce a breed of goat that produces a lot of milk as well as high quality meat. It can take many generations to improve an animal; it can often take longer than with plants, as animals generally take longer to become mature enough for reproduction.

Note that plants of the same species but with distinctively different characteristics are called *varieties*, whereas animals of the same species with distinct characteristics are called *breeds*. As they are still of the same species, different plant varieties can interbreed and different animal breeds can interbreed.

 **SCIENCE IN CONTEXT**   **TULIP MANIA**

Plants are also bred in horticulture, for gardens, houseplants, and cut flowers, to improve the colour, shape, and form of the flowers and leaves. This is because people like new things.

For example, tulips were introduced to Europe in the 1500s from Turkey. They were so exotic that they became a luxury item that all wealthy people had to have.

Plant breeders rapidly developed new varieties through selective breeding, such as flowers with different-coloured lines or specks on the petals.

At the peak of 'tulip mania' in the Netherlands in the 1630s, single tulip bulbs were being sold for more than 10 times the annual income of a skilled craftsman. Prices suddenly collapsed in 1637, mainly because people got bored with tulips.

Challenge Question: Although tulip breeders have produced plants with very dark purple flowers, they have not been able to use selective breeding to produce a tulip with a perfectly black flower. Suggest an explanation why they have not been able to.

Δ Fig. 17.6 A completely black flower is almost impossible to breed, but that doesn't stop people wanting to try to produce it because many people would pay a lot of money for something so rare.

QUESTIONS

1. Give two examples of selective breeding.
2. Describe the steps in the process of selective breeding.

End of topic checklist

Key terms

adaptation, artificial selection, continuous variation, discontinuous variation, mutation, natural selection, selective breeding, variation

SUPPLEMENT antibiotic-resistant bacteria

During your study of this topic you should have learned:

○ That variation is differences between individuals of the same species.

○ That continuous variation results in a range of phenotypes between two extremes; examples include body length.

○ That discontinuous variation results in a limited number of phenotypes with no intermediates; examples include ABO blood groups.

○ How to describe mutation as genetic change.

○ That mutation is the way in which new alleles are formed.

○ How to describe natural selection in terms of: genetic variation within populations; production of many offspring; struggle for survival, including competition for resources; a greater chance of reproduction by individuals that are better adapted to the environment than others; these individuals pass on their alleles to the next generation.

○ **SUPPLEMENT** How to describe the development of strains of antibiotic resistant bacteria as an example of natural selection.

○ How to describe selective breeding in terms of: selection by humans of individuals with desirable features; crossing these individuals to produce the next generation; selection of offspring showing the desirable features.

○ That selective breeding by artificial selection is carried out over many generations to improve crop plants and domesticated animals, and how to apply this to given contexts.

End of topic questions

1. In a class of students, a survey was carried out of the variation shown in some features.

   a) Of the 35 students, 28 were able to roll their tongue and 7 could not. There were no students who could partly roll their tongue. State which type of variation this shows. Explain your answer.

   b) State the other type of variation in humans, and give an example.

   c) Explain how this type of variation differs from the type shown by tongue-rolling.

2. **SUPPLEMENT** Aphids are common insect pests of crop plants. In countries where the climate is seasonal, they reproduce in different ways at different times of the year. During the summer, when there is a lot of food about, wingless female aphids reproduce asexually. As it gets cooler towards winter and plants start to die back, winged males and females reproduce sexually. Fertilised eggs are laid that survive through the winter, and hatch in the spring.

   a) Describe and explain two features of reproducing asexually that allow aphids to be successful during the summer.

   b) Describe and explain one feature of sexual reproduction that helps aphids to be successful in the following year.

3. Using an example of your own choice, explain how natural selection can result in some individuals of a population passing on more copies of their genes to the next generation than others.

4. **SUPPLEMENT** Describe using words or pictures the development of strains of bacteria that are resistant to many types of antibiotics.

Human activity can have a huge impact on food chains and food webs, and even lead to the extinction of whole species. The introduction of just 24 wild rabbits into Australia in 1859 resulted in a rapidly spreading population that has decimated huge areas of diverse environment, causing havoc to native animal and plant species. Various attempts to keep the rabbit population in Australia in check have been attempted, for example, by introducing the rabbit disease myxomatosis. Even so, scientists estimate that rabbits still cost Australian farmers more than $200 million annually in lost production, in addition to the ongoing damage that they wreak on the natural environment.

STARTING POINT

1. Where does all the energy in living organisms ultimately come from?

2. What are the differences between food chains and food webs?

3. How is carbon recycled in nature?

SYLLABUS SECTIONS COVERED

18.1 Energy flow

18.2 Food chains and food webs

18.3 Carbon cycle

18
Organisms and their environment

△ Plants make their food by using light energy from the Sun.

△ Fig. 18.1 Alligators can survive for months without food, although they are always on the lookout for a good meal.

Organisms and their environment

INTRODUCTION

All animals need to eat, to provide the fuel for respiration. Some animals such as the common shrew need to consume two or three times their body weight of insects, slugs and worms every day in order to survive. They live life quickly, being on the hunt for food for most of the time, especially at night. By contrast, alligators only need to feed about once a week, and can live for months without food. They live life much more slowly than shrews, waiting in ambush for prey to get close before attacking. Most animals eat on average somewhere between these extremes, although adult mayflies have no mouthparts and never eat. They live their brief lives of a few days using energy stored from earlier stages in their life cycle, as their only purpose is to reproduce, after which they die.

KNOWLEDGE CHECK

✓ Respiration is the release of energy from food molecules.
✓ During photosynthesis, plants transfer energy from light to stored chemical energy in sugars.
✓ Energy released by respiration is used for a range of purposes, including making new body tissue.
✓ Organisms that feed on one another can be displayed in a food chain that shows who eats what.
✓ Food chains within a habitat can be combined to produce a food web.
✓ Decomposers digest dead organic material, releasing some of the products of digestion into the environment.

LEARNING OBJECTIVES

✓ State that the Sun is the principal source of energy input to biological systems.
✓ Describe the flow of energy through living organisms, including light energy from the Sun and chemical energy in organisms, and its eventual transfer to the environment.
✓ Describe a food chain as showing the transfer of energy from one organism to the next, beginning with a producer.
✓ Construct and interpret simple food chains.
✓ Describe a food web as a network of interconnected food chains and interpret food webs.
✓ Describe a producer as an organism that makes its own organic nutrients, usually using energy from sunlight, through photosynthesis.
✓ Describe a consumer as an organism that gets its energy by feeding on other organisms.
✓ State that consumers may be classed as primary, secondary, tertiary, or quaternary according to their position in a food chain.
✓ Describe a herbivore as an animal that gets its energy by eating plants.
✓ Describe a carnivore as an animal that gets its energy by eating other animals.

- ✓ Describe a decomposer as an organism that gets its energy from dead or waste organic material.
- ✓ Use food chains and food webs to describe the impact humans have through overharvesting of food species and through introducing foreign species to a habitat.
- ✓ **SUPPLEMENT** Describe a trophic level as the position of an organism in a food chain and food web.
- ✓ **SUPPLEMENT** Identify the following as the trophic levels in food webs and food chains: producers, primary consumers, secondary consumers, tertiary consumers, and quaternary consumers.
- ✓ **SUPPLEMENT** Explain why the transfer of energy from one trophic level to another is often not efficient.
- ✓ **SUPPLEMENT** Explain, in terms of energy loss, why food chains usually have fewer than five trophic levels.
- ✓ **SUPPLEMENT** Explain why it is more energy efficient for humans to eat crop plants than to eat livestock that have been fed on crop plants.
- ✓ Describe the carbon cycle, limited to: photosynthesis, respiration, feeding, decomposition, formation of fossil fuels, and combustion.

ENERGY FLOW

Plants make their food (sugars) from carbon dioxide and water using light energy from the Sun (see photosynthesis in Section 6). In systems terminology, sunlight is the energy input for plants.

Most food chains on the surface of the Earth begin with photosynthesising plants. This means that the Sun is the principal (main) input of energy into biological systems, such as food chains and food webs. As the energy from sunlight is transferred to chemical energy in the living organisms, energy flows through food chains and webs. Eventually, all the energy will be transferred to the environment, for example, being stored as chemical energy in fossil fuels, or transferred as heat to their surroundings by living organisms as they respire.

SCIENCE IN CONTEXT **ENERGY INPUT FROM THE SUN**

As a result of the curvature of the Earth, the amount of light energy from the Sun that falls on every square metre is greatest at the equator, and decreases as you move towards the poles. The tilt of Earth's axis in relation to the Sun causes variation in the amount of sunlight energy received by high latitude regions at different times of the year, causing seasons.

These differences in energy received have major effects on the living organisms in each region. Parts of the world near the equator that receive sufficient rainfall, such as tropical rainforests, have a greater productivity of plants in a year than other regions. This greater productivity supplies more food for animals, leading to a greater productivity of animals – some of these areas are the most biodiverse on the planet.

The seasonal effects in high-latitude regions result in rapid plant growth in summer months and virtually no growth during the winter, although some of this effect is the result of lack of heat energy from the Sun as much as lack of light energy.

Challenge Question: Seasonal variations in energy input from the Sun are responsible for many aspects of animal and plant behaviour. Explain how animal migration and seed germination are timed to ensure the most effective use of the Sun's energy.

FOOD CHAINS AND FOOD WEBS
Food chains

You should be familiar with food chains from your earlier work. A **food chain** shows the transfer of energy from one organism to another, or 'who eats what' in a habitat. For example, in Fig.18.2, owls eat shrews, shrews eat grasshoppers, grasshoppers eat grass. (Remember, the arrows in a food chain show the direction of energy flow.)

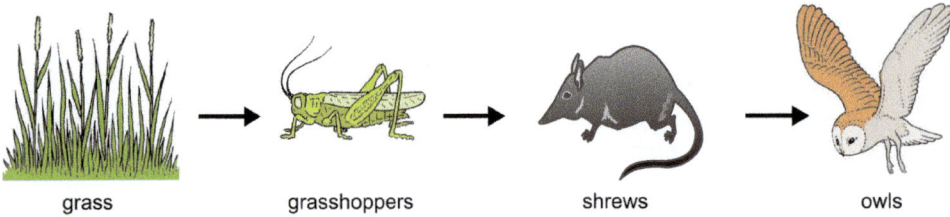

grass   grasshoppers   shrews   owls

△ Fig. 18.2 An example of a food chain.

Each stage in a food chain shows a separate level at which that species is feeding.

- Grass – this is the **producer** level. Producers are organisms that make their own organic nutrients (food). All food chains start with a producer. Although there are exceptions, producers are usually plants that use light energy in photosynthesis.
- Grasshoppers – these are the *primary consumers*, '**consumer**' because they eat the grass and 'primary' because they are the first eaters of other organisms in the food chain. This level may also be called **herbivores**, because they are animals that get their energy by eating plants.
- Shrews – these are consumers too, but they are specifically *secondary consumers* because they eat the primary consumers. They are also called **carnivores**, because they are animals that get their energy by eating other animals.
- Owls – these are also consumers, but they are specifically *tertiary consumers* because they eat the secondary consumers. They are also carnivores.

Animals at the top of the food chain may also be called the top consumers, or top **predators**. All animals are consumers, because they are organisms that get their energy by feeding on other organisms, in contrast to plants, which are all producers.

SUPPLEMENT

Each feeding level of a food chain is called a **trophic level**. So producers are one trophic level, primary consumers (or herbivores) are the trophic level that feeds on producers, and so on. If anything ate owls, they would be *quaternary consumers*, but food chains often don't reach that level. We also use the concept of trophic levels when describing the position of organisms in food webs and ecological pyramids, as you will see later in this section.

What isn't shown in a food chain is what happens to all the dead plant and animal material that isn't scavenged. This material decays as a result of the action of **decomposers**, such as fungi and bacteria. These organisms get their energy from the breakdown of dead or waste organic material. In the process, many of the broken down remains are released into the environment. Decomposers play an essential role in the environment, as you will see later when looking at nutrient cycles.

◁ Fig. 18.3 This fungus is growing through the dead tree and secreting enzymes that cause the wood to break down into simpler chemicals.

SCIENCE IN CONTEXT **OTHER PRODUCERS**

Not all producers are plants, and not all producers use light energy. There are some species of bacteria which produce their own food without the presence of light energy from the Sun. Instead they get the energy they need for the formation of sugars from chemical reactions. This whole process is known as *chemosynthesis*.

These bacteria are the source of food for food chains and webs that exist where there is no sunlight, such as deep in oceans and in underground caves. Be careful to avoid the statement that 'all life on Earth depends on the Sun', as this is an oversimplification and not totally accurate.

Challenge Question: If life does exist elsewhere in the Solar System, one place it may be found is on Europa, one of Jupiter's moons, below the icy surface of the water oceans that are thought to exist there. If there is life on Europa, scientists think it may rely on chemosynthesis rather than photosynthesis. Suggest why they think this.

Food webs

If we look more closely at food chains, it is rare to find an organism that is eaten by just one other species, or a predator that feeds on just one type of **prey**. It may also be the case that a predator may feed on different kinds of organism – an **omnivore**, for example, is a secondary or tertiary consumer when eating other animals, but a primary consumer when feeding on plants. So food chains within a habitat are linked together in a network to form a **food web**. A food web is a better description of the feeding relationships in a habitat and shows how living organisms are interconnected.

Food webs still usually group the organisms according to their feeding level. For example, in the simplified food web shown in Fig. 18.4, the rabbit, rat, mouse, seed-eating bird, and herbivorous insect are all primary consumers and are placed just above the producer level.

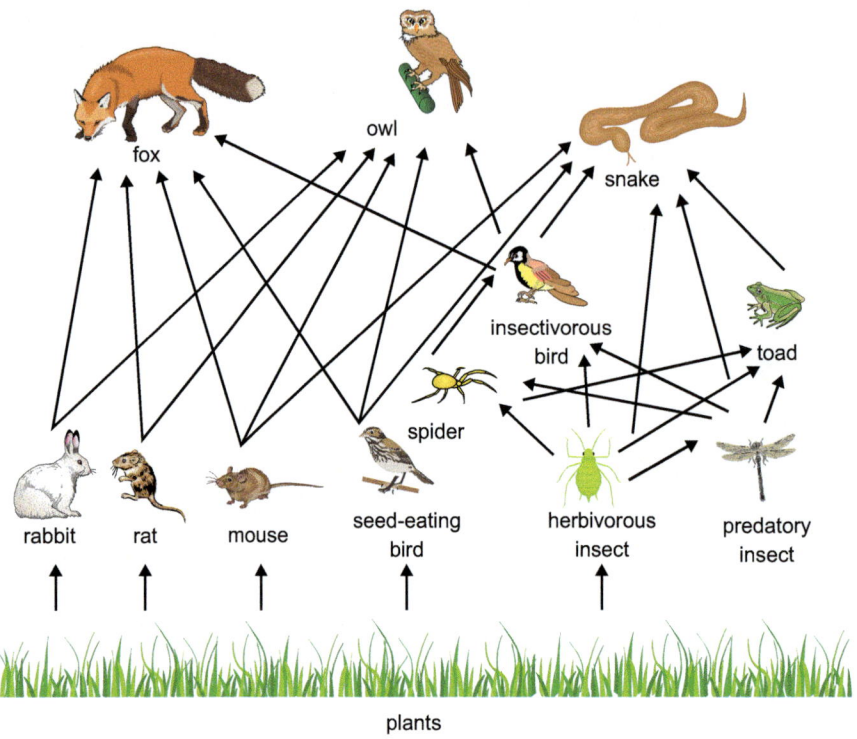

Δ Fig. 18.4 A simplified food web.

There are usually many more species in a habitat than shown in Fig. 18.4, and linking them all in one food web can get confusing. So food web diagrams may focus on the relationships between key organisms rather than all of them. For example, they may only include the most numerous species, or focus on the most vulnerable species. This can be helpful if you want to use the food web to predict what would happen if the food web were changed in some way, such as by human activity.

You could use the food web shown to predict what would happen if the plants were sprayed with a chemical that kills insects. This would kill the herbivorous insects and so reduce the amount of food available to all the animals that feed on them.

Interpreting human impact on food chains and food webs

We can use food chains and food webs to help us understand the wider impact on habitats that we have when we affect particular organisms, for example, by overharvesting particular food species. For example, along the Atlantic coast of the USA there has been overfishing of large shark species. These species are predators of smaller fish, such as skate and rays. These smaller fish feed on shellfish, including scallops.

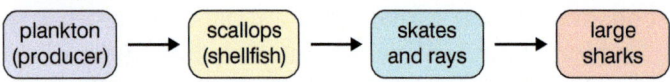

Δ Fig. 18.5 A marine food chain.

As the numbers of large sharks have decreased, there has been less **predation** of the smaller fish. So the numbers of skates and rays have greatly increased. This has had a major impact on the shellfish, with scallops becoming nearly extinct in several areas.

Many of the fish that we eat are predators, higher up in food chains, and so this effect of changing the **population** sizes of organisms lower in the food chains is being seen in many parts of the ocean.

We also affect food chains and webs when we introduce species from one area to another (introduce **foreign species**). This may happen intentionally, such as:

- to provide more food for humans (for example, introducing goats to provide meat and milk)
- to control a pest species (for example, introducing cane toads to Australia to control beetles that are pests of sugarcane plantations – see below)
- because the species is a pet (for example, cats).

Sometimes the introduction is accidental, such as the introduction of rats to some places because they were onboard the ships that transported humans to those places.

Goats have become a pest in many places, particularly on islands, because they eat much of the vegetation. This prevents new trees growing, and many of the local plants that are not adapted to being browsed like this also die out. Changing the plants that grow in the area will also change the animals that can live there.

Cats and rats have also become pests on islands because they eat many birds and their eggs. In New Zealand, many species of ground-nesting birds have become extinct because of these introductions.

Cane toads were introduced to Australia to control beetles that were attacking the sugarcane plantations in the northern regions. Unfortunately the toads didn't stay in the plantations, because they needed more shelter. So they moved out, and started eating small animals in other areas. This left less food for the predators of the local food web, including many species of small lizards. Population sizes of these lizards have decreased and some are at risk of extinction. The toad is now considered a pest in these areas.

Δ Fig. 18.6 Cane toads have glands in their skin that produce chemicals that are toxic to many animals. So there are few predators in Australia that can eat them.

SCIENCE IN CONTEXT **EXTINCTION IN HAWAII**

A deep hole on one of the Hawaiian islands provides a 10 000 year record of the effects of humans on the plants and animals that lived there. Before humans arrived, the only organisms must have arrived by chance on the wind or water. Only a limited number of species could travel the thousands of miles from the mainland to the islands. Since the time that they arrived, they evolved into a range of new species that were found nowhere else. The birds, in particular, evolved into a wide range of forms, including some that were too large to fly and behaved more like goats, grazing the plants.

Δ Fig. 18.7 The Laysan albatross is now protected when it nests on Midway Island in Hawaii, although cats are still a threat to its eggs.

About 900 years ago, the first bones of a rat appear in the deep hole. It probably arrived with people on a boat. More rat bones are found in the hole from since that time. Since 900 years ago, many island species have disappeared from the bone collection. Many species of birds, including the large species, became extinct. Only birds that nest on the islands and then leave for the rest of the year are still found in Hawaii. All the species of land snails, which were an important food for predators on the islands, became extinct, and this affected other species in the island food web.

The rats, and other species that humans brought to the island, including cats and goats, changed the island food web forever.

Challenge Question: New animals and plants have always arrived on islands – that's how islands became colonised in the first place. Discuss whether we should therefore be concerned by the arrival of new species brought by humans.

QUESTIONS

1. **SUPPLEMENT** Define the following terms: *producer, consumer, herbivore, carnivore, decomposer, trophic level*.

2. State the principal source of energy to biological systems. Explain your answer.

3. Describe the difference between a food chain and a food web.

4. Explain how food webs can be **a)** useful and **b)** difficult to construct.

SUPPLEMENT

Energy transfers in biological systems

Plants transfer the energy from light into stored chemical energy inside them. This energy is transferred to animals when they digest and assimilate plant food to make new substances in their own body tissues.

The energy that a plant receives from light, or that an animal gets in its food, is always greater than the amount of energy it stores in the substances in its tissues. This is because some of the energy that it takes in is transferred to the environment in various forms. This means that the transfer of energy from one trophic level to the next is never 100% efficient, and in fact, is only on average about 10% efficient. In other words, on average, 90% of the energy at one trophic level is not passed onto the next (i.e., it is 'lost'). This is the reason why ecological pyramids usually get smaller as you go up the trophic levels, and why food chains usually have fewer than five trophic levels. There are more details about this below.

Energy losses from plants

The amount of energy from sunlight that falls on the Earth's surface varies at different times of day and year, and varies in different parts of the world (with places near the equator receiving more light energy than places nearer the poles). On average, tropical areas receive between 3 and 5 kWh/m² per day (which is about the same energy as a one-bar electric heater left on for 3–5 hours).

Plants use only a tiny proportion of this for many reasons, as shown in Fig. 18.8. It has been estimated that most plants only transfer about 1–2% of the energy in the light that falls on them into chemical energy in their tissues (**biomass**). This is the energy potentially available to a herbivore that eats the plant.

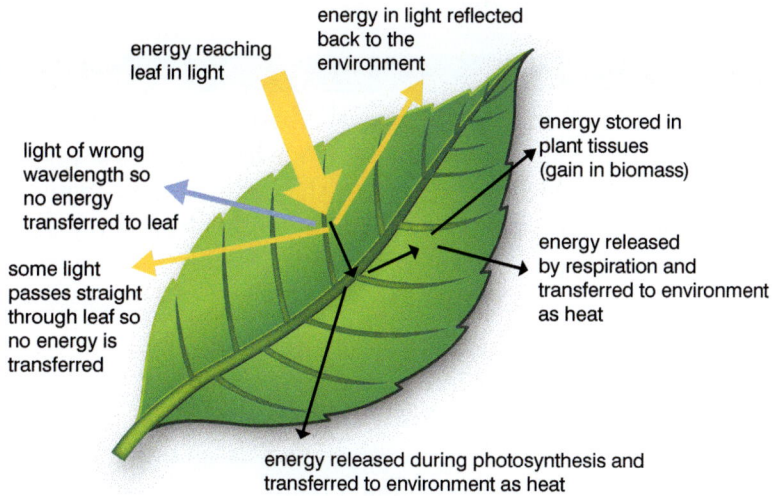

energy reaching leaf in light

energy in light reflected back to the environment

light of wrong wavelength so no energy transferred to leaf

energy stored in plant tissues (gain in biomass)

some light passes straight through leaf so no energy is transferred

energy released by respiration and transferred to environment as heat

energy released during photosynthesis and transferred to environment as heat

Δ Fig. 18.8 Energy gains and losses of a plant.

Energy losses in animals

When an animal eats, the food is digested in the alimentary canal and the soluble food molecules are absorbed into the body. The undigested and unabsorbed food in the alimentary canal is egested as faeces (see Section 7).

Absorbed food molecules may be used for different purposes in the body:

- to produce new tissue or gametes for reproduction
- as a source of energy for respiration
- converted to waste products in chemical reactions.

The energy stored in the food molecules may stay in the body stored in body tissue, or it may be transferred to the environment stored in waste materials such as urine.

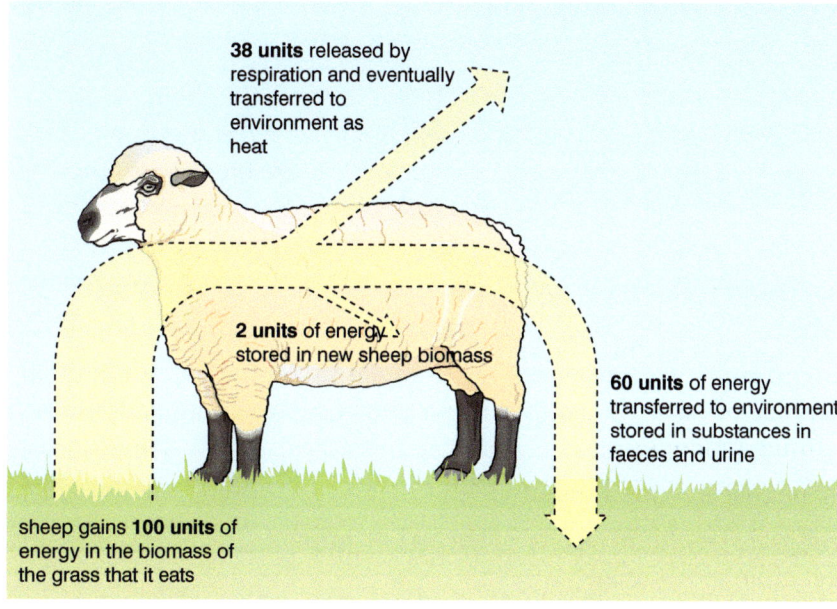

38 **units** released by respiration and eventually transferred to environment as heat

2 **units** of energy stored in new sheep biomass

60 **units** of energy transferred to environment stored in substances in faeces and urine

sheep gains **100 units** of energy in the biomass of the grass that it eats

Δ Fig. 18.9 The energy flow through a sheep.

When food molecules are broken down during respiration and other reactions, some of the energy released from the molecules is transferred to the environment as heat, through the processes of conduction, convection, and radiation. So only a small proportion of the energy stored in the animal's food is converted into energy stored in its body tissues as an increase in the animal's biomass.

QUESTIONS

1. Draw a flowchart to show the energy gains by, and losses from, a plant leaf.

2. Draw a flowchart to show the energy gains by, and losses from, a herbivore.

3. Explain why the amount of energy stored in an organism's tissues is always less than the amount of energy that it gained.

Developing practical skills

Some students were collecting data on the abundance of plants and snails on some school grounds so that they could construct ecological pyramids.

Devise and plan investigations

1. Draw a food chain for these organisms, and identify each trophic level.

Human food chains

Humans are *omnivores*, meaning we get our energy from eating both plant and animal tissues. This gives us choice of what we eat, but these choices have an impact on the energy efficiency of farming practices.

Consider two food chains that include humans:

- crop plant (e.g. wheat grain) → herbivore (e.g. sheep) → human
- crop plant (e.g. wheat grain) → human

If we consider these food chains in terms of energy flow, we can see that if humans eat the grain, there is much more energy available to them than if they eat the sheep that eat the grain. This is because energy will be lost from the sheep in the food chain. So it is energetically more efficient within a crop food chain for humans to be herbivores, and eat the crop plants, than to be carnivores, and eat the livestock that has been fed on crop plants. One way that this energy efficiency can be seen is that a certain area of crop land can produce food to feed far more people if it is used to grow crop plants for humans to eat than if it is used to grow plants to feed livestock.

In reality, we often eat animals that eat plants that either we cannot eat (e.g. grass), or that are too widely distributed for us to collect (e.g. tiny algae in oceans, which form the food of many fish that we eat).

THE CARBON CYCLE

Unlike energy, which is transferred through organisms and eventually to the environment in a way that is not useful to the organisms, nutrients continually transfer between the environment and organisms and back again, in what is described as **nutrient cycles**. One example is the **carbon cycle**.

Carbon is continually cycled through the living and non-living parts of the environment, in different forms at different stages of the carbon cycle. Carbon dioxide from the atmosphere is converted to complex carbon compounds in plants during photosynthesis. This is often called the 'fixing' of carbon by plants. Respiration in plants returns some of this fixed carbon back to the atmosphere as carbon dioxide. Carbon in the form of complex carbon compounds passes along food chains when animals feed on plants or other animals. At each stage of a food chain, some of this carbon is released as carbon dioxide back to the atmosphere as the result of respiration.

When organisms die, or produce waste, their bodies and waste decay as they are digested by decomposers. This is also known as **decomposition**. Carbon dioxide is released when the decomposers respire using the complex carbon compounds from the dead and waste material.

△ Fig. 18.10 Water excludes air from the ground, which prevents decay organisms from respiring. So dead plant material in waterlogged ground builds up over time, forming peat. Peat can be burnt as a fuel, although this is being discouraged so that peat bog habitats can be protected.

Combustion

If dead organic material is buried too quickly by sediment or water for decomposers to cause decay, and remains buried, then it may be converted to other complex carbon compounds. Peat is formed when mosses and other plants are buried in swampy ground for hundreds of years. Over many millions of years, where there were once huge forests growing in swampy regions, heat and pressure have turned the organic material into coal. Heat and pressure over many millions of years also produces oil and natural gas from the decaying bodies of tiny marine organisms that were buried in sediment at the bottom of oceans. Peat, coal, oil and natural gas are **fossil fuels**. We can release the carbon from the complex carbon compounds in fossil fuels into the air as carbon dioxide during **combustion**, when we burn them.

Make sure you understand what form carbon is in (carbon dioxide or complex carbon compounds such as carbohydrates) at each stage of the carbon cycle.

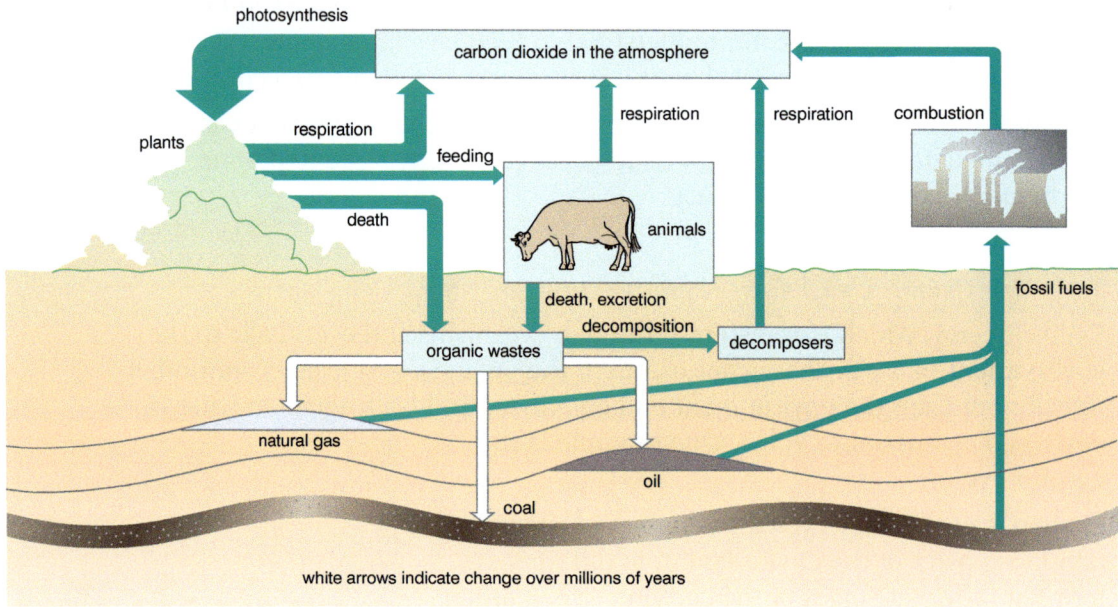

△ Fig. 18.11 A summary of the carbon cycle.

QUESTIONS

1. Describe the role of the following in the carbon cycle:

   a) respiration

   b) photosynthesis

   c) decomposition.

2. State in what form carbon is when it is in the following stages of the carbon cycle.

   a) Earth's atmosphere

   b) plant tissue

   c) fossil fuels.

End of topic checklist

Key terms

biomass, carbon cycle, carnivore, combustion, consumer, decomposer, decomposition, foreign species, food chain, food web, fossil fuel, herbivore, nutrient cycle, omnivore, population, predation, predator, prey, producer

SUPPLEMENT trophic level

During your study of this topic you should have learned:

○ That the Sun is the principal source of energy input to biological systems.

○ How to describe the flow of energy through living organisms, including light energy from the Sun and chemical energy in organisms, and its eventual transfer to the environment.

○ That a food chain is the transfer of energy from one organism to the next, beginning with a producer.

○ How to construct and interpret simple food chains.

○ That a food web is a network of interconnected food chains.

○ How to interpret food webs.

○ That a producer is an organism that makes its own organic nutrients, usually using energy from sunlight, through photosynthesis.

○ That a consumer is an organism that gets its energy by feeding on other organisms.

○ That consumers may be classed as primary, secondary, tertiary, or quaternary according to their position in a food chain.

○ That a herbivore is an animal that gets its energy by eating plants.

○ That a carnivore is an animal that gets its energy by eating other animals.

○ That a decomposer is an organism that gets its energy from dead or waste organic material.

○ How to use food chains and food webs to describe the impact humans have through overharvesting of food species and through introducing foreign species to a habitat.

○ **SUPPLEMENT** How to identify the following as the trophic levels in food webs and food chains: producers, primary consumers, secondary consumers, tertiary consumers, and quaternary consumers.

End of topic checklist continued

○ **SUPPLEMENT** How to explain why the transfer of energy from one trophic level to another is often not efficient.

○ **SUPPLEMENT** How to explain, in terms of energy loss, why food chains usually have fewer than five trophic levels.

○ **SUPPLEMENT** How to explain why it is more energy efficient for humans to eat crop plants than to eat livestock that have been fed on crop plants.

○ How to describe the carbon cycle, limited to: photosynthesis, respiration, feeding, decomposition, formation of fossil fuels, and combustion.

End of topic questions

1. The photograph below shows African lions eating a dead zebra. Before the lions killed the zebra, the zebra had been feeding on grass.

   a) State whether the lion is a carnivore or a herbivore. Explain your answer.

   b) State at which trophic level of a food chain the zebra feed.

   c) Draw a food chain for the organisms described above.

   d) Lions also feed on the herbivores gazelle and wildebeest. Use all the organisms in this question to draw a food web for the African grassland.

2. In a tropical forest, the layer of dead leaves (called the leaf litter) on the forest floor is usually very thin at all times of the year. In temperate woodlands (where there are seasons of summer and winter), many trees drop their leaves in the autumn and grow new ones in the spring.

   a) Tropical trees drop a few leaves at a time at any time of year. State what happens to the leaves on the ground. Explain your answer as fully as possible.

   b) The leaf litter in a temperate woodland is deep all through winter, when it may be cold enough for snow, until spring arrives. Then the leaf litter disappears. Suggest an explanation for these observations.

3. Use the food web in Fig. 18.4 to predict what would happen to the following species if all the herbivorous insects were killed by a chemical. Explain your answers.

   a) Predatory insects.

   b) Insectivorous birds.

   c) Mice.

   d) Snakes.

4. **SUPPLEMENT** One food chain from a British woodland is made up of four organisms:

   oak tree leaves → caterpillars → treecreeper bird → hawk

   If on average only 10% of the input energy is passed on at each trophic level what percentage of the original energy stored in the oak leaves is available to the hawk?

5. **SUPPLEMENT** Explain as fully as you can why a food chain is unlikely to include more than five trophic levels.

6. **SUPPLEMENT** Food chains in northern regions on Earth may be much shorter than food chains in tropical rainforests. Thinking only in terms of energy, suggest an explanation for this difference.

7. **SUPPLEMENT** A farmer feeds wheat grain grown in his fields to his chickens, to fatten them up for meat. Evaluate this in terms of energy efficiency.

8. The graph below shows the change in carbon dioxide concentration above a forest over two days, and the light intensity just above the top of the trees.

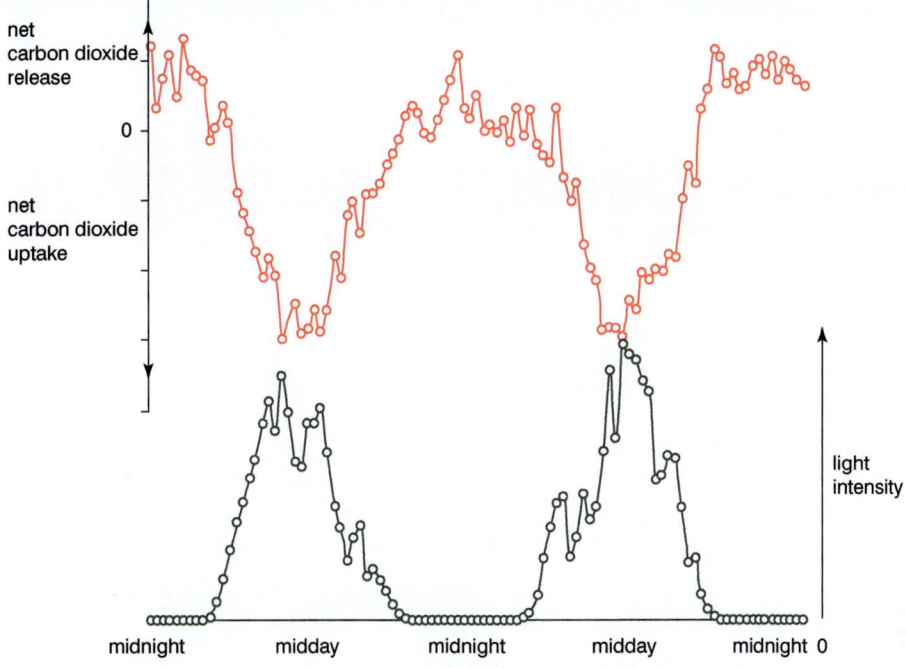

a) Explain the changes in light intensity shown in the graph.

b) Explain the changes in carbon dioxide concentration shown in the graph. (Remember that there are more organisms than just the trees in the forest.)

There is probably no place on Earth that is not affected by human activity. An estimated 40% of the land surface is cultivated to produce food, either from crops or animals. Changing land use, so that it produces food or provides places for us to live and work, destroys habitats for other organisms. The waste we produce causes pollution of land, water, and air unless it is properly controlled. As the human population grows, we must find ways of reducing our impact on ecosystems, so that we conserve resources and the many organisms with which we share the Earth.

STARTING POINT

1. Why are some animal species becoming endangered?

2. How can we conserve endangered animal species?

SYLLABUS SECTIONS COVERED

19.1 Habitat destruction

19.2 Conservation

19
Human influences on ecosystems

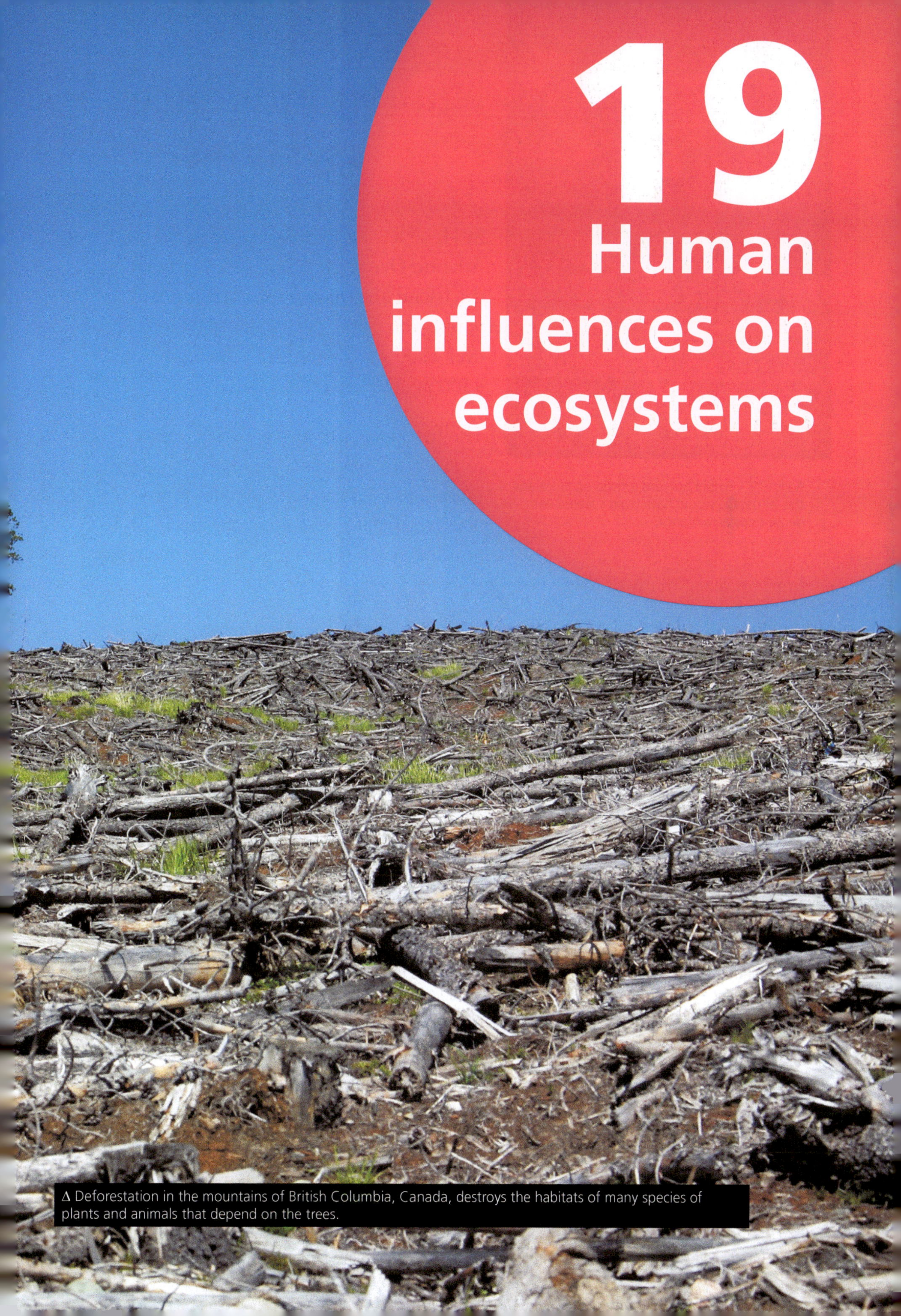

△ Deforestation in the mountains of British Columbia, Canada, destroys the habitats of many species of plants and animals that depend on the trees.

Human influences on ecosystems

△ Fig. 19.1 Before humans arrived on St Helena this landscape would have been covered with dense tropical rainforest.

INTRODUCTION

St Helena is an isolated island in the Atlantic Ocean. The first people to reach the island arrived in 1502. The human population of the island slowly increased to over 1000 people in the 1700s, and around 4500 people live there now.

The first people to arrive on St Helena found many plants and animals that were unique to the island. As people cleared the dense tropical forest, to make space for building and for growing crops and keeping herd animals, many of these unique species became extinct.

The introduction of animals that didn't naturally live there, such as cats, goats, and rats, also had a devastating effect on wildlife. Cats catch and kill small animals and birds, and rats steal and eat eggs from bird nests. Many of the lower areas near the sea are now completely bare of vegetation as a result of grazing by goats.

KNOWLEDGE CHECK

✓ Earth's atmosphere is affected by human activity.
✓ Humans can impact species through overharvesting and introducing foreign species.

LEARNING OBJECTIVES

✓ Describe an ecosystem as a unit containing the community of organisms and their environment interacting together.
✓ Describe biodiversity as the number of different species that live in an area.
✓ Describe the reasons for habitat destruction, including: increased area for housing, crop plant production, and livestock production; extraction of natural resources; and freshwater and marine pollution (a detailed description of eutrophication is not required).
✓ State the undesirable effects of deforestation as an example of habitat destruction, to include: reducing biodiversity, extinction, loss of soil, flooding, and increase of carbon dioxide in the atmosphere.
✓ **SUPPLEMENT** Explain the undesirable effects of deforestation as an example of habitat destruction, to include: reducing biodiversity, extinction, loss of soil, flooding, and increase of carbon dioxide in the atmosphere.
✓ Outline why organisms become endangered or extinct, including: climate change, habitat destruction, hunting, overharvesting, pollution, and introduced species.
✓ Describe how endangered species can be conserved, limited to: monitoring and protecting species and habitats; education; captive breeding programmes; and seed banks.

HABITAT DESTRUCTION

Biodiversity refers to the number of different species that live in an area. An ecosystem is a unit scientists use to describe a community of organisms and their environment interacting together. Different types of ecosystem can be complex webs of interactions between many different species. Some **habitats**, such as tropical rainforest, naturally have a high biodiversity. Other habitats, such as in the Arctic and Antarctic, may have a much lower biodiversity, but this does not mean they are unimportant, as species live there that do not live anywhere else. Many people are very concerned about the effect that habitat destruction around the world is having on biodiversity.

As you saw in Section 18, food chains and food webs can be altered when humans deliberately, or accidentally, introduce new species into an area and/or cause the disappearance of existing species. The negative impacts on a habitat may involve the populations of some species decreasing. Some may even become **extinct**. **Extinction** occurs when there are no individuals of that species left at all.

We change and destroy habitats not only when we grow large areas of crops, or grassland for grazing livestock. We also use large areas of ground for building houses, factories, and offices, and the road networks that link them.

Habitat destruction also occurs when we clear land to extract natural resources such as limestone rocks, coal, and minerals (e.g. iron and copper). **Pollution** from the mining activities can damage the environment even further by adding poisonous chemicals to the water, or by creating large areas of waste.

△ Fig. 19.2 This open-cast mine for copper-containing rocks has cleared an area the size of a mountain.

Habitat destruction can also happen in freshwater and marine environments as a result of pollution caused by human activity, for example:

- oil spills from container ships or from oil rigs can poison marine animals and clog up the feathers of sea birds
- chemical fertilisers and pesticides that get into in river water can cause the death of many organisms
- warming of the oceans as a result of **climate change** caused by the *enhanced greenhouse effect* (see later) can make it unsuitable for the growth of corals, which is resulting in the large-scale damage of many coral reefs around the world.

△ Fig. 19.3 This river kingfisher has been badly affected by an oil spill.

Deforestation

Deforestation is the permanent destruction of large areas of forest and woodland habitats. It happens, for example, in areas that provide quality wood for furniture, such as the tropical hardwood forests of Malaysia, or to create farming or grazing land (all over the world).

Deforestation can result in many kinds of damage to the environment and the organisms that live there. Biodiversity will be reduced, as species will no longer survive and may become extinct. Soil is lost which can lead to increased risk of flooding, and carbon dioxide levels will increase in the atmosphere.

SUPPLEMENT

Effects of deforestation

Loss of plant species due to deforestation will result in a loss of animal species in the same community, because the animals use the plants for food or shelter. Many tropical rainforests are areas of high biodiversity, where many organisms live. They also contain many species found nowhere else because they have evolved together in an energy-rich and relatively unchanging environment. Destruction of tropical rainforests, such as in the Amazon Basin, is causing not only a reduction in biodiversity in those areas, but also a high rate of extinction of species that live nowhere else.

Forests act as a major carbon store because carbon dioxide is taken up from the atmosphere during photosynthesis and used to produce the chemical compounds that make up trees. When forests are cleared, and the trees are either burnt or left to rot, this carbon is released quickly as carbon dioxide. This rapidly increases the proportion of carbon dioxide compared with oxygen in the air surrounding the forest. On the scale of deforestation in the Amazon Basin, the amount of carbon dioxide released into the atmosphere is so great that the increase cannot be brought back into balance as a result of photosynthesis.

Deforestation also has an effect on the natural cycling of water. Trees, like other plants, draw ground water up through their roots and release it into the atmosphere by transpiration (see Section 8). As forest trees are removed, not only is there less transpiration, but the amount of water that can be held within plants decreases, which in turn means more rainwater is flowing into rivers, increasing the risk of flooding (see below).

Removing the protective cover of vegetation from the soil can also result in **soil erosion**. This is where the soil is washed away by rain. The top layers of soil are the ones that contain the most nutrients, from the decay of dead vegetation, so soil erosion removes essential nutrients from the land. Soil nutrients are also lost by **leaching**, which is the soaking away of soluble nutrients in soil water because there are few plant roots in the soil to absorb the nutrients and lock them away in plant tissue. This loss of nutrients from the soil is permanent, and makes it very difficult for forest trees to regrow in the area, even if the land is not cultivated.

As soil is washed away, it enters rivers, reducing the flow of water or completely blocking them. In addition, we have already seen that the removal of vegetation means more water is entering rivers. The result of more water flowing into blocked rivers can be severe flooding in the surrounding areas.

△ Fig. 19.4 This satellite image of a river estuary in Madagascar shows large amounts of soil in the water (orange). This is a result of deforestation near the river.

CONSERVATION

Conservation refers to the protection of species and their habitats, so that the species can continue to survive and reproduce successfully. If the number of organisms in a population falls too low, the organisms may have difficulty in finding mates for reproduction and the population will die out. If this happens to all the populations of these organisms, the species will become extinct.

The International Union for the Conservation of Nature (IUCN) Red List shows which species are **endangered** (at risk of extinction). Regular monitoring of endangered species is used to help conservationists prioritise which species, and which habitats, to protect.

Organisms can become endangered (or extinct) as a result of several human activities.

- Climate change may occur as a result of changes in the atmosphere from combustion of fossil fuels and from methane. If conditions change too much, the organisms may not be adapted to the new conditions and so will die, or move to areas to which they are better suited. For example, plants and animals adapted to living in cold places may need to move nearer to the polar regions or higher up mountains, if the climate gets warmer.
- The destruction of habitats, such as by deforestation or pollution, destroys the places where organisms live. If the destruction is on a large scale, there may be nowhere else left for the organism to live.
- Some species have been hunted to extinction, or to near extinction. For example, the dodo was hunted to extinction in 1681, and the last known thylacine, or Tasmanian tiger, died in captivity in 1936.
- **Overharvesting** is when a species that could be managed sustainably is overexploited and so many are taken that the population falls to dangerous levels. The term can apply to both animal and plant species (unlike 'hunting' which usually just refers to killing animals).
- Pollution can poison organisms or change the conditions in the habitat so that they can no longer live there.
- Introducing species to an area can cause extinction by affecting the food web. For example, the introduced species may become a predator of the local species – as happened with the cane toad in Australia or rats in Hawaii (see Section 18). The introduced species will also affect local species if it becomes a competitor for food or space.

△ Fig. 19.5 This skin of a Siberian tiger was taken from poachers by an anti-poaching force. Tiger numbers in the wild are now so small that all species may become extinct in the next decade or so.

Endangered species may be conserved by breeding them in safe places. For example, animals may be kept in zoos, wildlife parks, or nature reserves and bred in **captive breeding programmes** to increase their numbers. Botanic gardens are places where plants that are at risk of extinction in the wild are kept and conserved. Endangered plant species may be conserved by growing them or by keeping their seeds in **seed banks**. However, if the habitats they lived in are not conserved as well, then there is little point in trying to return the species to the wild.

SCIENCE IN CONTEXT

THE HAWAIIAN GOOSE

A successful example of conservation is the survival of the Hawaiian goose. When James Cook visited the island of Hawaii in 1778 there was an estimated population of around 25 000. As a result of hunting the geese for food, and the introduction by people of predators such as cats, there were only about 30 birds by 1952.

A few breeding pairs were brought back to the Slimbridge reserve in the UK in the 1950s, where they bred successfully. Since then more birds have been bred in zoos and wildfowl sanctuaries around the world. Some have also been successfully re-introduced to national parks on an island near Hawaii where they are protected.

△ Fig. 19.6 A Hawaiian goose.

Challenge Question: The Hawaiian goose was reintroduced onto an island near Hawaii. Suggest why it wasn't reintroduced to Hawaii itself.

Education to change people's attitudes plays an important part in conserving endangered species. For example, people used to travel to Africa to hunt and kill large animals such as elephants and lions, but now they visit as tourists to photograph them instead. Elephants are still at risk of being shot, either for their ivory tusks or because they damage crops of local people. Changing attitudes of people who might buy ivory products will reduce the market for ivory, and placing bee hives near crops (elephants can hear the bees and tend to avoid going close to them) could help to protect crops without harming the elephants.

QUESTIONS

1. Explain the meaning of these terms:

   a) *endangered*

   b) *captive breeding programme*

   c) *seed bank*.

2. Identify three ways in which human activities may cause a species to become endangered.

End of topic checklist

Key terms

biodiversity, captive breeding programme, climate change, conservation, deforestation, endangered, extinction, habitat, leaching, overharvest, pollution, seed bank, soil erosion

During your study of this topic you should have learned:

- ○ That an ecosystem is a unit containing a community of organisms and their environment interacting together.

- ○ That biodiversity is the number of different species that live in an area.

- ○ That the reasons for habitat destruction include: increased area for housing, crop plant production, and livestock production; extraction of natural resources; and freshwater and marine pollution.

- ○ **SUPPLEMENT** How to explain the undesirable effects of deforestation as an example of habitat destruction: reducing biodiversity, extinction, loss of soil, flooding, and increase of carbon dioxide in the atmosphere.

- ○ How to explain why organisms become endangered or extinct: climate change, habitat destruction, hunting, overharvesting, pollution, and introduced species.

- ○ That endangered species can be conserved by methods including: monitoring and protecting species and habitats; education; captive breeding programmes; and seed banks.

End of topic questions

1. What are the correct definitions for "ecosystem" and "community"?

	Ecosystem	Community
A	A population and its environment interacting together.	A community and their environments interacting together.
B	All of the individuals of different species in an ecosystem.	All of the populations of different species in an ecosystem.
C	All of the populations of different species in an ecosystem.	A community and their environments interacting together.
D	A community and their environments interacting together.	All of the populations of different species in an ecosystem.

2. **a)** Describe biodiversity.

   b) Describe three reasons for habitat destruction.

3. In the Amazon Basin large areas of rainforest have been cut down to make space for growing livestock.

   a) Give one reason why some people have decided to keep livestock instead of leaving the rainforest to grow.

   b) Suggest one advantage of this change, and explain your answer.

   c) Give two disadvantages of this change.

4. **SUPPLEMENT** If rainforest is cleared on a large scale and then left to recover, it is rare that the same species of plants and animals return to the area, even after many years. Explain why this happens.

Practice questions for Sections 17–19

Note: practice questions, sample answers and comments have been written by the authors. The marks awarded for these questions indicate the level of detail required in the answers. In examinations, the way marks are awarded may be different. References to assessment and/or assessment preparation are the publisher's interpretation of the syllabus requirements and may not fully reflect the approach of Cambridge Assessment International Education.

Example answer

Question 1

The food web on the right shows the relationship of some of the organisms on a rocky shore.

a) In the food web, identify which organisms are:

   i) producers   (2)

     algae ✔ ①

     and seaweed ✔ ①

   ii) primary consumers   (3)

     topshell ✔ ①

     limpet ✔ ①

     periwinkle ✔ ①

   iii) secondary consumers.   (3)

     dog whelk ✔ ①

     gull ✔ ①

b) In an ecosystem, the number of crabs is severely reduced.

   i) Suggest **one** reason for the reduction of an organism in an ecosystem.   (1)

     A disease killed them ✔ ①

   ii) Predict the impact on dog whelks, limpets, and gulls in the food web. Explain your answers.   (3)

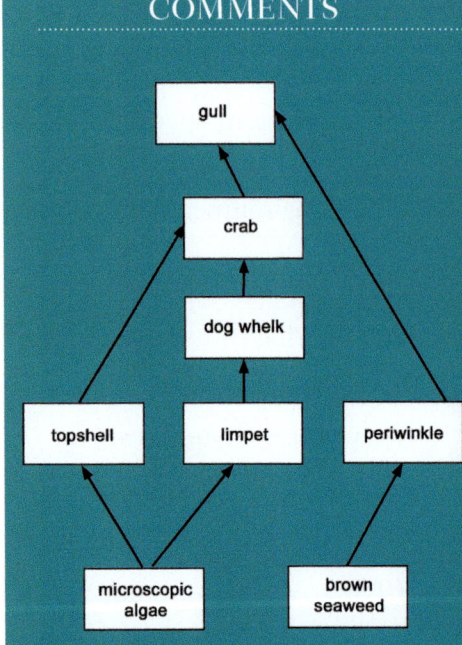

COMMENTS

a) **i)** These are partly correct, but it is best to always give the full name – microscopic algae and brown seaweed

  ii) Correct answer.

  iii) These are both correct, but crab is also a carnivore.

b) **i)** Correct. Also correct would be introducing foreign species; overharvesting food crops; habitat destruction or hunting.

  ii) The first part is correct and makes good use of key vocabulary.

c) Both suggestions are sensible.

Dog whelks – increase because of less predation from crabs ✔ ①

Limpets – decrease because of the increase in dog whelks ✔ ①

Gulls – decrease because they have less food ✗ ①

c) A student investigated the distribution of brown seaweed and periwinkles down a rocky shore. Her results are shown in the table.

Distance below high water on seashore/metres	Distribution of organisms	
	Observed distribution of brown seaweed	Density of periwinkles/ mean number of periwinkles per m²
0	absent	0
10	rare	16
20	occasional	52
30	abundant	156
40	abundant	128
50	occasional	44
60	rare	12

Suggest **two** reasons for the pattern of distribution of the periwinkles. (2)

periwinkles feed on brown seaweed ✔ ①

periwinkles get shelter from the seaweed ✔ ①

(Total 14 marks)

⑫/⑭

Question 2

Which feature shows discontinuous variation in humans? (1)

    A ABO blood group

    B hair length

    C lung volume

    D shoe size

(Total 1 mark)

Question 3

What is the name of a group of organisms of the same species living in the same area? (1)

    A community

    B ecosystem

    C habitat

    D population

(Total 1 mark)

Question 4

Carbon is cycled in nature.

a) **i)** Which atmospheric gas contains carbon? (1)

    ii) Which process uses carbon fixation to fix atmospheric carbon into glucose? (1)

    iii) Describe the process of decomposition. (2)

    iv) Which process allows carbon in plants to become carbon in animals? (1)

    v) Which two processes return carbon dioxide to the atmosphere? (2)

b) State **one** process, not mentioned in part **(a)**, by which carbon dioxide enters the air. (1)

(Total 8 marks)

Question 5

The graph shows human population size and estimated extinctions of species.

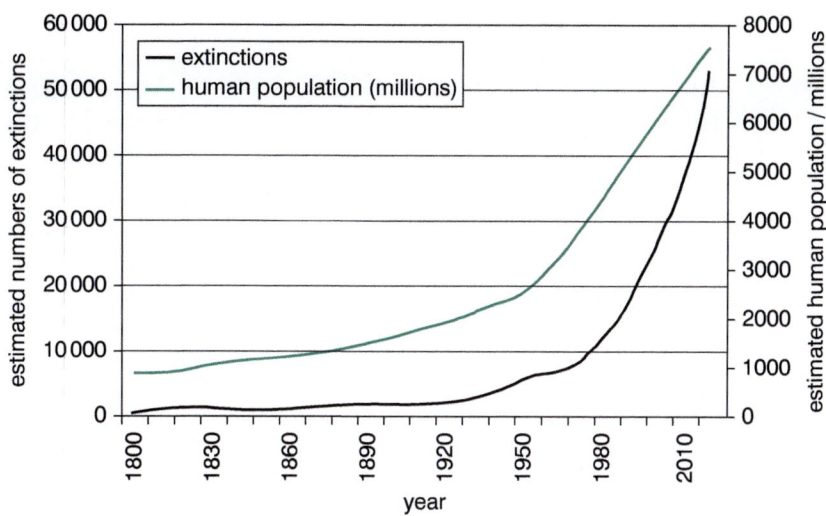

a) Compare the patterns in human population growth and species extinctions. (2)

b) Suggest why the numbers of species extinctions are estimated and **not** actual numbers. (2)

c) Explain why habitat destruction can lead to extinction. (2)

d) Give one other cause of extinction and explain your answer. (3)

(Total 9 marks)

Question 6

In an ecosystem, the following measurements were made.

Organism	Number in ecosystem	Total biomass of organisms/g
Oak trees (producers)	1	500 000
Aphids (primary consumers)	100 000	100
Ladybirds (secondary consumers)	200	10

a) Draw a food chain to illustrate the feeding relationships of the three organisms. (2)

b) In observations of the ecosystem, ladybirds were seen to be fed on by spiders, and spiders fed on by blackbirds.

   i) Draw the complete food chain that includes these feeding relationships. (1)

   ii) SUPPLEMENT Explain fully why food chains longer than this are rare. (5)

(Total 8 marks)

Question 7

The table gives information on how the land area covered by forest has changed from 1990 to 2005.

Country	Area covered by forest/millions of hectares		Area of forest lost from 1990 to 2005/%
	1990	**2005**	
Bolivia	109.9	58.7	46.6
Brazil	851.5	477.7	
Colombia	113.9	60.7	
French Guiana	9.0	8.1	
Peru	125.5	68.7	
Suriname	16.3	14.8	
Venezuela	91.2	47.7	

Data from http://rainforests.mongabay.com/deforestation_alpha.html

a) Calculate the area of forest lost for each country, as a percentage of the area in 1990. The first one has been done for you. (6)

b) During the time period 1990 to 2005, identify in which country there is, in terms of percentage lost:

   i) the greatest deforestation (1)

   ii) the least deforestation. (1)

c) Give two reasons for deforestation. (2)

d) i) State what would happen to biodiversity after deforestation occurs. (1)

   ii) Describe two other undesirable effects of deforestation. (2)

(Total 13 marks)

Developing experimental skills

The information in this section is based on the Cambridge International syllabus. You should always refer to the appropriate syllabus document for the year of examination to confirm the details and for more information. The syllabus document is available on the Cambridge International website at www.cambridgeinternational.org.

INTRODUCTION

As part of your biology study of the Cambridge IGCSE Co-ordinated Sciences course, you will develop practical skills and have to carry out investigative work in science.

This section provides guidance on carrying out an investigation.

The experimental and investigative skills are divided as follows:

1. Apparatus and techniques.

2. Plan an experiment.

3. Make observations and record measurements.

4. Interpret observations and data.

5. Evaluate methods.

1. APPARATUS AND TECHNIQUES

Skill: Demonstrate knowledge of how to select and safely use techniques, apparatus and materials (including following a sequence of instructions where appropriate).

Questions to ask:

How shall I use the equipment and chemicals safely to minimise the risks – what are my safety precautions?

✓ When writing a Risk Assessment, investigators need to be careful to check that they have matched the hazard with the concentration of a chemical used. Many acids, for instance, are corrosive in higher concentrations, but are likely to be irritants or of low hazard in the concentration used when working in biology experiments. Eye protection is likely to be needed when handling and using many laboratory chemicals.

✓ Don't forget to consider the hazards associated with all the chemicals and biological materials, even if these are very low.

✓ You should be able to justify the precautions taken when carrying out an investigation.

How much detail should I give in my description?

✓ You need to give enough detail so that someone else who has not done the experiment would be able to carry it out to reproduce your results.

How should I use the equipment to give me the precision I need?

✓ You should know how to read the scales on the measuring equipment you are using.

✓ You need to show that you are aware of the precision needed.

△ Fig. 20.1 The volume of liquid in a burette must be read to the bottom of the meniscus. The volume in this measuring cylinder is 202 cm³, not 204 cm³.

EXAMPLE 1

This is an extract from a student's notebook. It describes how she carried out an experiment to investigate the production of carbon dioxide by yeast at different temperatures.

What are my safety precautions?

a) Chemicals

I have looked up the hazards associated with the chemical I am using:
Glucose (solutions from 0.05 to 0.25 M): LOW HAZARD

Although it is only a low hazard, it is still best to wear eye protection when using the solutions, especially as some of the liquids will be hot. It is also important to handle all chemicals carefully, and wipe up any spills of liquid.

COMMENT

A data source has been used to look up the chemical hazards.

b) Organisms

I found my information on yeast from chemical safety data sheets: Dried yeast may cause eye, skin, and respiratory tract irritation. It is expected to be a low hazard for usual handling. If the dust is inhaled, you should 'remove from exposure and move to fresh air immediately'. I will handle the powdered yeast carefully when making up my suspension, trying to avoid making any dust.

COMMENT

A data source has been used to look up the biological hazards.

c) Equipment

I must be careful when using the water bath not to get water near the electrical sockets.

I need to handle the glassware (conical flask, gas syringe and glass tubing) carefully. In particular, I need to protect my hands with a towel (or glove) when linking together the glass delivery tubes from the rubber bung in the conical flask to the gas syringe with rubber tubing. The tubing needs to be lubricated with water and I need to hold my hands close together to limit the movement of glass if a break occurs.

Some sensible precautions have been suggested.

How much detail should I give in my description?

The student's method is given below:

1 *100 cm³ of 0.25 M glucose solution was transferred to each of six conical flasks.*
2 *The conical flask was transferred to a water bath at 20 °C. It was left for a few minutes to reach the temperature.*
3 *1.00 g of yeast was then added, and the mixture swirled to mix in the yeast. The stop-clock was started.*
4 *The bung, on which I had placed tubing connecting it to the gas syringe, was placed in the conical flask.*
5 *Every minute, the volume of carbon dioxide in the gas syringe was recorded.*
6 *The average volume of carbon dioxide produced per minute was calculated.*
7 *The experiment was repeated three times and the average rate of carbon dioxide production calculated.*
8 *The investigation was then carried out at 10 °C, 30 °C, 40 °C, 50 °C and 60 °C.*
9 *A graph was drawn of the average rate of carbon dioxide production against temperature.*

The method is detailed and well written. As it is the effect of temperature that is being investigated, it is important for the sugar solution to reach the required temperature before the yeast is added, and this has been included in the written method.

2. PLAN AN EXPERIMENT

Skill: Plan experiments and investigations.

Questions to ask

What do I already know about the area of biology I am investigating, and how can I use this knowledge and understanding to help me with my plan?

✓ Think about what you have already learned and any investigations you have already done that are relevant to this investigation.

✓ List the factors might affect the process you are investigating.

What is the best method or technique to use?

✓ Think about whether you can use or adapt a method that you have already used.

✓ A method, and the measuring instruments, must be able to produce **valid** measurements. A measurement is valid if it measures what it is supposed to be measuring.

You will make a decision as to which technique to use based on:

- the accuracy and precision of the results required

- the simplicity or difficulty of the techniques available, or the equipment required – is this expensive, for instance?
- the scale – for example using standard laboratory equipment or on a microscale, which may give results in a shorter time period
- the time available to do the investigation
- health and safety considerations.

What am I going to measure?

✓ The factor you are investigating is called the **independent variable**. A **dependent variable** depends on the value of the independent variable that you select.

✓ You need to choose a range of measurements that will be enough to allow you to plot a graph of your results and so find out the pattern in your results.

✓ You should be able to justify the range you have chosen, rather than a lower or higher range.

How am I going to control the other variables?

✓ These are **control (constant) variables**. Some of these may be difficult to control. This may be especially difficult if you are carrying out an ecology investigation in the field, where varying factors such as light and temperature are impossible to control.

✓ You must decide how you are going to control any other variables in the investigation and so ensure that any conclusions you draw are valid.

What equipment is suitable and will give me the accuracy and precision I need?

✓ The **accuracy** of a measurement is how close it is to its true value.

✓ **Precision** is related to the smallest scale division on the measuring instrument that you are using, e.g. when measuring the distance moved by the bubble in Method 3 in Example 2 below, a ruler marked in millimetres will give greater precision that one divided into centimetres only.

✓ A set of precise measurements also refers to measurements that have very little spread about the mean value.

✓ You need to be sensible about selecting your devices and make a judgement about the degree of precision. Think about what is the least precise variable you are measuring and choose suitable measuring devices. There is no point having instruments that are much more precise than the precision you can measure the least precise variable to.

What are the potential hazards of the equipment, chemicals, organisms, and techniques I will be using and how can I reduce the risks associated with these hazards?

✓ You can find information on the hazards of most chemicals from manufacturers' chemical data sheets or online resources. Teachers must always check that all the hazards have been identified.

✓ You should be able to suggest appropriate safety precautions when presented with details of a biology investigation.

EXAMPLE 2

You are going to plan an investigation into carbon dioxide production by yeast under different conditions. You have been told that yeast is a unicellular organism that respires by breaking down glucose to release energy, and that chemical reactions involved in this process are controlled by the yeast's enzymes.

What do I already know?

You already know that respiration produces carbon dioxide. As sugar is a reactant in the process, conditions that might affect the production of carbon dioxide include the *concentration of sugar*: the *type of sugar* may also be a factor.

The chemical reactions involved in this process are controlled by the yeast's enzymes, so you might expect this process to be also affected by *temperature*.

Some conditions that might be expected to affect the process are therefore:

✓ concentration of sugar

✓ type of sugar

✓ temperature.

You need to think about how to measure the effect these factors, or one of these factors, has on carbon dioxide production.

All of these factors are independent variables. The amount of carbon dioxide produced is the dependent variable because it is affected by the independent variables: concentration of sugar, type of sugar, and temperature.

What is the best method or technique to use?

An investigator needs to set up an experiment so that they can measure the carbon dioxide produced by the respiring yeast.

Several methods are available to produce valid measurements.

Method 1

A simple way is to add some yeast to a sugar solution in a conical flask, as shown in the diagram and count the bubbles produced in given time periods, e.g. every minute, or over a period of time, e.g. one hour.

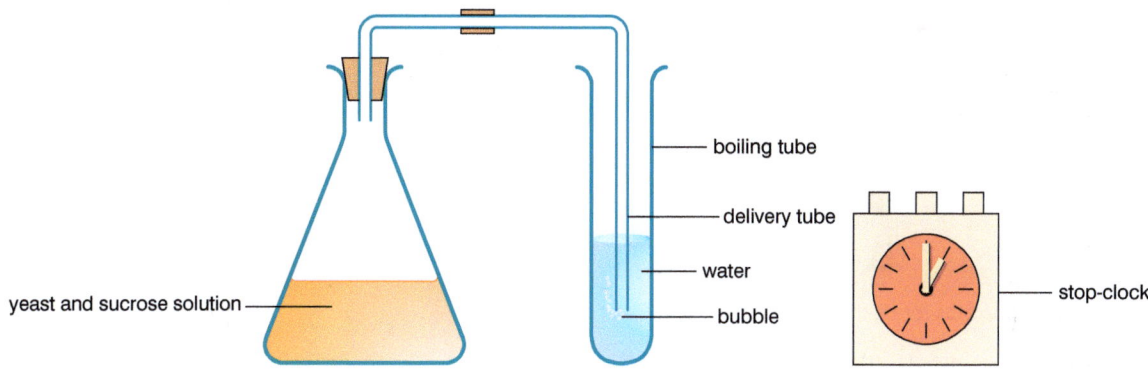

△ Fig. 20.2 Apparatus for method 1.

Method 2

An alternative method is to set up the yeast and sugar solution as before, but this time connect the glass tube to a gas syringe. This time, you can measure the volume of carbon dioxide produced in given time periods, e.g. every minute, or over a period of time, e.g. one hour.

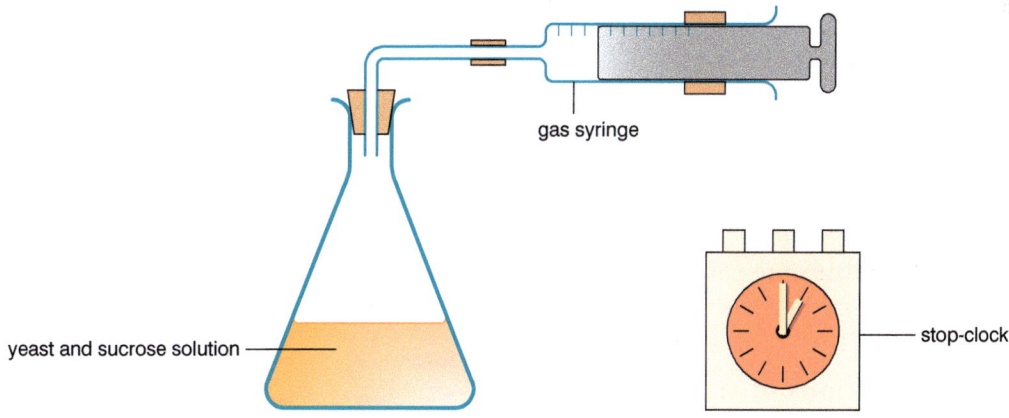

△ Fig. 20.3 Apparatus for method 2.

Method 3

Another method works on a smaller scale. The yeast and sugar solution is placed in a syringe, which is connected to a pipette that contains a bubble of water. The movement of the bubble is measured in given time periods, e.g. every minute, or over a period of time, e.g. one hour. It is less suitable for the temperature investigation, however, as it would be inadvisable to immerse the syringes in water baths.

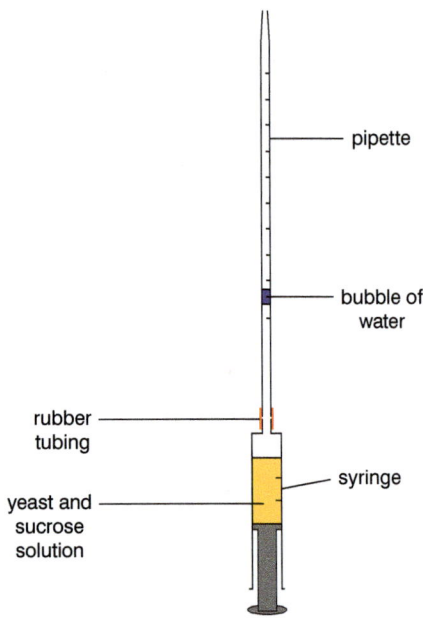

Δ Fig. 20.4 Apparatus for method 3.

Choosing the method:

✓ Accuracy and precision: methods 2 and 3 would be preferred when accurate and precise results are needed. Method 1 cannot be used to actually measure volume of carbon dioxide, but might be best for a preliminary test or a quick comparison.

✓ Microscale: method 3 might be most suitable here.

✓ Time available: method 3 might be most suitable, as smaller volumes are involved and the investigation could be carried out more quickly.

✓ Health and safety considerations are less relevant here.

What am I going to measure?

You need to make a measure of the amount of carbon dioxide produced using different concentrations of sugar, different types of sugar, or at different temperatures.

The **independent variables** are the factors under investigation: the concentration of sugar, the type of sugar, and the temperature.

The **dependent variables** are the measurements made: the number of bubbles per minute, the volume of carbon dioxide per minute, or the distance moved by the bubble (which is a measure of the volume of carbon dioxide).

Different concentrations of sugar

A sensible range of sugar concentrations to use might be based on concentrations that yeast might encounter in nature. An investigator might choose to use concentrations ranging from 0 M to 0.25 M (at intervals of 0 M, 0.05 M, 0.10 M, 0.15 M, 0.20 M and 0.25 M), as a preliminary test would show that these concentrations would give suitable results in the time period allocated. Concentrations chosen would be no higher as these might have an osmotic effect: water could be drawn from the yeast cells by osmosis, preventing them from functioning properly.

The investigator would plot a graph using the results from the six different concentrations to look for a pattern in those results.

Different types of sugar

Here, an investigator might decide to use sugars commonly found in nature. These include fructose, glucose, lactose, maltose, and sucrose.

Different temperatures

Respiration in yeast is a series of enzyme-controlled reactions. In most cases, enzymes work best around 40 °C, and cease to function above around 60 °C. An investigator might decide, therefore, to measure carbon dioxide production at five different temperatures (10 °C, 20 °C, 30 °C, 40 °C, 50 °C and 60 °C). Again, this would be sufficient to allow the investigator to plot a graph of results and so find any pattern in the results.

How am I going to control the other variables?

The investigator must ensure that any differences in carbon dioxide production must be the result of, in the different investigations, sugar concentration, type of sugar, and temperature, and not the result of some other factor. In other words, it must produce valid measurements.

So, the investigator must decide what other factors could affect the experiment and try to keep these constant. These are the control (constant) variables.

Factor under investigation/ independent variable	Factors to be kept constant / constant variables				
	Yeast concentration	Sugar concentration	Type of sugar	Temperature	Length of investigation
Sugar concentration	yes	no – vary	yes	yes	yes
Type of sugar	yes	yes	no – vary	yes	yes
Temperature	yes	yes	yes	no – vary	yes

Δ Table 20.1 Variables in Example 2.

EXAMPLE 3

In this experiment, the production of heat from the respiration of germinating seeds is being investigated. Some seeds have been placed in an insulated flask (thermos flask) and their temperature measured over a period of time.

But a rise in temperature alone would not be sufficient to demonstrate that it is the respiring seeds that are causing this rise. Some other factor could be involved. So an almost identical experiment is set up, but this time with seeds that have been killed. So any change in the temperature must be the result of the living seeds' respiration. Both sets of seeds are also sterilised with disinfectant so that any temperature rise cannot be explained by the growth of microorganisms on the seeds.

The almost identical experiment, with the killed seeds, is an example of an **experimental control**. Experimental controls are a way of minimising the effects of any other possible variables so that the control can be compared with the experiment where the independent variable has been changed.

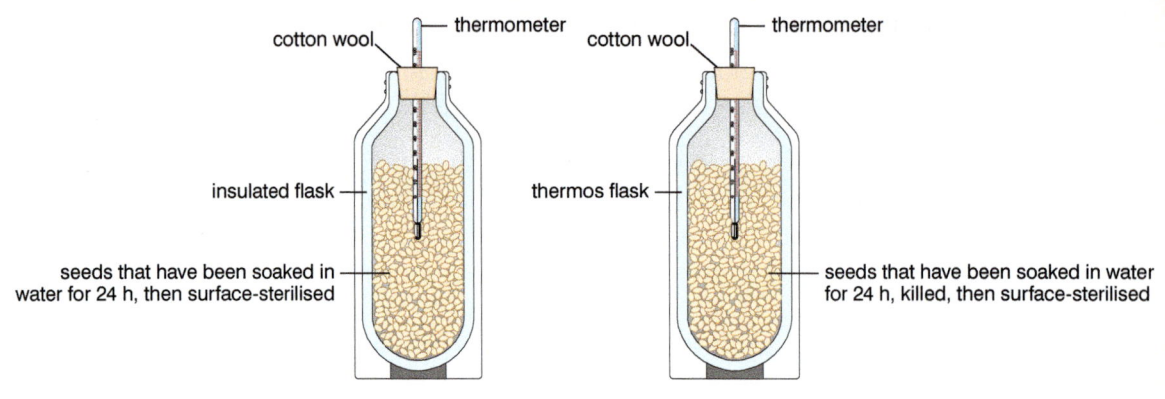

△ Fig. 20.5 Apparatus for Example 3.

What equipment is suitable and will give me the accuracy and precision I need?

You now know what you will need to measure and so can decide on your measuring devices.

Referring back to Example 2:

Measurement	Quantity required	Equipment
Mass of yeast	1.00 g	balance measuring up to two decimal places
Volume of sugar solution	100 cm³	100 cm³ volumetric flask
Temperature	10–60 °C	thermometer, with 1 °C precision
Time	one minute intervals	stop-clock (1 s precision)

△ Table 20.2 Suitable equipment for experiment.

You will need to be sensible about selecting your equipment and make a judgement about the degree of accuracy. The accuracies of different pieces equipment need to be comparable. It would be not appropriate to measure the yeast that was put in every conical flask accurately without measuring the volume of sugar solution poured onto each flask to a similar level of accuracy.

What are the potential hazards of the equipment and how can I reduce the risks?

In Example 2, the chemical hazards are as follows:

Fructose solution	LOW HAZARD
Glucose solution	LOW HAZARD
Lactose solution	LOW HAZARD
Maltose solution	LOW HAZARD
Sucrose solution	LOW HAZARD
Yeast, dried	LOW HAZARD

These indicate that there are no specific hazards the investigator needs to be aware of. However, when handling *any* chemicals, it would be sensible to wear eye protection.

In terms of the equipment and technique, the major hazards are:

✓ spilling hot liquids (at 50 °C and 60 °C)

✓ breaking glass when connecting tubing together

✓ possible contact between water and electrical sockets.

3. MAKE OBSERVATIONS AND RECORD MEASUREMENTS

Skill: Make and record observations, measurements and estimates.

Questions to ask:

How many different measurements or observations do I need to take?

✓ Sufficient readings have been taken to ensure that the data are consistent.

✓ It is usual to repeat an experiment to get more than one measurement. If an investigator takes just one measurement, this may not be typical of what would normally happen when the experiment was carried out.

✓ When repeat readings are consistent they are said to be **repeatable**.

Do I need to repeat any measurements or observations that are anomalous?

✓ An **anomalous result** or **outlier** is a result that is not consistent with other results.

✓ You want to be sure a single result is accurate. So you will need to repeat the experiment until you get close agreement in the results you obtain.

✓ If an investigator has made repeat measurements, they would normally use these to calculate the arithmetical mean (or just mean or average) of these data to give a more accurate result. You calculate the mean by adding together all the measurements, and dividing by the number of measurements. Be careful though – anomalous results should not be included when taking averages.

✓ Anomalous results might be the consequence of an error made in measurement. But sometimes outliers are genuine results. If you think an outlier has been introduced by careless practical work, you should omit it when calculating the mean. But you should examine possible reasons carefully before just leaving it out.

✓ You are taking a number of readings in order to see a changing pattern; for example, measuring the volume of gas produced in a reaction every 10 seconds for 2 minutes (so 12 different readings). It is likely that you will plot your results onto a graph and then draw a **line of best fit**.

✓ You can often pick an anomalous reading out from a results table (or a graph if all the data points have been plotted, as well as the mean, to show the range of data). It may be a good idea to repeat this part of the practical again, but it is not necessary if the results show good consistency.

✓ If you are confident that you can draw a line of best fit through most of the points, it is not necessary to repeat any measurements that are obviously inaccurate. If, however, the pattern is not clear enough to draw a graph, then readings will need to be repeated.

How should I record my measurements or observations – is a table the best way? What headings and units should I use?

✓ A table is often the best way to record results.

✓ Headings should be clear.

✓ If a table contains numerical data, do not forget to include units – data are meaningless without them.

✓ The units should be the same as those that are on the measuring equipment you are using.

✓ Sometimes you are recording observations that are not quantities. Putting observations in a table with headings is a good way of presenting this information.

EXAMPLE 4

How many different measurements or observations do I need to take?

A student cut a number of cylinders of tissue from a potato and weighed them, and recorded the mass of each cylinder. Six dishes were set up, with each dish containing a different concentration of sucrose solution. Four potato cylinders were placed into each dish. After one hour, the cylinders were removed, blotted dry, and reweighed. The student then calculated the percentage change in mass for each cylinder. The final results after calculating the percentage change are shown below.

REMEMBER

Percentage change is calculated using the formula:
percentage change = (actual change ÷ original value) × 100

Concentration of sucrose/M	Percentage change in mass of potato cylinders/%				Average percentage change in mass/%	Texture of potato cylinders (qualitative)
	Experiment 1	Experiment 2	Experiment 3	Experiment 4		
0.0	+31.4	+33.7	+31.2	+32.5	+42.9	firm
0.2	+20.9	+33.4	+22.8	+21.3	+21.7	firm
0.4	−2.7	−1.8	−1.9	−2.4	−2.2	slightly soft
0.6	−13.9	−12.8	−13.7	−13.6	−13.5	soft
0.8	−20.2	−19.7	−19.3	−20.4	−19.9	floppy
1.0	−19.9	−20.3	−21.1	−20.3	−20.4	very floppy

△ Table 20.3 Results for Example 4.

In this table of results:

✓ each repeat is recorded in its own column

✓ the description of each measurement is clear

✓ the units are given

✓ the data are recorded to the same number of decimal places, and decimal points are aligned

✓ calculations of means are recorded to the appropriate number of significant figures.

The student has recorded four calculations for each concentration investigated. This is sufficient to see if the data are consistent.

Do I need to repeat any measurements or observations that are anomalous?

The result from the cylinder in Experiment 2 in a concentration of 0.2 M sucrose is not consistent with the other results for this concentration. It is an anomalous result and is highlighted in the table. The student has not included this result in the calculation of the mean for this concentration. However, there is no need to repeat the measurement as there are already three other results showing good consistency.

How should I record my measurements or observations?

Here are some results of food tests carried out by an investigator:

Food substance tested	Colour change on heating with Benedict's solution
10% glucose solution	blue → green → yellow → orange → red
Biscuit	blue → greenish blue
Grape	blue → green → yellow → orange → red
Honey	blue → green → yellow → orange → red-brown
Potato	remained blue

Δ Table 20.4 Results of some food tests.

Note the clear table headings. The right hand column doesn't simply say 'colour' but refers to colour *change*.

COMMENT

Don't forget that some investigations might benefit from including both numerical data and observations, e.g., in the osmosis experiment in Example 4, it was helpful to include information on the firmness of the potato cylinders at the end of the experiment.

4. INTERPRET OBSERVATIONS AND DATA

Skill: Interpret and evaluate experimental observations and data.

Questions to ask:

What is the best way to show the pattern in my results? What type of chart or graph should I use?

✓ Graphs are usually the best way of demonstrating trends in data.

✓ Pie charts are generally used to show percentage or proportionality, for example, comparing the different food types in a diet.

✓ A bar chart is used when one of the variables is a **categoric** (or *discrete*) **variable**, for example, when one of the variables is the type of leaf, or species of organism. Bar charts should be made up of bars of equal width that do **not** touch.

✓ Histograms may look like bar charts, but they are used for *continuous* data, for example, when showing the number of students in a class in different height ranges (such as 150–159 cm, 160–169 cm, and so on). Unlike in bar charts, the bars in histograms **should** touch.

✓ A line graph is used when both variables are continuous, for example, time and temperature, time and volume.

✓ Scatter graphs can be used to show the degree of *correlation* between two variables.

✓ Sometimes a line of best fit is added to a scatter graph, but usually the points are left without a line.

When drawing bar charts, histograms or line graphs:

✓ choose scales that take up most of the graph paper (at least half of the graph grid in each direction)

✓ make sure the axes are linear and allow points to be plotted accurately – for example, one big square = 5 or 10 units; but not 3 units – and each square on an axis should represent the same quantity

✓ label the axes with the variables (ideally with the independent variable on the x-axis)

✓ make sure the axes have units

✓ if more than one set of data is plotted, use a key to distinguish the different data sets

✓ points on a line graph should be drawn with a sharp pencil, and be clearly marked as crosses (or encircled dots) of a suitable size (not too large, or too small).

If I use a line graph, should I join the points, and how?

✓ A best-fit line, straight or curved, should be drawn if a trend can be seen and there is good reason to believe that intermediate values can be predicted.

✓ A best-fit line or curve should show the 'average' trend of your plotted points. It actually does not need to go exactly through any of your plotted points – but if there are many points not exactly on the line, there should be a roughly equal distribution of them either side of the line.

✓ Remember there may be some points that don't fall on the curve – these may be incorrect or anomalous results. Any points that are clearly anomalous should be ignored when drawing your line.

✓ A graph will often make it obvious which results are anomalous.

✓ If you are uncertain of the values in between, then the points can be joined point-to-point.

Do I have to calculate anything from my results?

✓ It will be common to calculate means from data.

✓ Sometimes it is helpful make other calculations, before plotting a graph (see Example 4). Other types of calculation include:

- the volume of water taken up by a plant from the distance moved by a bubble in a potometer (volume = $\pi r^2 \times$ distance)

- the rate of a reaction, e.g. involving enzymes (rate of reaction = $1 \div$ time taken by reaction; see Example 9).

✓ Investigators also look for numerical trends in data, for example, the doubling of a reaction rate every 10 °C or the doubling of numbers of microorganisms every 20 minutes.

✓ Sometimes you will have to make some calculations before you can make any conclusions.

Can I write a conclusion from my analysis of the results, and what biological knowledge and understanding can be used to explain the conclusion?

✓ You need to use your biological knowledge and understanding to explain your conclusion.

✓ It is important to be able to add some explanation which refers to relevant scientific ideas in order to justify your conclusion.

EXAMPLE 6

What is the best way to show the pattern in my results?

A student did an experiment to compare the loss of water from leaves of three different species of tree: hazel, lime, and oak.

For each species, he measured the mass of 10 leaves of similar size and hung the leaves on a line. After three hours, he removed the leaves and measured the masses of the leaves again and calculated the average loss of water in grams per hour.

Species	Average loss of water/g per hour
Apple	0.30
Hazel	0.05
Oak	0.01

△ Table 20.5 Results of Example 6.

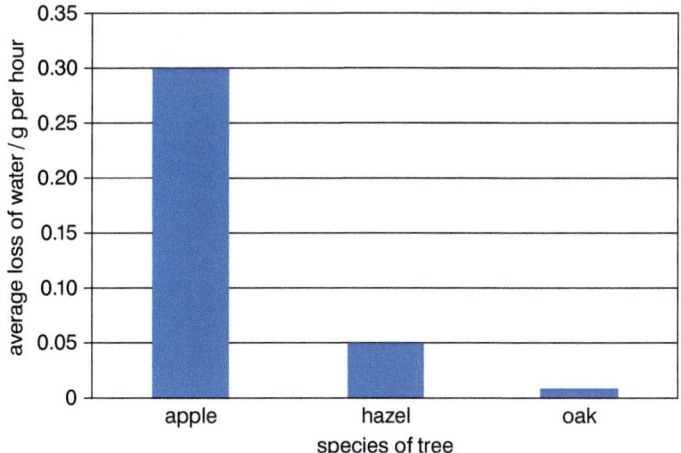

△ Fig. 20.6 Bar chart showing water loss from different leaves.

A bar chart is used to display the data in this instance, as the type of leaf is a categoric variable.

EXAMPLE 7

What is the best way to show the pattern in my results?

A student investigated the effect of different light intensities on photosynthesis in pondweed.

She measured the oxygen collected over a number of days.

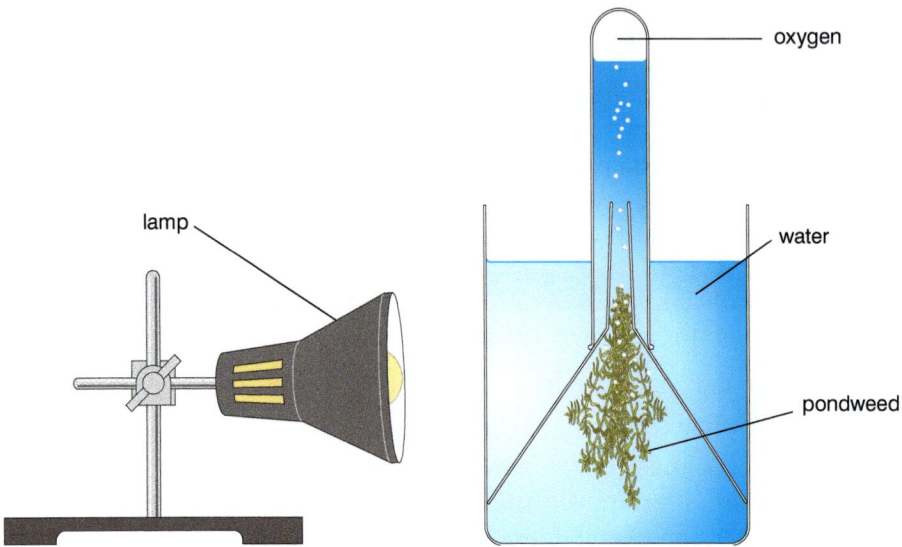

△ Fig. 20.7 Apparatus for Example 7.

A line graph is needed as both the volume of gas and time are continuous variables.

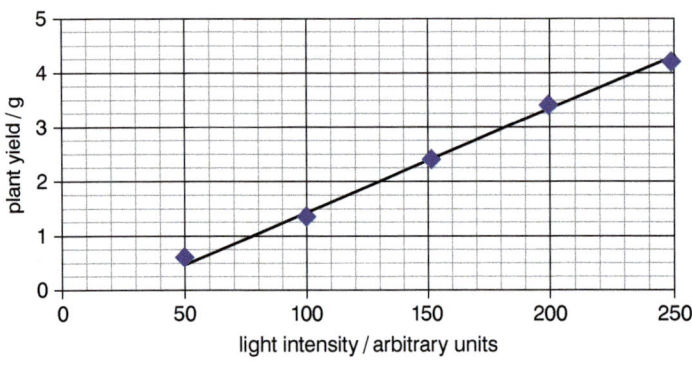

△ Fig. 20.8 Experimental results of Example 7.

If I use a line graph, should I join the points, and how?

In this case the data fit a straight line most closely. You need to look at the shape that the points make to help you decide how to join them.

EXAMPLE 8

What is the best way to show the pattern in my results?

A student investigated the effect of applying different amounts of fertiliser to 18 different tomato plants in the school garden. She measured the mass of tomatoes produced by each plant during the same growing period. She used a scatter graph to display the results she collected.

A scatter graph is most appropriate for displaying these data because each point represents one plant and is unrelated to another point. The results show a trend that should be described in the conclusion.

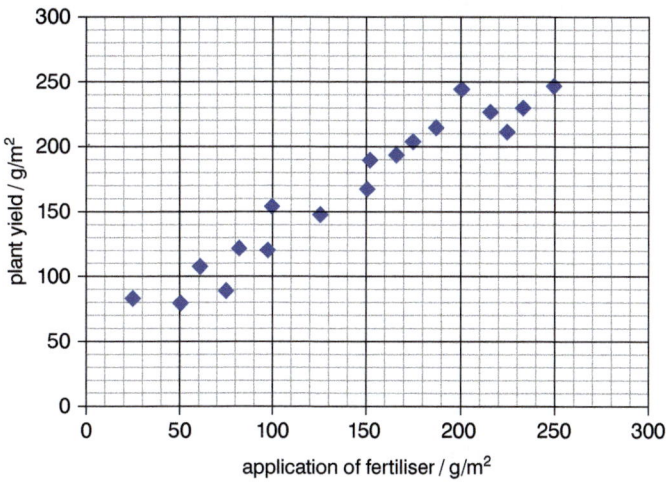

△ Fig. 20.9 Experimental results for Example 8.

EXAMPLE 9

If I use a line graph, should I join the points, and how?

In an investigation on the activity of an enzyme at different temperatures, an investigator obtained a set of results that shows how activity varies.

Photographic film (as used widely before the introduction of digital photography) is made from a sheet of plastic coated with light-sensitive silver particles bonded by the protein gelatin. When the gelatin is broken down by a protease enzyme, the silver particles fall off, and the film becomes clear.

Photographic film was cut into five strips of equal size, and each strip was placed into a test tube. An identical volume of protease was added to the tubes, and each tube was kept at a different temperature.

The amount of time taken for each strip of film to become clear was measured and recorded.

Temperature/°C	Average time taken for breakdown of gelatin/s
4	3450
13	667
25	175
30	130
40	133
50	7140

△ Table 20.6 Table of results for Example 9.

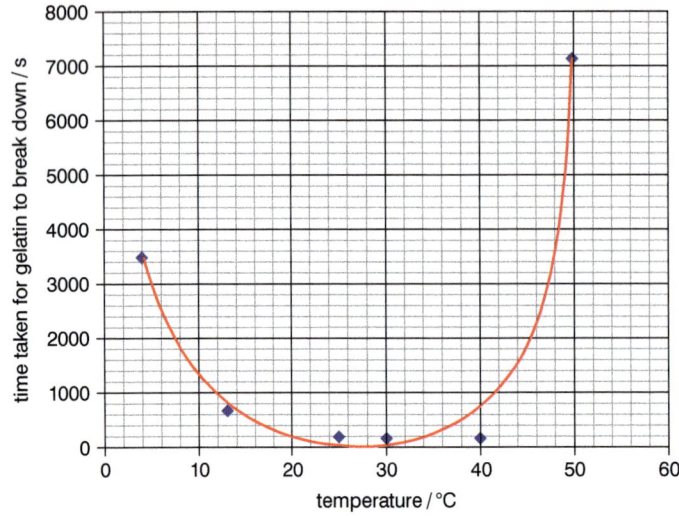

△ Fig. 20.10 Results for Example 9 with line of best fit plotted.

Do I have to calculate anything from my results?

The trend of how enzyme activity is affected by temperature is not well illustrated by the graph. It is shown better if the *rate of reaction* is calculated and plotted against temperature. The rate is the inverse of the time taken to break down the gelatin, i.e. 1 ÷ the time taken.

Temperature/°C	Average time taken for breakdown of gelatin/s	Rate of breakdown of gelatin (= 1/time)/s^{-1}
4	3450	0.00029
13	667	0.00150
25	175	0.00571
30	130	0.00769
40	133	0.00752
50	7140	0.00014

△ Table 20.7 Table of results with rate column added.

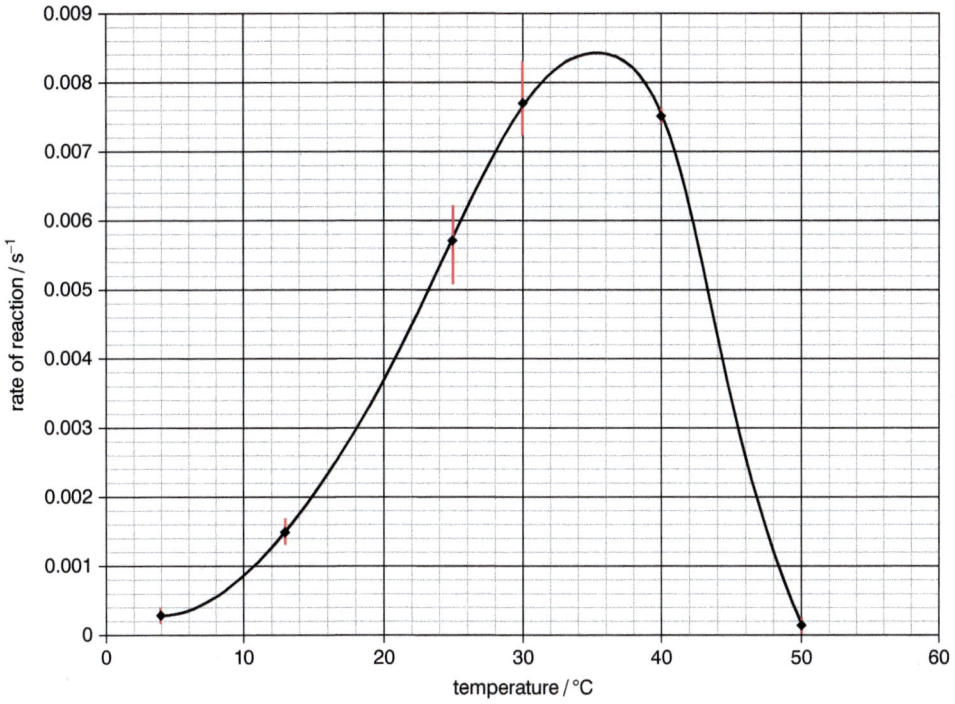

△ Fig. 20.11 Rate of reaction for Example 9.

Note that on this graph, range bars have been drawn to show how the data points are arranged around the mean.

Rates can also be calculated by looking at the change in steepness/gradient of a line graph, or part of a graph.

Can I write a conclusion from my analysis of the results?

A student wrote:

> Enzymes work best at a particular temperature. My graph suggests that the optimum temperature for protease is round 35 °C. At lower temperatures, enzymes work slowly because the molecules have less energy and move around more slowly, so there are fewer successful collisions between enzyme and substrate (gelatin) molecules.
>
> Enzymes work by a lock and key mechanism, with the substrate fitting into the enzyme's active site. At temperatures that are too high, the shape of an enzyme will be changed so that it will not work. This change is permanent and the enzyme is said to be denatured.

COMMENT

This is a good, concise conclusion and links the data to the mechanism of enzyme action.

5. EVALUATE METHODS

Skill: Evaluate methods and suggest possible improvements.

Questions to ask:

Do any of my results stand out as being inaccurate or anomalous?

✓ You need to look for any anomalous results or outliers that do not fit the pattern.

✓ You can often pick this out from a results table (or a graph if all the data points have been plotted, as well as the mean, to show the range of data).

What reasons can I give for any inaccurate results?

✓ When thinking about this it is important to be specific. Simply saying 'experimental error' is too vague.

✓ It is often possible to look at the practical technique and suggest explanations for anomalous results.

✓ When you carry out the experiment you will have a better idea of which possible sources of error are more likely.

✓ Try to give a specific source of error and avoid statements such as 'the measurements must have been wrong'.

Your conclusion will be based on your findings, but must take into consideration any uncertainty in these introduced by any possible sources of error. You should discuss where these have come from in your evaluation.

Error is a difference between a measurement you make, and its true value.

The two types of errors are:

✓ random error

✓ systematic error.

With **random error**, measurements vary in an unpredictable way. This can occur when the instrument you are using to measure lacks sufficient precision to indicate differences in readings. It can also occur when it is difficult to make a measurement. If two investigators measure the height of a plant, for instance, they might choose different points on the plant to make their measurements.

With **systematic error**, readings vary in a particular way. They are either consistently too high or too low. One reason could be down to the way you are making a reading, for example taking a burette reading at the wrong point on the meniscus, or not being directly in front of an instrument when reading from it.

What an investigator *should not* discuss in an evaluation are problems introduced by using faulty equipment, or by using the equipment inappropriately. These errors can, or could have been, eliminated, by:

✓ checking equipment

✓ practising techniques before the investigation, and taking care and patience when carrying out the practical.

Overall, was the method or technique I used good enough?

✓ If your results were good enough to provide a confident answer to the problem you were investigating the method probably was good enough.

✓ If you realise your results are not accurate when you compare your conclusion with the actual answer, it may be that you have a **systematic error** (an error that has been made in obtaining all the results). A systematic error would indicate an overall problem with the experimental method.

✓ If your results do not show a convincing pattern, then it might be fair to assume that your method or technique was not good enough and there may have been a **random error** (i.e. measurements vary in an unpredictable way).

If I were to do the investigation again, what would I change or improve upon?

✓ Having identified possible errors it is important to say how these could be overcome. Again you should try and be specific.

✓ When suggesting improvements, do not just say 'do it more accurately next time' or 'measure the volumes more accurately next time'.

✓ For example, if you were measuring small volumes, you could improve the method by using a burette to measure the volumes rather than a measuring cylinder.

✓ Sometimes investigations can be improved by extending the range (e.g. temperature, time, pH, etc.) over which they are carried out. However, sometimes an investigation can be improved by measuring more values within a smaller range, e.g. when trying to identify the optimum temperature of an enzyme.

EXAMPLE 10

Overall, was the method or technique I used accurate enough?

In this example, the pH of yoghurt was monitored during its production using a data logger. The data logger gave very precise, consistent readings. Part of the reason for this is that it had first been set, or *calibrated*, using solutions of known pH (called buffers). Although the line is not a completely smooth curve, there are no clearly anomalous results, so there do not appear to be any random errors. If the data logger has just been calibrated it is unlikely there are any systematic errors either. So these results do seem to be accurate enough.

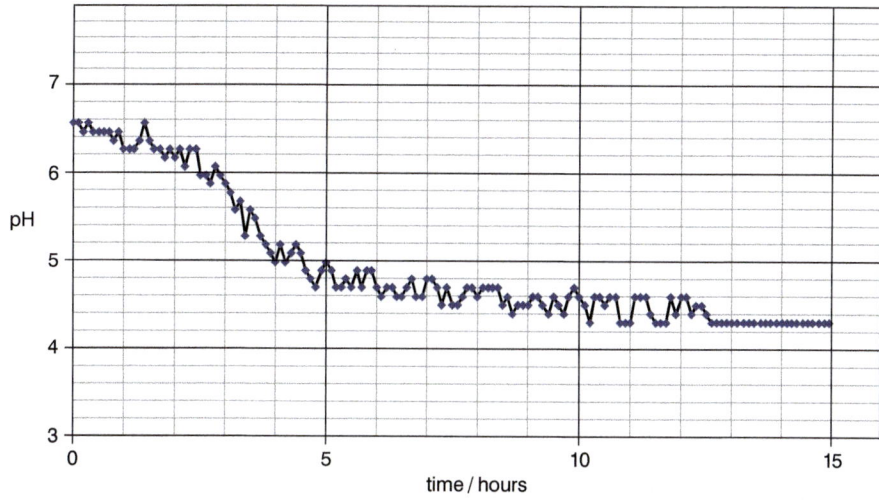

Δ Fig. 20.12 Experimental results for Example 10.

However, in Example 4, there is a clearly anomalous result. One possible reason for the anomalous result might be that the potato cylinder had not been blotted dry properly before its mass was measured. However, apart from the anomalous result, all the other repeats show good consistency, and there is a clear pattern as the sucrose concentration changes. This suggests that, overall, the method gave accurate results.

EXAMPLE 11

A student was asked to investigate the effect of exercise on heart rate. The student chose to ask 5 members of their class to jump up and down on the spot, using a metronome to keep time. The exercise lasted for 1 minute and then they measured their pulse rate for 1 minute, by finding their pulse in their neck. They then repeated the experiment, using faster metronome speeds.

Was the method or technique [I] used valid?

✓ Yes, it measures pulse rates after different types of exercise, but it may have some uncertainties.

What key variables were controlled?

✓ Type of exercise, length of exercise, tempo of exercise, and time of pulse rate were all controlled.

What possible sources of errors / uncertainty are there?

✓ 5 repeats is reasonable, but 5 different people will have different fitness levels, which could impact the results.

✓ Different people may need different rest breaks between repeats, which could also affect results.

✓ Different people may do the exercise slightly differently, which could also affect results.

✓ Pulse rate can be difficult to measure if people have to find their own pulse, so there may be a large uncertainty.

If I were to do the investigation again, what would I change or improve on?

✓ You could ask the same person, or people from the same sports team to do the repeats, so the sample group is more similar.

✓ You could measure the pulse rate, every minute for 5 minutes, to see how long it takes for it to return to normal, so that the next repeat is not impacted by the previous exercise.

✓ Use a sports watch or data logger to measure the pulse rate, as this is more reliable that a person measuring their own pulse rate.

Preparing for assessment

The information in this section is taken from the biology section of the Cambridge IGCSE Co-ordinated Sciences syllabus (0654) for examination from 2025. You should always refer to the appropriate syllabus document for the year of examination to confirm the details and for more information. The syllabus document is available on the Cambridge International website at www. cambridgeinternational.org.

INTRODUCTION

Examinations will test how good your understanding of scientific ideas is, how well you can apply your understanding to new situations and how well you can analyse and interpret information you have been given. The assessments are opportunities to show how well you can do these.

You will need to:

✓ have a good knowledge and understanding of science

✓ be able to apply this knowledge and understanding to familiar and new situations

✓ be able to interpret and evaluate evidence that you have just been given.

You need to be able to do these things under exam conditions.

OVERVIEW

Ensure you are familiar with the structure of your examinations. Consult the relevant syllabus for the year you are entering your examinations for details of the different papers and the weighting of each, including the papers to test practical skills. Your teacher will advise you of which papers you will be taking.

To prepare for assessment, you should know how to perform calculations, draw graphs and describe, explain and interpret biological ideas and information. You should also know how to handle content that may be unfamiliar to you, by knowing how to use your data-handling skills to apply biological principles and ideas in unfamiliar situations.

EXAMINATION TECHNIQUES

To help you to work to your best abilities in exams, there are a few simple steps to follow.

Check your understanding of the question

✓ **Read the introduction to each question carefully before moving on to the questions themselves**.

✓ Look in detail at any **diagrams, graphs** or **tables**.

✓ Underline or circle the **key words** in the question.

✓ **Make sure you answer the question that is being asked** rather than the one you wish had been asked!

✓ Make sure you understand the meaning of the '**command words**' in the questions.

Remember that any information you are given is there to help you to answer the question.

COMMAND WORDS

✓ **Analyse:** examine in detail to show meaning, identify elements and the relationship between them.

✓ **Calculate:** work out from given facts, figures, or information.

✓ **Comment:** give an informed opinion.

✓ **Compare:** identify/comment on similarities and/or differences.

✓ **Consider:** review and respond to given information.

✓ **Contrast:** identify/comment on differences.

✓ **Deduce:** conclude from available information.

✓ **Define:** give precise meaning.

✓ **Demonstrate:** show how or give an example.

✓ **Describe:** state the points of a topic / give characteristics and main features.

✓ **Determine:** establish an answer using the information available.

✓ **Discuss:** write about issue(s) or topic(s) in depth in a structured way.

✓ **Evaluate:** judge or calculate the quality, importance, amount, or value of something.

✓ **Examine:** investigate closely, in detail.

✓ **Explain:** set out purposes or reasons / make the relationships between things clear / provide why and/or how and support with relevant evidence.

✓ **Give:** produce an answer from a given source or recall/memory.

✓ **Identify:** name/select/recognise.

✓ **Justify:** support a case with evidence/argument.

✓ **Outline:** set out main points.

✓ **Predict:** suggest what may happen based on available information.

✓ **Show (that):** provide structured evidence that leads to a given result.

✓ **Sketch:** make a simple freehand drawing showing the key features, taking care over proportions.

✓ **State:** express in clear terms.

✓ **Suggest:** apply knowledge and understanding to situations where there are a range of valid responses to make proposals / put forward considerations.

Check the number of marks for each question

✓ Look at the **number of marks** allocated to each question.

✓ Look at the **space provided** to guide you as to the length of your answer. However, be aware there may be more space than you need.

✓ Make sure you include at least as many points in your answer as there are marks.

✓ Write neatly and try to keep within the space provided.

REMEMBER

Beware of continually writing too much because it may mean that you are not really answering the questions. Do not repeat the question in your answer.

Use your time effectively

✓ Don't spend so long on some questions that you don't have time to finish the paper.

✓ If you are really stuck on a question, leave it, finish the rest of the paper, and come back to it at the end.

✓ Even if you eventually have to guess at an answer, this gives you a better chance of getting it right than if you leave it blank.

ANSWERING QUESTIONS
Multiple choice questions

✓ Select your answer from the options given. Multiple choice questions are not as easy as some people think. Make sure you **consider all the options** before making your choice.

Short and long-answer questions

✓ In short-answer questions, **try to be concise in your answers.**

✓ Always aim to use **clear scientific language**.

✓ Present the information in a logical sequence.

✓ Do not be afraid to also use **labelled diagrams** or **flow charts** if it helps you to show your answer more clearly.

Questions with calculations

✓ **In calculations always show your working**. Even if your final answer is incorrect, part of your attempt may be correct.

✓ Write down your answers to as many **significant figures** as you are asked, or are used in the numbers in the question (and no more).

✓ Don't round off too early in calculations with many steps – it's always better to give too many significant figures at an early stage than too few.

✓ Use the correct **units**. In some questions the units will be mentioned, e.g. calculate the mass in grams; or the units may also be given on the answer line. If numbers you are working with are very large, you may need to make a conversion, e.g. convert joules into kilojoules or millimetres into metres.

Finishing your exam

✓ When you've finished your exam, **check through** your paper to make sure you've answered all the questions.

✓ Check that you haven't missed any questions at the end of the paper or turned over two pages at once and missed questions.

✓ Cover over your answers and read through the questions again, and check that your answers are as good as you can make them.

REMEMBER

For questions on investigative work, it is important that you understand the methods used by scientists when carrying out investigative work.

More information on carrying out practical work and developing your investigative skills is given in the previous section.

Glossary

A

absorption Movement of a substance into part of an organism, for example the movement of digested food molecules from the intestines into the blood.

acrosome A 'bag' of enzymes at the front of a sperm cell that makes it possible for the sperm nucleus to enter the egg cell.

active The opposite of passive, needing energy from respiration.

active immunity Defence against a pathogen by antibody production in the body; gained after an infection by a pathogen or by vaccination; memory cells are produced.

active site The space in an enzyme into which the substrate molecule fits.

active transport The movement of particles through a cell membrane from a region of lower concentration to a region of higher concentration (i.e. against a concentration gradient), using energy from respiration.

adaptation Features that organisms have which help them survive in their environment, or features of specialised cells, tissues, or organs that are adapted for a particular function.

adrenal gland Endocrine gland that secretes the hormone adrenaline.

adrenaline The hormone secreted by the adrenal glands that prepares the body for action.

aerobic respiration The chemical reaction in cells that uses oxygen to break down nutrient molecules (e.g. glucose) to release energy; produces carbon dioxide and water.

AIDS Acquired immunodeficiency syndrome, caused by the human immunodeficiency virus (HIV).

alimentary canal The tubular part of the digestive system, from mouth to anus.

allele An alternative form of a gene, producing one form of the characteristic that the gene codes for.

alveoli (singular *alveolus*) Tiny air sacs in the lungs where gases diffuse between the air in the lungs and the blood.

amino acid The basic unit of a protein.

amylase An enzyme that breaks down starch to the simple reducing sugars.

anaerobic respiration The chemical reactions in cells that break down nutrients to release energy without using oxygen.

animal Multi-celled organism with cells that have a nucleus, but do not have a cell wall.

anther Part of the stamen (the male part of the flower) that produces pollen.

antibiotic resistance Resistance in bacteria to the effect of an antibiotic that normally kills them.

antibiotics Drugs used to kill bacteria but which do not kill viruses.

antibodies Proteins that bind to antigens, leading to direct destruction of pathogens or marking of pathogens for destruction by phagocytes; specific antibodies have complementary shapes which fit specific antigens.

antigens Chemicals on the surface of cells, including pathogens, that have specific shapes.

anus Opening at the end of the large intestine through which faeces are egested.

aorta The largest artery, which receives blood from the left ventricle of the heart.

arteriole Small blood vessel that connects an artery to a capillary.

artery Blood vessel that carries blood away from the heart.

artificial selection See *selective breeding*.

asexual reproduction Process resulting in the production of genetically identical offspring from one parent; production of young without fertilisation.

assimilation The uptake and use of nutrients by cells.

atrium (plural *atria*) One of two chambers of the heart that receives blood from veins and pumps it into the ventricles.

B

bacteria Single-celled organisms that do not have a nucleus.

balanced diet The intake of food that supplies all the protein, fat, carbohydrate, vitamins and minerals that the body needs in the right amounts.

Benedict's solution Solution that changes colour in the presence of reducing sugars; used to test for their presence in a food sample.

biconcave A shape in which the middle is pressed inwards, making a red blood cell thinner in the middle than at the edges.

bile Liquid produced by the liver and stored in the gall bladder, which is highly alkaline and emulsifies fats and oils.

biodiversity The number of different species that live in an area.

biological catalyst A catalyst of metabolic reactions inside living organisms; an enzyme.

biomass The mass of living organisms.

biuret test The test used to indicate the presence of protein in a food sample.

bronchioles The tiny tubes in the lungs that carry air between the bronchi and the alveoli.

bronchus (plural *bronchi*) The division of the trachea as it joins to the lungs.

C

cancer A disease caused by uncontrolled division of cells.

capillary Smallest blood vessel, found within every tissue, which exchanges substances with the cells.

captive breeding programme Where animals are bred in zoos and wildlife parks to help protect them from extinction.

carbohydrate A molecule, such as starch or glycogen, made up of simple sugars.

carbon cycle How the element carbon cycles in different forms between living organisms and the environment.

carnivore An animal that gets its energy by eating only other animals.

carpel The female reproductive structure in flowers that contains one or more ovaries and their stigmas and styles.

catalyst A substance that increases the rate of a chemical reaction but is not itself changed by the reaction, such as an enzyme.

cell division A division of a cell into daughter cells.

cell membrane The structure surrounding cells that controls what enters and leaves the cell.

cellulose A carbohydrate of which plant cell walls are formed.

cell wall A layer of cellulose that surrounds plant cells, giving them support and shape. Bacterial cells also have a cell wall outside the cell membrane.

central nervous system (CNS) The brain and spinal cord.

cervix Part of female reproductive system, connecting the vagina and the uterus.

chemical digestion The breakdown of large insoluble food molecules into small soluble molecules that can be absorbed.

chlorophyll The green chemical in chloroplasts that captures light energy for photosynthesis.

chloroplast A cell structure found in plant cells that can capture energy from light for use in photosynthesis.

chromosome Structure found in a cell nucleus, made of DNA containing genetic information in the form of genes.

cilia Tiny hairs that project from a cell surface and that move things across the cell surface.

ciliated cell A cell that has cilia on its surface.

circulatory system The organ system that transports substances around the body (also known as the *blood system*).

climate change A change to wind, rainfall, and temperature as a result of increased greenhouse gases.

clone An individual that is genetically identical to other individuals.

clot A clump of semi-solid blood formed at the site of an injury to prevent blood loss and the entry of pathogens.

colon The main part of the large intestine, where water is absorbed.

combustion Burning, such as that of fossil fuels.

community All of the populations of different species in an ecosystem.

competition The interaction between organisms trying to get the same limited resource or resources.

complementary Having shapes that fit together, like the active site of an enzyme and its substrate.

concentration gradient The difference in the concentration of a substance between two areas; diffusion occurs down the concentration gradient: molecules move from the area of high concentration to the area of low concentration.

condom A rubber sheath placed over the erect penis to act as a mechanical barrier against sperm and sexually transmitted infections.

conservation The protection of species or habitats, to prevent their destruction.

consumer An organism that gets its energy by feeding on other organisms.

continuous variation Variation in a phenotypic feature that can have any value in a range; e.g. body length and body mass.

coordination The linking together of activities of the nervous system and its associated sense organs, muscles, and glands.

core temperature The temperature around the major organs of the heart and liver.

coronary artery Blood vessel that carries blood to the heart muscle.

coronary heart disease Diseases such as angina, high blood pressure, or heart attack, caused by the partial or complete blockage of a coronary artery, such as by cholesterol.

correlation When two factors vary in a linked way.

cotyledon Food store found in the seeds of flowering plants.

cuticle A waxy layer that covers a leaf, particularly the upper surface, to reduce water loss from the leaf.

cytoplasm The jelly-like liquid inside the cell that contains the cell structures and where many chemical reactions take place.

D

daughter cell A cell produced by division of a parent cell.

death rate The number of deaths within a particular time.

decay See *decomposition*.

decomposer An organism that gets its energy from dead or waste organic material.

decomposition (also called *decay*) The breakdown of dead or waste organic material; caused by the action of organisms called decomposers (certain bacteria or fungi).

deforestation The destruction of large areas of forest or woodland.

dehydration Loss of too much water.

denature A change in the structure of an enzyme, which may be the result of high temperature or pH change; this stops an enzyme working.

deoxygenated Lacking in oxygen.

deoxyribonucleic acid See *DNA*.

diaphragm (respiratory system) A sheet of tissue situated below the lungs that controls breathing.

diffusion The net movement of particles from a region of their higher concentration to a region of their lower concentration (i.e. down a concentration gradient) as a result of their random movement.

digestion The breakdown of food: physical digestion breaks food down into smaller pieces; chemical digestion breaks down large insoluble molecules into small soluble molecules.

digestive enzyme An enzyme found in the digestive system.

diploid A cell or nucleus containing two sets of chromosomes.

discontinuous variation Variation in a phenotypic feature with a limited number of phenotypes with no intermediates; e.g. ABO blood groups, seed shape and seed colour in peas.

DNA (deoxyribonucleic acid) The chemical that forms chromosomes and contains genetic information.

dominant An allele that is expressed if it is present in the genotype.

double circulation A circulatory system as in humans where the blood passes through the heart to be pumped first to the lungs, then returned to the heart to be pumped to the rest of the body.

double helix The twisted ladder shape formed by a molecule of DNA.

drug A substance that, when taken into the body, modifies or affects chemical reactions in the body.

duodenum First part of the small intestine where some food is digested.

E

ecosystem A unit containing the community of organisms and their environment, interacting together.

effective collision A collision between molecules that results in a chemical reaction.

effector An organ that responds to the nervous system, such as a muscle or a gland.

egestion The removal of undigested food from the body as faeces (compare with *excretion*).

egg cell The female gamete (sex cell).

electrical impulse Signal carried by neurones.

electrocardiogram (ECG) A recording of the activity of the heart.

embryo A ball of cells formed from a male gamete and a female gamete.

emulsify To break up large droplets of a lipid in an aqueous solution into smaller droplets.

endangered When a species is at high risk of becoming extinct.

endocrine gland An organ that produces a hormone.

enzyme A protein that acts as a biological catalyst, increasing the rate of reactions in the body.

enzyme–substrate complex The structure formed when an enzyme combines with its substrate.

epidermis The layer of cells on the outer surface of a body or organ, such as a leaf.

ethanol emulsion test The test used to show the presence of lipids (fats or oils) in a food sample.

evaporation When particles in a liquid (e.g. water) gain enough energy to move fast enough and become a gas (as in water vapour).

excess More than is needed.

excretion The removal of the waste products of metabolism and substances in excess of requirements.

expiration Breathing out (exhalation).

extinction When all individuals of a species have died.

F

faeces The undigested material that remains after digestion of food.

fat A solid lipid.

fatty acid One of the basic units of a lipid, along with glycerol.

femidom The female equivalent of a condom, which prevents sperm and sexually transmitted infections entering the woman's body.

fertilisation The fusion (joining) of the nuclei of two gametes (sex cells); in plants, when a pollen nucleus fuses with a nucleus in an ovule; in animals, when a nucleus from a sperm fuses with the nucleus in an egg cell.

fetus The name given to the developing baby in the uterus; the stage after an embryo.

fibre Plant material that is difficult to digest and keeps the food in the alimentary canal soft and bulky, aiding peristalsis. Also called *roughage*.

filament Part of the stamen (the male part of a flower) that supports the anther.

flaccid When a plant cell has lost water and the cytoplasm does not push against the cell wall to give the plant support. This will cause a plant to wilt.

flagellum Large cell structure that moves back and forth to move the cell, such as the 'tail' of a sperm cell.

food chain A diagram that shows the transfer of energy from one organism to the next, beginning with a producer.

food web A network of interconnected food chains.

foreign species a species that has been introduced from a different habitat (often invasive).

fossil fuel Fuel formed over a long period of time from organic material, such as peat, coal, oil, or natural gas.

G

gall bladder Where bile from the liver is stored.

gamete A sex cell.

gas exchange The exchange of gases between the air and the body across a gas exchange surface.

gas exchange surface The site of gas exchange, such as the alveoli in the lungs or the cell surfaces within a leaf.

gene A length of DNA that codes for a protein.

genetic diagram A diagram that displays how different combinations of alleles may be inherited by offspring from their parents, e.g. a Punnett square.

genotype The genetic make-up of an organism in terms of the alleles present.

germination The start of plant growth from a seed, which only occurs when there is the right amount of oxygen and water and an appropriate temperature.

glucagon A hormone secreted by the pancreas that increases blood glucose concentration.

glucose A simple sugar.

glycerol One of the basic units of a lipid, along with fatty acids.

glycogen Large carbohydrate that is stored in liver cells for use as an energy source.

growth A permanent increase in size and dry mass.

guard cells Cells that are found in pairs on the surface of a leaf, can swell or shrink to open and close stomata.

H

habitat The place where an organism lives.

haemoglobin The red chemical in red blood cells that combines reversibly with oxygen.

haploid A cell or nucleus containing a single set of chromosomes, as in gametes.

heart The organ of the circulatory system that pumps blood.

heart rate The number of heart beats in a given time, e.g. beats per minute.

herbivore An animal that gets its energy by eating only plants.

heterozygous Having two different alleles of a particular gene.

homeostasis The maintenance of a constant internal environment.

homozygous Having two identical alleles of a particular gene.

hormonal system The system of the body consisting of endocrine glands which secrete hormones to alter the activity of target organs.

hormone A chemical substance produced by an endocrine gland and carried by the blood, which alters the activity of one or more specific target organs.

host An organism that is attacked by a pathogen.

human immunodeficiency virus (HIV) A pathogen transmitted by sexual contact, by blood, through the placenta, or in breast milk, which may lead to AIDS.

humidity A measure of the concentration of water molecules in the air.

hydrochloric acid Acid found in gastric juice in stomach that kills harmful microorganisms in food and provides an acidic pH for optimum enzyme activity.

hydrogencarbonate indicator Solution that is red-orange in neutral pH, yellow in acidic solutions, purple in alkaline solutions.

hygiene Keeping things clean, which helps to reduce the risk of transmission of a pathogen.

ileum Main part of the small intestine, where digested food is absorbed.

immune Cannot catch a transmissible disease.

immune system The system of the body that protects the body against infection; includes white blood cells.

impulse Electrical signal carried along a neurone.

ingestion The taking of food and drink into the alimentary canal.

inheritance The transmission of genetic information from generation to generation.

insoluble A substance that does not dissolve.

inspiration Breathing in (inhalation).

insulin A hormone secreted by the pancreas that decreases blood glucose concentration.

intercostal muscles Muscles between the ribs that help with breathing.

iodine solution A solution that changes colour in the presence of starch, used to test for its presence.

kinetic energy The energy stored by moving molecules.

L

lactic acid Product of anaerobic respiration in muscles.

large intestine The final part of the alimentary canal, made up of the colon, rectum, and anus.

larynx The 'voice box' at the top of the trachea, which produces sounds when air moves through it, e.g. when speaking.

leaching The loss of dissolved mineral nutrients in soil water as it soaks deep into the ground beyond the reach of plant roots.

lipase An enzyme that breaks down fats and oils to fatty acids and glycerol.

lipid Molecule made from fatty acids and glycerol.

liver Large organ used for storage and production of useful molecules.

livestock Animals grown as a source of food.

lumen The space inside a blood vessel through which blood flows.

lungs Organs in the human body where gas exchange takes place.

lymphocyte A type of white blood cell that makes antibodies to attack a pathogen.

M

magnification How much a microscope increases the observed size of a structure compared with its actual size. Calculated either by multiplying the eyepiece lens magnification with the objective lens magnification, or by using the formula: magnification = image size ÷ actual size.

mechanical digestion See *physical digestion*.

meiosis A reduction division in which the chromosome number is halved from diploid to haploid resulting in genetically different cells; involved in the production of gametes.

memory cell Cells produced by lymphocyte blood cells that responds to a second infection by the same pathogen by the production of many more antibodies.

menstrual cycle The repeating sequence of events in a woman's reproductive organs; each cycle of ovulation (ripening and release of an egg) and menstruation (shedding of the unneeded uterus lining) takes about 28 days.

metabolism All the reactions that occur inside the body that keep an organism alive.

mineral ions Nutrients that plants and animals need in small amounts, such as nitrates that are needed for making amino acids.

mitochondria Cell structures in which respiration takes place (singular *mitochondrion*).

mitosis Nuclear division giving rise to genetically identical cells; used for growth, repair of damaged cells, replacement of cells and asexual reproduction.

monohybrid cross The inheritance of characteristics produced by the alleles of a single gene.

motile Capable of moving by itself.

motor neurone A nerve cell that carries electrical impulses from the central nervous system to an effector.

movement An action by an organism or part of an organism causing a change of position or place.

MRSA (methicillin-resistant *Staphylococcus aureus*) A type of antibiotic-resistant bacterium.

mucus Slimy liquid that is produced in the lining the trachea, bronchi and bronchioles.

multicellular An organism that has a body which contains many cells.

mutation A genetic change in an organism; the way in which new alleles are formed.

 N

natural selection The process by which individual organisms that are better adapted to the environment than others have a greater chance of survival and reproduction, passing on their alleles to the next generation.

nectar Sweet liquid produced in flowers to attract insects for pollination.

negative feedback Where a change in a stimulus causes the body to respond to reverse that change, e.g. in controlling body temperature.

nerve A bundle of nerve cells (neurones).

nervous system The system of the body that uses neurones to coordinate and regulate body functions.

net movement The sum of all the movements in different directions, e.g. the overall movement of all the particles being considered in diffusion or osmosis.

neurone A nerve cell.

nucleus Structure inside plant and animal cells that contains the genetic material (DNA).

nutrient cycle How a nutrient cycles between living organisms and the environment.

nutrition The taking in of materials for energy, growth, and development.

 O

oesophagus Part of the alimentary canal, between the mouth and stomach.

oestrogen A female sexual hormone secreted by the ovaries that produces secondary sexual characteristics in girls.

oil A liquid lipid.

omnivore An animal that eats both plants and animals.

optimum pH The pH at which an enzyme works best.

optimum temperature The temperature at which an enzyme works best.

organ A collection of tissues that together have a particular function (e.g. the kidney).

organism An individual living thing.

organ system A group of organs that work together to carry out a particular function (e.g. the digestive system).

osmosis The net movement of water molecules from a dilute solution (region of higher water potential) to a concentrated solution (region of lower water potential), through a partially permeable membrane.

ovary (in plants and humans) A structure that contains egg cells.

overharvest When a species that could be managed sustainably is overexploited and so many are taken that the population falls to dangerous levels.

oviduct A tube that carries the egg released from the ovary to the uterus; where fertilisation occurs.

ovulation When an egg is released by an ovary.

ovule Female structure inside the ovary of a flower that contains one egg cell.

oxygenated Containing a lot of oxygen.

oxygen debt The need for oxygen after anaerobic respiration to break down lactic acid.

P

palisade mesophyll cells Cells in the upper part of a leaf that contain the most chloroplasts and carry out most of the photosynthesis in a leaf.

pancreas Organ that secretes digestive enzymes and the hormones insulin and glucagon.

partially permeable The condition of a membrane that lets some substances pass through but not others.

passive The opposite of *active*; happening without the need for additional energy.

pathogen A disease-causing organism.

pedigree diagram A diagram that shows the inheritance of phenotypes within different generations of a family.

penis External male reproductive organ; contains the urethra which carries urine and sperm outside the body.

peripheral nervous system (PNS) The nerves outside of the brain and spinal cord.

peristalsis The rhythmic muscular contractions of the alimentary canal that move food from mouth to anus.

pest An organism that is causing a problem.

pesticide A chemical used to kill pests.

petal Modified leaves that surround a flower; often large and colourful to attract insects for pollination.

phagocyte A type of white blood cell that engulfs and destroys pathogens.

phagocytosis To flow and engulf, as when phagocytes engulf pathogens.

phenotype The observable features of an organism.

phloem The plant tissue that transports sucrose and amino acids through the vascular bundles.

photosynthesis The process carried out in plant cells that makes sugars by combining carbon dioxide and water molecules using energy from light.

physical digestion The breakdown of food into smaller pieces without chemical change to the food molecules; this increases the surface area of food for the action of enzymes in chemical digestion.

plant Multi-celled organism that obtains its nutrition from photosynthesis, made of cells that each contain a nucleus and cell wall.

plasma The liquid, watery part of blood that carries dissolved food molecules, urea, hormones, carbon dioxide, and other substances around the body, and also helps to distribute heat.

plasmid A small circle of genetic material found in some bacteria in addition to the main circular DNA.

plasmolysed A plant cell in which the cell membrane is pulling away from the cell wall due to a lack of water.

plasmolysis When the contents of a plant cell shrink due to a lack of water and the cell membrane pulls away from the cell wall.

platelet Fragment of a much larger cell that causes blood clots to form at sites of damage in blood vessels.

pollen grain A male structure in plants that contains the male gamete.

pollination The transfer of pollen grains from an anther to a stigma; necessary before fertilisation can take place.

pollution The adding of substances to the environment that cause harm.

population A group of organisms of one species, living in the same area, at the same time.

predation The killing of prey animals by predators for food.

predator An animal that kills and eats other animals for food.

prey An animal killed for food by a predator.

producer An organism that makes its own organic nutrients, usually using energy from sunlight, through photosynthesis.

product A molecule that is formed during a reaction.

prostate gland Part of the male reproductive system that helps make semen by producing some of liquid that sperm swim in.

protease An enzyme that breaks down protein to amino acids.

protein A large molecule that is made of many amino acids joined together.

protein synthesis The process in which proteins are made inside a cell.

pulmonary artery Blood vessel that carries blood from the heart to the lungs.

pulmonary vein Blood vessel that carries blood from the lungs to the heart.

Punnett square A form of genetic diagram used to predict the possible genotypes of the offspring of a particular cross.

pure-breeding A group of individuals identical for one or more phenotypic characteristics that, when crossed, always produce offspring with the same phenotypes.

R

receptor An organ or cell that can respond to an external stimulus like light or heat, and transmit a signal to a sensory nerve.

receptor organ An organ that receives information about the environment (such as eye or ear) and responds by stimulating sensory neurones.

recessive An allele that is only expressed when there is no dominant allele of the gene present in the genotype.

rectum Part of the large intestine where undigested food material is compacted to form faeces.

red blood cell Blood cell that contains the red protein haemoglobin, which carries oxygen.

reducing sugar A sugar, such as glucose, that gives a positive result when tested with Benedict's solution.

reduction division See *meiosis*.

reflex An automatic response to a stimulus.

reflex arc The pathway of neurones that an electrical impulse follows during a reflex response.

relay neurone A nerve cell in the central nervous system that forms a connection between other neurones.

reproduction The processes that make more of the same kind of organism.

respiration The chemical reactions in cells that break down nutrient molecules and release energy for metabolism.

response A change in an organism as a result of detecting a stimulus.

ribosome Very small cell structures found in plant, animal, and bacterial cells.

ribs Part of the skeleton surrounding the heart and lungs; move to aid the ventilation of the lungs.

rickets Deficiency disease which causes softening of bones; caused by a lack of calcium and vitamin D.

root cortex cells Unspecialised cells found in plant roots between the epidermis and the transport vessels.

root hair cell A cell in the epidermis of roots that has a long extension of cytoplasm, where uptake of substances from soil water occurs.

roughage See *fibre*.

S

salivary glands Glands in the mouth that produce saliva to aid digestion of food.

scrotum Sac supporting the testes.

scurvy Deficiency disease in which gums bleed and wounds do not heal properly; caused by a lack of vitamin C.

secondary sexual characteristic A feature that develops at puberty as a result of sexual hormones.

secretion The release of chemicals that have been made inside the cell into the fluid outside the cell.

seed The structure formed from an ovule that contains the plant embryo and food stores.

seed bank A large collection of many different species of seed stored for use in the future.

selective breeding (also called *artificial selection*) The breeding together of individual organisms that have desirable features; carried out over many generations to improve crop plants and domesticated animals.

sense organ An organ that responds to a stimulus by causing an electrical impulse in a sensory neurone.

sensitivity The ability to detect and respond to changes in the internal or external environment.

sensory neurone A nerve cell that carries electrical impulses from a receptor to the central nervous system.

sepal Leaf-like structure that protects the flower in the bud.

septum Wall dividing the left and right sides of the heart.

set point The value around which the normal range fluctuates, e.g. internal body temperature or blood glucose concentration.

sex chromosome A chromosome that affects the sex of the individual; for humans, the X and Y chromosomes.

sexual reproduction Process involving the fusion (joining) of the nuclei of two gametes to form a zygote and the production of offspring that are genetically different from each other.

sexually transmitted infection (STI) An infection that is transmitted through sexual contact.

simple sugar A basic sugar unit (e.g. glucose) that can join together with other sugar units to make large carbohydrates such as starch and glycogen.

single circulation A circulatory system as in fish where the blood is pumped by the heart to the gills then to the rest of the body before returning to the heart.

sink The parts of plants that use or store sucrose or amino acids.

small intestine Part of the alimentary canal, made up of the duodenum and ileum, where nutrients are absorbed.

soil erosion The washing away of soil as a result of wind and rainfall when there is little vegetation to hold on to the soil.

soluble Dissolves easily in a solvent, such as water.

solute A substance that can dissolve in a liquid (the solvent).

solution A liquid (solvent) containing a dissolved substance (solute) or substances.

solvent A liquid that a substance (the solute) is able to dissolve in.

source The parts of plants that release sucrose or amino acids.

specialisation When a cell develops special features that help it work in a particular way.

specific Limited, usually to one or a few. For example, enzymes are specific because they only work with one or a few similarly shaped substrates.

sperm cell Male gamete (sex cell) in animals.

sperm duct Tube that carries sperm from a testis to the urethra in the penis.

spinal cord Part of the central nervous system found inside the spinal column of the skeleton.

spongy mesophyll The layer of cells in the lower part of the leaf in which there are many air spaces, so increasing the internal surface area to volume ratio.

spore Small structure produced by ferns and fungi for dispersal and growth into new individuals.

stamen The male reproductive structure in flowers, made up of the anther and filament.

starch A complex carbohydrate made from many glucose units.

stigma The female reproductive structure in flowers to which pollen grains attach in pollination.

stimulus A change in the internal or external environment that produces a response by an organism.

stomach Part of the alimentary canal where acid and protease enzymes are secreted.

stomata (singular *stoma*) Tiny holes in the surface of a leaf (mostly the lower epidermis), which allow gases to diffuse into and out of the leaf.

style The female reproductive structure that supports the stigma in a flower.

substrate A molecule that fits into the active site of an enzyme molecule at the start of a reaction.

sucrose Common sugar produced by plants.

synthesis The building of larger molecules from smaller ones, such as the formation of proteins from amino acids.

T

target organ An organ that is affected by a hormone.

testis (plural *testes*) The site of sperm production; secretes the male sexual hormone testosterone.

testosterone The male sexual hormone that is secreted by the testes.

tissue A group of similar specialised cells that work together to carry out a particular function.

toxic Poisonous.

trachea The tube leading from the mouth to the bronchi.

translocation The movement of sucrose and amino acids, through the phloem tissue of a plant, from sources to sinks.

transmissible disease A disease in which the pathogen can be passed from one host to another.

transpiration Loss of water vapour from the leaves of a plant.

trophic level The position of an organism in a food chain, food web, or ecological pyramid.

turgid Plant cells that have a full vacuole and the cytoplasm pushes against the cell wall, giving the plant structure and support.

turgor pressure The pressure of the cytoplasm inside a plant cell against the cell wall.

U

urea A waste product produced in the liver and transported by blood plasma.

urethra The tube carrying urine from the bladder to the outside of the body.

uterus Where a baby develops inside a mother.

V

vaccination Putting weakened pathogens or their antigens into the body to stimulate an immune response by lymphocytes which produce antibodies; memory cells are also produced for long-term immunity.

vaccine A harmless version of a pathogen used in immunisation.

vacuole A large sac found in the middle of many plant cells, containing cell sap.

vagina Part of female reproductive system; tube leading to the cervix and uterus.

valve Flaps in the heart, and in veins, that prevent the flow of blood in the wrong direction.

variation Differences between individuals of the same species.

vascular bundle Tissue that forms the veins in plant roots, stems, and leaves, containing xylem vessels and phloem cells.

vasoconstriction The narrowing (constriction) of blood vessels.

vasodilation The widening (dilation) of blood vessels.

vein (*animal*) A blood vessel that carries blood towards the heart. (*plant*) See *vascular bundle*.

vena cava The largest human vein that delivers blood from the body to the right atrium.

ventilation Moving air into and out of the lungs (breathing).

ventricle One of two chambers of the heart that receive blood from the atria and pump it out through arteries.

vitamin A nutrient needed by the body in tiny amounts to remain healthy, such as vitamins A, C, and D.

vitamin C Vitamin needed for healthy skin.

 W

waste product A product of a chemical reaction that is not needed, such as oxygen in photosynthesis.

water potential The potential for a solution to take up more water molecules.

water potential gradient The difference in water potential between two regions, e.g. in a plant.

white blood cell Blood cell involved in phagocytosis or antibody production.

wilt When a plant droops because its cells are no longer turgid.

 X

xylem vessel A tube formed from dead cells in the vascular bundles of a plant, which transports water and mineral ions from the roots to the leaves and other parts of the plant; it also provides support.

 Z

zygote A fertilised egg, formed from the fusion of a male gamete and female gamete.

Answers

All answers, including answers to practice questions, have been written by the authors. In examinations, the way marks are awarded may be different. These are the answers to the questions in each topic. Answers to end of topic and practice questions are available from the Collins website.

SECTION 1 CHARACTERISTICS OF LIVING ORGANISMS

Characteristics of living organisms

Page 12

1. **a)** Any suitable answers for human, such as:
   movement – walking; respiration – combination of oxygen with glucose to release energy, carbon dioxide and water; sensitivity – vision; growth – increase in height; reproduction – having a baby; excretion – producing urine; nutrition – eating food.

   b) Any suitable answers for a specific animal, such as: movement – crawling; respiration – combination of oxygen with glucose to release energy, carbon dioxide, and water; sensitivity – smell; growth – increase in length; reproduction – producing young; excretion – losing carbon dioxide through respiratory surface; nutrition – eating food.

   c) Any suitable answers for a plant, such as:
   movement – growing towards light; respiration – combination of oxygen with glucose to release energy, carbon dioxide, and water; sensitivity – detecting direction of light; growth – increase in height; reproduction – producing seeds;

   excretion – diffusion of waste products out of leaf for photosynthesis (oxygen) and respiration (carbon dioxide); nutrition – taking in nutrients from soil and making glucose by photosynthesis.

2. Movement – to reach best place to get food or other conditions favourable for growth;
   respiration – to release energy from food that can be used for all life processes;
   sensitivity – to detect changes in the environment;
   growth – to increase in size until large/mature enough for reproduction;
   reproduction – to pass genes on to next generation;
   excretion – to remove harmful substances from body;
   nutrition – to take in substances needed by the body for growth and reproduction.

SECTION 2 CELLS

Cell structure

Page 23

1. **a)** Drawing should be drawn with thin, clear pencil lines, no crossing out, to show the outline of the cell in the photograph and the central shape.

   b) Diagram should be labelled to show nucleus, cytoplasm, and cell membrane.

2. **a)** cell wall, large vacuole, chloroplast

   b) cell wall, circular DNA, plasmids

3. **a)** chloroplast

   b) large vacuole

   c) cell wall

4. Mitochondria release energy during aerobic respiration. Ribosomes are where new proteins are formed.

5. Because they are too small to be seen properly with a light microscope.

Page 26

1. a) Lining some tubes in animal organs, such as the respiratory tract of humans; the cilia on the outside of the cells help move substances along inside the tubes.

b) Throughout the body, conduct electrical impulses around the body.

c) In blood; carry oxygen around attached to haemoglobin inside the cell.

d) Near the tips of plant roots; have long cell extensions to increase surface area for absorption of substances into the root.

e) In plant leaves; carry out most of the photosynthesis in plants.

2. Sperm cells are small, and have a nucleus that can enter the egg cell for fertilisation. Egg cells are large and contain a nucleus and a lot of cytoplasm to provide nutrients for the fertilised cell during the early stages of division.

Page 27

1. a) Any suitable two, such as: muscle tissue, nervous tissue, bone tissue.

b) Any suitable two, such as: heart, liver, brain.

c) Any suitable two, such as: nervous system, digestive system, circulatory system.

2. a) Any suitable two, such as: palisade tissue, epidermis/epidermal tissue.

b) Any suitable two, such as: leaf, root.

Size of specimens

Page 28

1. 2 mm. The magnification is image size ÷ actual size = 10 ÷ 0.5 = ×20. So second image size = magnification × actual size = 20 × 0.1 = 2 mm.

2. If you are not using a suitable magnification for the specimen you are looking at, you may not be able to see what you want to. (It is most useful to start by focusing at a lower magnification and then moving up to the magnification you want to use.)

3. actual size = image size ÷ magnification = 2.5 mm ÷ 100

4. 0.025 mm = 25 µm

SECTION 3 MOVEMENT INTO AND OUT OF CELLS

Diffusion

Page 40

1. Any answer that means the same as the following:

net movement – the sum of movement in all the different directions possible;

diffusion – the net movement of particles from a region of their higher concentration to a region of their lower concentration, as a result of their random movement.

2. Passive, because no energy is provided by the cell for it to happen.

3. Only particles that are small enough to pass through the membrane can diffuse. Larger molecules cannot diffuse through the membrane.

Osmosis

Page 43

1. It is a solvent and many substances can dissolve in it.

2. In humans, substances dissolve in blood so they can be transported around the body.

3. a) Water molecules move from a higher concentration of water to a region of lower concentration of water, through a partially permeable membrane.

b) Osmosis is the net movement of water from a region of high water potential to a region of low water potential through a partially permeable membrane. It is the net movement of water down a water potential gradient.

4. a) Both involve the passive movement of molecules as the result of a concentration gradient.

b) Osmosis only considers the movement of water molecules; diffusion considers the solute molecules.

5. The strong cell wall prevents more water entering a plant cell than there is space for in the cell (i.e. when the cell is full of water). The cell wall gives cells that are full of water a specific shape and a rigidity, and this helps to support the plant, keeping it upright.

Page 45

1. a) A plant cell that is not full of water.

b) A plant cell that is full of water.

c) The removal of water from a plant cell so that the cell membrane surrounding the cytoplasm pulls away from the cell wall.

d) The pressure on the cell wall caused by the water in the cytoplasm that prevents more water entering the plant cell.

2. The water potential of the cells inside the plant root is lower than the water potential of the soil water surrounding the root, so water moves down the water potential gradient into the root.

3. Diagram should show water molecules leaving the red blood cell as a result of osmosis and entering the solution. Labels should indicate a water potential gradient from the cell to the solution and indicate that the loss of water results in the cell shrinking.

Active transport

Page 46

1. The movement of particles through a cell membrane from a region of lower concentration to a region of higher concentration (i.e. against a concentration gradient), using energy from respiration.

2. Uptake of nitrate ions by root cells in plants which plants need but are in higher concentration inside plant cells than in soil water; uptake of glucose from digested food in the small intestine by cells of the villi in humans, to ensure all the glucose is absorbed.

SECTION 4 BIOLOGICAL MOLECULES

Biological molecules

Page 53

1. a) fatty acids and glycerol
  b) simple sugars
  c) amino acids

2. Protein is formed from amino acids, carbohydrates from simple sugars; carbohydrates are often made from one kind of simple sugar, proteins from many different kinds of amino acids.

Page 55

1. a) i) An orange-red precipitate would form, because glucose is a reducing sugar.

ii) The solution wouldn't change colour as there is no starch present.

b) i) There would be no change in colour because sucrose and the starch in wheat flour are not reducing sugars.

ii) The solution would turn blue-black because of the starch in flour.

2. Crush the seeds using a mortar and pestle, then:

a) mix part with ethanol, pour off the liquid and add to water – if the mixture turns cloudy, then fat is present;

b) mix part with water to form a solution, add a few drops of biuret solution – if protein is present, a blue ring forms at the surface, which disappears to form a purple solution.

SECTION 5 ENZYMES

Enzymes

Page 61

1. A substance that increases the rate of reaction but remains unchanged at the end of the reaction.

2. A substance that is found in living organisms that acts as a catalyst.

3. Without enzymes, the metabolic reactions of a cell would happen too slowly for life processes to continue.

Page 62

1. A substrate is a molecule that an enzyme joins with at the start of a reaction. Substrate molecules are changed to product molecules during a reaction.

2. When the substrate is joined to the active site, this makes it easier for the bonds inside the substrate to be rearranged to form the products.

3. The part of an enzyme into which a substrate fits closely during a reaction.

4. Only a substrate with a shape that is complementary to the shape of the active site can fit it into it. So an enzyme can only work with a particular shape of substrate.

Page 65 (top)

1. As temperature increases, the rate of the reaction will increase, up to a maximum point (the optimum), after which it decreases rapidly as the enzyme is denatured.

2. The optimum pH for pepsin is around pH 2, which is very acidic like the contents of the stomach. The optimum for trypsin is around pH 8, which is more alkaline like the contents of the small intestine. Each enzyme has an optimum pH that matches the environment in which they work, so that they act most efficiently there.

Page 65 (bottom)

1. a) The cooler a substance is, the less kinetic energy its molecules have. So collisions between the enzymes and substrate molecules are less frequent and it takes longer for the substrate to fit into the active site. Therefore the cooler the temperature, the slower the rate of reaction.

b) As temperature increases, the atoms within the molecules of the enzyme vibrate more. This changes the shape of the active site, making it more difficult for the substrate to fit into the active site and so slowing down the rate of reaction. Eventually, the atoms vibrate so much that the shape of the active site is destroyed and the enzyme is denatured.

2. At a pH above and below the optimum of pH 2, the shape of the active site is changed as the interactions between the amino acids in the enzyme are affected by the pH. This makes it more difficult for the substrate to fit into the active site, so the rate of reaction slows down.

SECTION 6 PLANT NUTRITION

Photosynthesis

Page 72 (top)

1. carbon dioxide + water $\xrightarrow[\text{light energy}]{\text{chlorophyll}}$ glucose + oxygen

2. Light provides the energy needed for photosynthesis.

3. a)
$$6CO_2 + 6H_2O \xrightarrow[\text{light energy}]{\text{chlorophyll}} C_6H_{12}O_6 + 6O_2$$

b) Labels should show: CO_2 from air, H_2O from soil water, $C_6H_{12}O_6$ used in cells for respiration or converted to other chemicals for use in cells, O_2 released into air if not needed in respiration.

4. Most organisms other than plants get their energy in chemical form from the food that they eat. That energy was originally transferred from light to energy in chemicals during photosynthesis in a plant cell and then transferred along the food chain.

Page 72 (bottom)

1. Plants make their own foods and need to convert the carbohydrates made by photosynthesis into other substances, such as proteins, which contain additional elements.

2. a) Nitrate ions are needed to make amino acids and proteins.

b) Magnesium ions are needed to make chlorophyll, which is the green substance in plants.

Page 74

1. Test the leaf of a variegated plant for starch. Starch is only produced in the green parts of the leaf, where there is chlorophyll, so only the green parts of the leaf photosynthesise. The pale parts of the leaf, without chlorophyll, act as the control.

2. Heat in a water bath, keeping the ethanol away from open flames such as from a Bunsen burner, because ethanol gives off flammable fumes.

3. Place one de-starched plant in an atmosphere with no (or limited) carbon dioxide (due to the presence of potassium hydroxide) and one in an atmosphere high in carbon dioxide (for example, due to carbon dioxide given off in a reaction between marble chips and dilute acid). Shine light on the plants. After several hours, test one leaf from each plant. Only the leaf in high carbon dioxide will have produced significant amounts of starch as a result of photosynthesis.

Page 76

1. Carbon dioxide is soluble and acidic, so when more gas is being produced, such as during respiration, the solution becomes more acidic. When carbon dioxide is removed from the solution, such as during photosynthesis, the solution becomes less acidic.

Page 77

1. a) As light intensity increases, so rate of photosynthesis increases.

b) As carbon dioxide concentration increases, so rate of photosynthesis increases.

c) As temperature increases, the rate of photosynthesis increases up to a maximum, after which it decreases rapidly.

2. a) As light intensity increases, more energy is supplied to drive the process of photosynthesis.

b) As carbon dioxide concentration increases, so there is more reactant for the process.

c) As temperature increases, up to the maximum the particles in the reaction including enzymes are moving faster and collide into each other more. Above the maximum the rate of photosynthesis decreases because the enzymes that control the process start to become denatured.

Leaf structure

Page 79

1. Any four from: epidermis, spongy mesophyll, palisade mesophyll, phloem, xylem.

2. Thin broad leaves, chlorophyll/chloroplasts in cells, veins containing xylem tissue that transports water and mineral ions to the leaves and phloem tissue that takes products of photosynthesis to other parts of the plant, transparent epidermal cells, palisade cells tightly packed in a single layer near top of leaf, stomata to allow gases into and out of leaf, spongy mesophyll layer with large internal surface.

3. A large surface area helps to maximise the rate of diffusion, in this case diffusion of carbon dioxide into

cells for photosynthesis and oxygen out of cells so that it can be released into the air.

4. It allows as much light as possible to pass through the epidermal cells to reach the palisade (and spongy mesophyll) cells below, where there are chloroplasts.

SECTION 7 HUMAN NUTRITION

Diet

Page 89

1. carbohydrates, proteins, and fats/oils

2. carbohydrates from pasta, rice, potato, bread, wheat flour; proteins from meat, pulses, milk products, nuts; fats/oils from vegetable oils, butter, full-fat milk products, red meat

3. vitamins, minerals, water, and fibre

4. Vitamins and minerals are needed for maintaining the health of skin, blood, bones, etc. Water is needed e.g. as a solvent and to maintain the water potential of cells. Fibre is needed to help digested food to move easily through the alimentary canal.

5. A diet that contains all the groups of food, in the correct proportions.

Digestive system

Page 91

1. Sketch should show the following labels correctly attached to organs shown on the diagram:

- mouth, where food is broken down by physical digestion (chewing) and amylase enzyme starts digestion of starch in food
- oesophagus moves food from mouth to stomach by peristalsis
- stomach, where churning mixes food with protease enzymes and

acid to start digestion of protein molecules

- small intestine, where alkaline bile neutralises the acid chyme and enzymes from pancreas complete digestion of proteins, lipids, and carbohydrates, and where digested food molecules are absorbed into the body
- large intestine, where water is absorbed from undigested food
- rectum, where faeces are held until they are egested through the anus
- liver, where bile is made and where some food molecules are assimilated
- gall bladder, where bile is stored until needed
- pancreas, where proteases, lipases and amylase are secreted, which pass to the small intestine.

2. Egestion is the removal of undigested food from the alimentary canal – food that has never crossed the intestine wall into the body. Excretion is the removal of waste substances that have been produced inside the body.

Page 92

1. Chemical digestion uses chemicals (enzymes) to help break down large food molecules into smaller ones. Physical digestion is the chewing by the teeth to break large pieces of food into smaller ones before swallowing, or the breaking up of large fat droplets into smaller ones by bile – there is no chemical change to the food molecules.

2. Bile helps to emulsify fats in food, breaking them up into much smaller droplets and creating a much larger

surface area for lipase enzymes to act on.

Chemical digestion

Page 94

1. The digestive enzymes break down food molecules that are too large to cross the wall of the small intestine into smaller ones that can be absorbed across cell membranes and so enter the body. If we did not have enzymes, we would not be able to absorb many nutrients from our food.

2. **a)** amylase

   b) fatty acids and glycerol

3. **a)** The acid increases stomach acidity, providing the right conditions for enzymes that digest food in the stomach.

   b) Bile neutralises the acidity of food from the stomach, providing the right conditions for enzymes that digest food in the small intestine. It also emulsifies fats, providing a larger surface area for lipase enzymes to work on.

SECTION 8 TRANSPORT IN PLANTS

Xylem and phloem

Page 104

1. In vascular bundles that form veins throughout the roots, stems and leaves.

2. Phloem cells carry dissolved food materials, such as sucrose and amino acids.

Water uptake

Page 105

1. It enters through the root hair cells, moves through the root cortex cells

to the xylem in the centre of the root. It moves through the xylem up the stem and into the leaves. In the leaves, it moves out of the xylem into the mesophyll cells.

2. **a)** osmosis, **b)** active transport

Transpiration

Page 108

1. Evaporation from the surfaces of a plant, particularly from the stomata of a leaf into the air.

2. Diagram should include annotations like the following, at the appropriate point: water molecules evaporate from surfaces of spongy mesophyll cells into air spaces; water molecules from air spaces move into and out through stomata into the air – diffusion (net movement) usually from inside leaf to outside; osmosis causes water molecules to move from xylem into neighbouring cells until they reach a palisade cell or a spongy mesophyll cell; transpiration is the evaporation of water from a leaf.

3. Closing stomata reduces diffusion of water molecules out of the leaf. At night, carbon dioxide is not needed for photosynthesis, so keeping stomata open would lose water unnecessarily.

4. **a)** When temperature is higher, particles move faster, so water molecules will diffuse out of the leaf more quickly.

   b) When air humidity is lower, there is a lower concentration of water molecules in the air outside the leaf. This increases the concentration gradient between the inside of the leaf and the outside air. This means the rate of diffusion will be faster.

Translocation

Page 109

1. phloem

2. sucrose, amino acids

3. A source is a part of a plant where a substance is formed e.g. sucrose is made in leaf cells. A sink is a part of a plant where the substance leaves or is converted into something else, e.g. cells in the root or fruit may be sinks for sucrose.

SECTION 9 TRANSPORT IN ANIMALS

Circulatory systems

Page 128

1. to pump blood around the body

2. valves in the heart and veins

3. Blood passes twice through the heart for every full circulation around the body.

4. The blood being pumped twice for each complete circulation allows higher blood pressure to be maintained around the body. Also, pressure in the two circulations can be different, so the higher blood pressure needed to get blood through most of the body doesn't damage the delicate capillaries in lung tissue.

Heart

Page 130

1. left atrium, right atrium, left ventricle, right ventricle

2. Arteries carry blood away from the heart, whereas veins carry blood towards the heart.

3. vena cava, right atrium, atrioventricular valve, right ventricle, semilunar valve, pulmonary artery, pulmonary vein, left atrium, atrioventricular valve, left ventricle, semilunar valve, aorta

Page 132

1. taking a pulse count, listening to the heart, taking an ECG

2. Resting heart rate varies widely due to many factors, including age, health, and fitness, so a single value for the average is too limited.

3. As level of physical activity increases, so heart rate increases.

4. Heart rate increases with exercise so the blood can circulate faster round the body, delivering oxygen and glucose to muscle cells for the increased rate of respiration to generate the energy needed for contraction. It also removes waste carbon dioxide from muscle tissue more rapidly to prevent it building up and affecting cells.

Page 134

1. to supply the oxygen and glucose needed for the heart muscle cells to respire and to remove waste carbon dioxide

2. Any four from: smoking, diet containing a lot of saturated fat, stress, genetic factors, lack of exercise.

3. Eat a diet that is relatively low in saturated fat, don't smoke, try to reduce stress, and make sure they get enough exercise.

Blood vessels

Page 137

1. Arteries are large vessels with thick, elastic muscular walls; capillaries are tiny blood vessels with very thin walls that are often only one cell thick;

veins are large vessels with a large lumen, and valves to prevent backflow of blood.

2. a) aorta, b) pulmonary veins

3. The walls stretch as each surge of blood enters then slowly recoil as the blood flows through, evening out the pressure so that the change in pressure is reduced.

Blood

Page 140

1.

Blood component	Function
plasma	carries dissolved substances, such as carbon dioxide, glucose, urea, and hormones; also transfers heat energy from warmer to cooler parts of the body
red blood cells	carry oxygen
white blood cells	protect against infection
platelets	cause blood clots to form when a blood vessel is damaged

2. Phagocytes engulf pathogens inside the body and destroy them. Lymphocytes produce antibodies that attack pathogens.

3. Any cut that damages a blood vessel can create an easy route of infection into the body. So forming a blood clot where there is damage, as quickly as possible, helps to reduce the risk of infection. The platelets do this by converting soluble fibrinogen to insoluble fibrin, which forms a mesh to trap blood cells to produce the clot.

SECTION 10 DISEASES AND IMMUNITY

Diseases and immunity

Page 149

1. A disease caused by a pathogen that can be passed from one host to another.

2. Named viruses: cold virus, Covid virus, HIV virus.
Named bacteria: salmonella, E. coli, Clostridium

Page 152

1. Direct: any one from blood (HIV, hepatitis B, hepatitis C) or semen (HIV, syphilis, gonorrhoea, other STI). Other examples may be correct.

   Indirect: water droplets in air (colds, flu); drinking water (cholera, typhoid, dysentery); contaminated surfaces (athlete's foot, food poisoning pathogens); insect bite (malaria, dengue fever).

2. Any transmissible disease with a suitable description and explanation of how to control its spread.

Page 153

1. A barrier that physically prevents entry into the body, e.g. nose hairs that filter air, or skin.

2. A chemical that destroys the pathogen, e.g. lysozymes in mucus or acid in stomach.

3. White blood cells attack pathogens, either by engulfing and destroying them, or by producing antibodies that destroy pathogens.

Page 156

1. a) Proteins made by lymphocytes that help defend against infection.

   b) An organism or virus that causes a disease.

2. Part of the response of lymphocytes to an infection is to release memory cells into the blood. On a second infection, the memory cells respond by causing large quantities of antibodies to be released quickly. This destroys the pathogens before they cause disease.

SECTION 11 GAS EXCHANGE IN HUMANS

Gas exchange in humans

Page 165

1. trachea, bronchi, bronchioles, alveoli.

2. The lungs are adapted by having: many alveoli to provide a large surface area; thin alveoli walls to reduce the length of the diffusion pathway; good ventilation with air and a good blood supply to maintain concentration gradients for oxygen and carbon dioxide.

3. Sketch similar to Fig. 11.3, with annotations showing: thin lining of alveolar wall and wall of capillary allows rapid diffusion; high concentration gradients for gases between blood and air in alveolus due to continuous blood flow through capillary and ventilation of alveolus (lungs); large surface area of alveoli, maximising area of contact between capillary and alveoli, maximising area over which diffusion can occur.

Page 168

1. The percentage of oxygen is less in expired air than inspired air. The percentage of carbon dioxide is greater in expired air than inspired air. The percentage of water vapour is higher in expired air than inspired air.

2. As level of exercise increases, rate and depth of breathing increase.

3. There is less oxygen in expired air than inspired air because oxygen is used for respiration. There is more carbon dioxide in expired air than inspired air because the body produces carbon dioxide in respiration. There is more water vapour in expired air than inspired air because water molecules evaporate from the moist surface of the alveoli due to the warmth of the body.

4. More exercise means more carbon dioxide is produced from an increased rate of respiration. Carbon dioxide is a soluble acidic gas so could cause the body tissues and blood to become more acidic. A change in pH can affect many enzymes and so affect the rate at which life processes are carried out in the body. Slowing down the rate of life processes may harm the body.

SECTION 12 RESPIRATION

Aerobic respiration

Page 176–177

1. a), **b)** and **c)**:

glucose (*from digested food from alimentary canal*) + oxygen (*from air via lungs*) ⟶ carbon dioxide (*excreted through lungs*) + water (*used in cells or excreted through kidneys*) (+ energy (*transferred to other chemicals in cell processes*))

 d) glucose replaced by fats from hump, and very little water excreted through kidneys

2. inside cells

3. Any three from: muscle cells for contraction; synthesis of new molecules, such as proteins, for growth; active transport across cell membranes; passage of nerve impulses; maintenance of core body temperature.

4. $C_6H_{12}O_6 + 6O_2 \longrightarrow 6CO_2 + 6H_2O$

Anaerobic respiration

Page 178

1. During vigorous exercise, they may not be able to get enough oxygen from the blood for all the energy they need for contracting. So the additional energy comes from anaerobic respiration.

SECTION 13 COORDINATION AND RESPONSE

Coordination and response

Page 186

1. The ability to detect and respond to changes in the external environment and internal conditions of the body.

2. A receptor detects a change in the environment, which is the stimulus. This causes a response from an effector in the body.

3. muscles, glands

4. Sensory neurones have long dendrons and axons that link the sense organ with the central nervous system.

Relay neurones are short neurones with many dendrites, found in the central nervous system, that link sensory neurones to motor neurones or other relay neurones.

Motor neurones have many dendrites to link with relay neurones and end on the effector, such as a muscle.

Page 188

1. A reflex action is an automatic, often rapid, response to a stimulus.

2. Reflex actions are usually very fast, which makes it possible to respond to a stimulus very quickly. Reflex actions are usually important in survival, e.g. to protect you from touching something very hot, or blinking to protect the eye if something comes toward it.

Hormones

Page 189

1. **a)** A chemical messenger in the body, produced by a gland and carried by the blood, that alters the activity of one or more target organs.

   b) A gland that secretes hormones.

   c) An organ that contains cells that are affected by hormones.

2. When faced with attack, or when suddenly frightened.

3. It prepares the body for action, for example, by increasing the amount of oxygen and glucose delivered to muscle cells for rapid respiration.

Homeostasis

Page 193

1. The maintenance of a constant internal environment, i.e. keeping the conditions inside the body within limits that allow cells to work efficiently.

2. Control of core body temperature (other answers possible).

3. Skin blood vessels (arterioles) dilate when the core body temperature is too high. This allows blood to reach the skin surface more easily and the heat energy it carries to be transferred to the environment more rapidly. These arterioles constrict when the core body temperature falls too low. This reduces blood flow near the skin's surface, so heat energy cannot be transferred as easily to the skin surface and so cannot be transferred to the environment as quickly. This keeps more heat energy within the body.

4. When a change in a stimulus causes a control centre to trigger the opposite change in response, so keeping a condition within limits (returning it to its set point).

5.

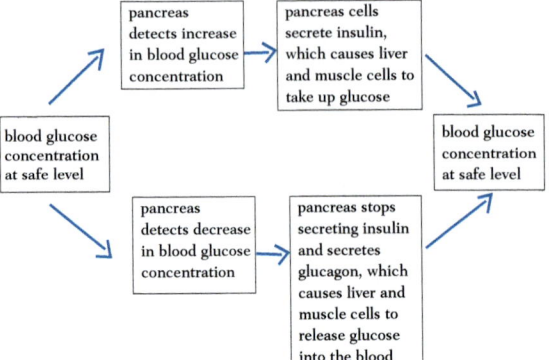

SECTION 14 DRUGS

Drugs

Page 202

1. **a)** A substance that, when taken into the body, affects chemical reactions in the body.

   b) A chemical that kills bacteria or prevents them from growing.

   c) When one type of bacteria is no longer killed or affected by an antibiotic.

2. If the infections are caused by bacteria, they can be treated with antibiotics, but if they are caused by viruses, they cannot be treated with antibiotics.

3. Using antibiotics only when essential so that bacteria are not exposed to antibiotics more often than necessary, as that is what increases the likelihood of the development of resistance.

SECTION 15 REPRODUCTION

Asexual reproduction

Page 210

1. Reproduction without the fusion of gametes, resulting in the production of genetically identical offspring from one parent.

2. Binary fission is where the genetic material is copied and the cell splits in half. Only one parent cell is involved and there is no fusion of gametes before division.

3. Advantages: no need for fertilisation, so reproduction faster and easier; if conditions remain stable, all new individuals will grow as well as the parent plant. Disadvantages: no genetic variation between plants, so if conditions change/the parent plant is susceptible to a particular disease, then all plants will do badly; this increases the risk that the plants in that area will all die.

Sexual reproduction

Page 211

1. a) The fusion of the nuclei of a male gamete and a female gamete to produce a zygote.

   b) The production of genetically different offspring from two parents as a result of fertilisation.

2. a) a sex cell or gamete such as a sperm, male gamete in a pollen grain, or an egg cell

   b) a zygote, or fertilised egg cell

3. Advantage: produces individuals with new variations of genetic material that increase the chance of survival when conditions change.

   Disadvantage: variation in offspring may also result in many offspring being less well adapted to environmental conditions than parent plants and so producing a lower yield.

Sexual reproduction in plants

Page 213

1. The carpel, made up of: the stigma, where pollen grains attach; style, which supports the stigma; ovary, which surrounds and protects the ovule, inside which is the female gamete.

2. The stamen, made up of: an anther that contains pollen grains, inside which are the male gametes; filament, which holds the anther above the flower to help with shedding of pollen.

Page 216

1. Pollination is the transfer of pollen from a stamen to a stigma. Fertilisation is the fusion of the male gamete with the female gamete to form a zygote.

2. Any three differences from: wind-pollinated flowers are usually small, no colour (white), make large amounts of lightweight pollen; insect-pollinated flowers are usually large, may be brightly coloured, produce nectar and sometimes scent, make (relatively) small amounts of larger pollen grains.

3. a) Can make less pollen OR less waste of pollen as insects more likely to deliver pollen to flower than random distribution in wind.

   b) If the insect species die out, the plant will not get pollinated.

Page 219

1. When the embryo in a seed starts to grow, splitting the seed coat and increasing in size and complexity.

2. a) Seeds need a supply of oxygen for aerobic respiration to release energy needed for growth, although they may be able to start germination using anaerobic respiration.

   b) Seeds need water for germination and will not germinate in dry soil – water is needed, for example, to transport substances needed by the growing embryo plant.

c) Seeds need warmth for germination because this allows the chemical reactions involved in growth to happen at a fast enough rate, although the amount of warmth they need may depend on where they naturally grow. Seeds from plants that live in colder areas may need a period of deep cold before they will germinate. Seeds from plants that live in areas prone to fire may not germinate until after a fire.

Sexual reproduction in humans

Page 221 (top)

1. a), b) and c) Sketch should be similar to Fig. 15.21. Labels and annotations as follows:
   - testes, where sperm (male gametes) are produced
   - sperm duct, which carries sperm to urethra
   - prostate gland and seminal vesicles, which produce liquid in which sperm swim
   - penis, which when erect delivers sperm into vagina of female
   - urethra, the tube that carries sperm from sperm ducts to outside the body.

2. a), b) and c) Sketch should be similar to Fig. 15.22. Labels and annotations as follows:
   - ovaries, where egg cells form
   - oviducts, which carry the eggs to the uterus and where fertilisation by sperm takes place
   - uterus, where embryo implants into lining and fetus develops
   - cervix, base of uterus where sperm are deposited during sexual intercourse
   - vagina, where penis is inserted during sexual intercourse.

3.

	Egg cell	Sperm cell
size	very large, 0.2 mm diameter	very small, 45 μm long
numbers	thousands in ovary but usually only one released each month	>100 million produced each day
mobility	unable to move on its own	motile: self-propelling with tail

Page 221 (bottom)

1. a) In an oviduct.
   b) male – sperm; female – egg
2. The nucleus of the sperm cell fuses with the nucleus of the egg cell forming a zygote (fertilized egg).

Sexual hormones in humans

Page 222

1. testosterone
2. oestrogen
3. To make sexual reproduction possible, and to show that the individual is sexually mature.
4. a) The release of an egg from an ovary.
   b) The changes that happen in the female reproductive system over about 28 days, including the development and breakdown of the uterus lining and ovulation.
5. a) ovaries contain a developing egg during the first half of the menstrual cycle and release it around day 14 in ovulation.
   b) The uterine lining thickens through the menstrual cycle, and is then released during the first

week of the cycle, also known as menstruation.

Sexually transmitted infections

Page 223

1. An infection that is transmitted through sexual contact.

2. Through sexual contact; via blood through cuts or sharing of needles; across the placenta from mother to fetus; through milk from mother to baby when breast-feeding.

3. The virus attacks the immune system, reducing the ability of the body to fight off other infections. This leads to AIDS.

4. a) Do not have sexual contact with a partner infected with a sexually transmitted infection; use barrier methods such as condoms or femidoms during intercourse.

   b) If one person does not pass the pathogen on to another person during sexual intercourse, that person cannot pass the pathogen on to other sexual partners, so the chain of transmission is broken.

SECTION 16 INHERITANCE

Chromosomes, genes and proteins

Page 232 (top)

1. gene, chromosome, nucleus, cell

2. A gene codes for a protein, and so a particular characteristic; an allele is one form of the gene, coding for a variation in the protein or characteristic. Any suitable example, e.g. gene for eye colour, allele for blue eye colour or brown eye colour.

3. The passing on of genetic information from one generation to the next.

Page 232 (bottom)

1. 24

Meiosis

Page 233

1. a) mitosis

   b) It produces cells that are genetically identical.

2.

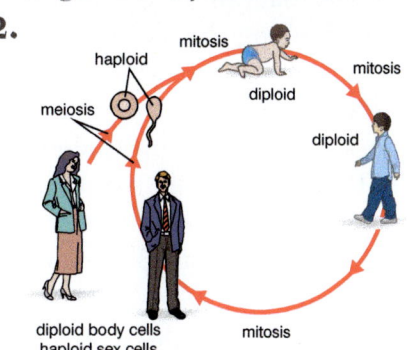

3. a) meiosis, b) mitosis

4. Meiosis produces non-identical cells, so there is variety in the gamete cells. When the gamete cells fuse, this will mean that the offspring will vary from each other.

Monohybrid inheritance

Page 235

1. a) A dominant allele is an allele that is expressed in the phenotype if it is present at all in the genotype.

   b) A recessive allele is an allele that is only expressed in the phenotype when there is no dominant allele of the gene present in the genotype.

   c) Having two identical alleles of a particular gene.

   d) Having two different alleles of a particular gene.

2. a) 2 b) 1 c) 2

Page 237–238

1. The inheritance of a characteristic controlled by one gene.

2. genotype (the alleles in the chromosomes) BB, phenotype (what the organism looks like) brown; genotype Bb, phenotype brown

(because the allele for brown colour is dominant); genotype bb, phenotype black (because the organism doesn't have the allele for brown colour)

3. a) The answer may be presented as a full layout diagram or a Punnett square, showing the adult genotypes and phenotypes (male BB brown and female bb black), the possible gametes produced (male B and B, female b and b), genotypes and phenotypes of possible offspring (all Bb brown).

   b) None of the offspring will be black, so the probability is 0.

Page 240

1. XX

2. XY

3. At each fertilisation there is a 0.5 probability that the X egg will be fertilised by an X sperm or a Y sperm. So the chance of the child being born male is 0.5, or 1 in 2, or 50%.

Page 242

1. three

2. two

3. two

4. The alleles for freckles must be dominant. This is because I has no freckles but both of her parents, C and D, do. This means I must be homozygous recessive and C and D must be both heterozygous. If having no freckles was dominant, then if I had no freckles, then at least one of C or D would also have had to have no freckles.

SECTION 17 VARIATION AND SELECTION

Variation

Page 256 (top)

1. All the differences between individuals of the same species.

2. Discontinuous, e.g. ABO blood group or seed shape in peas; continuous, e.g. body length or body mass.

Page 256 (bottom)

1. A genetic change that may produce a new allele that results in a different form of a characteristic.

Selection

Page 258

1. Answer along the lines of: the influence of the environment on the inheritance of a characteristic, such that some variations of the characteristic cause the individual organisms to be more successful at producing offspring than others and so pass on the alleles for that characteristic to their offspring.

2. a) If the individuals in a population are not all the same (there is variation), natural selection will favour some over others.

   b) If there is competition between variations, those that are better adapted will be more likely to survive to produce offspring and pass on their alleles.

   c) Individuals with particular variations of adaptive features will be more likely to survive and reproduce than individuals with other variations. So they are more likely to produce offspring that carry their alleles, and so those alleles will become more common in the next generation.

3. As the better-adapted individuals contribute more offspring to the next generation than those that are less well adapted, more individuals will have the alleles for the better adaptations. This means that more

individuals in the next generation will be better suited to their environment.

4. Diagram should show the following: person infected with bacteria > bacteria grow in number inside patient > treatment of patient with antibiotics kills off least-resistant bacteria but most-resistant bacteria survive > some of these bacteria escape into the environment from the patient and infect another person > the same antibiotic cannot be used on that patient as the bacteria are resistant.

Page 260

1. Any two suitable examples, e.g. increased meat production in sheep, large eggs produced by chickens, unusual flower colours, crops that produce large amounts of grain.

2. Parent organisms with desirable characteristics are selected and bred together. Offspring with the best combination of characteristics are selected and then bred together. The process is repeated over many generations to produce a number of individuals that all exhibit the desired characteristics.

SECTION 18 ORGANISMS AND THEIR ENVIRONMENT

Food chains and food webs

Page 272

1. Producer: an organism that makes its own organic nutrients, usually using energy from sunlight, through photosynthesis.

   Consumer: an organism that gets its energy by feeding on other organisms.

   Herbivore: an animal that gets its energy by eating plants.

Carnivore: an animal that gets its energy by eating other animals.

Decomposer: an organism that gets its energy from dead or waste organic material.

Trophic level: the position of an organism within a food chain, food web or ecological pyramid.

2. The Sun provides light energy, transferred as chemical energy to build plant tissue, which is then transferred as chemical energy through all other organisms in the ecosystem.

3. A food chain only shows the relationship between one producer, one herbivore that eats the producer, one carnivore that eats one herbivore, and so on.

   A food web shows the feeding relationships between all, or many of, the organisms living in an area.

4. a) Food webs help us to understand the relationship between organisms in an area, and can help us predict what might happen to the organisms as a result of a change to the ecosystem.

   b) It can be difficult to organise the information in a food web because some organisms feed at many trophic levels, and it may not be possible to include all organisms (e.g. decomposers) on a food web because of space for the drawing.

Page 274

1. Energy in light from Sun (gain) > some reflected, some passes straight through, some wrong wavelength (losses) > energy from light transferred to chemical substances during photosynthesis > energy transferred to environment from photosynthetic reactions and from respiration as heat (losses) > energy stored in plant biomass.

2. Energy stored in food (gain) > energy in undigested food lost transferred to environment in faeces (loss) > energy stored in absorbed food molecules transferred to energy in waste products such as urea in urine and transferred to the environment (loss) > energy released in respiration transferred as heat to environment (loss) > energy stored in animal biomass.

3. Not all the energy gained is stored in new tissue in the organism because a lot of the energy is lost from the organism in waste or as heat from respiration.

The carbon cycle

Page 276

1. a) Respiration releases carbon dioxide into the atmosphere from the breakdown of complex carbon compounds inside organisms.

b) Photosynthesis fixes/converts carbon dioxide from the atmosphere into complex carbon compounds in plant tissue.

c) Decomposition decays/breaks down dead plant and animal tissue by decomposers, releasing carbon dioxide into the atmosphere during their respiration.

2. a) carbon dioxide
b) complex carbon compounds
c) complex carbon compounds

SECTION 19 HUMAN INFLUENCES ON ECOSYSTEMS

Habitat destruction

Page 287

1. a) Any two such as: using land for building, extracting natural resources, pollution by chemicals, using land for growing crops, livestock production.

b) Any two such as: pollution by chemicals including oil or discarded plastic waste, warming of oceans.

2. a) The destruction/cutting down of large areas of forest and woodland.

b) The washing away of soil by heavy rainfall.

c) When there are no individuals of that species left alive.

3. a) There will be a reduction in biodiversity because of the loss of plant species as well as a loss of animal species because they use the plants for food or shelter.

b) Soil is washed away and nutrients leached from soil by increased water flow through ground, so decreasing soil fertility.

c) Increases atmospheric carbon dioxide concentration as less carbon dioxide taken from air through photosynthesis and stored as wood.

Conservation

Page 289

1. a) At risk of extinction/dying out.

b) Breeding animals in captivity, such as in zoos or wildlife parks.

c) A collection of seeds of many plant species stored for use in the future.

2. Any three from: climate change, hunting, overharvesting, habitat destruction, pollution, introduction of foreign species.

Index

Acknowledgements

The publishers wish to thank the following for permission to reproduce photographs. Every eff ort has been made to trace copyright holders and to obtain their permission for the use of copyright materials. The publishers will gladly receive any information enabling them to rectify any error or omission at the fi rst opportunity:

(t = top, c = centre, b = bottom, r = right, l = left)

Cover and title page image Ann Paganuzzi, p4 Jubal Harshaw/Shutterstock, p5t SciePro/Shutterstock, p5b worldlandscape/Shutterstock, p6t wavebreakmedia/Shutterstock, pp8-9 stockphoto-graf/Shutterstock, p10 SciePro/Shutterstock, p11t Alex Stemmer/Shutterstock, p11b Shutterstock/Image Point Fr, pp16-17 Jubal Harshaw/Shutterstock, p18 TinyDevil/Shutterstock, p19 Jose Luis Calvo/Shutterstock, p20 Kallayanee Naloka/Shutterstock, p21tl Jose Luis Calvo/Shutterstock, p21tr July Store/Shutterstock, p21bl DENNIS KUNKEL MICROSCOPY/ Science Photo Library, p21br Designua/Shutterstock, p22 Melba Photo Agency/Alamy Stock Photo, p24 buccaneer/Alamy Stock Photo, pp34-35 ZanozaRu/Shutterstock, p36 Menno van der Haven/Shutterstock, p37cr Dr Andrew Lambert/Science Photo Library, p37bl GIPhoto Stock/Science Photo Library, p39 Picsfi ve/Shutterstock, p42l David Cook/blueshiftstudios/Alamy Stock Photo, p42r David Cook/ blueshiftstudios/Alamy Stock Photo, p44tl J.C.Revy, ISM/Science Photo Library, p44tr J.C.Revy, ISM/Science Photo Library, p44b Barbol/ Shutterstock, pp50-51 DavidBautista/Shutterstock, p53 Martin Shields/Alamy Stock Photo, p54tl Andrew Lambert Photography/Science Photo Library, p54br Andrew Lambert Photography/Science Photo Library, p55 Andrew Lambert Photography/Science Photo Library, pp58-59 dinsor/Shutterstock, p60 F8 Studio/Shutterstock, p65 Martyn F Chillmaid/Science Photo Library, pp68-69 Anest/Shutterstock, p70 MarcelClements/Shutterstock, p73c SciencePhotos/Alamy Stock Photo, p76 A Krotov/Shutterstock, p77 Triff/Shutterstock, p78 Dr Keith Wheeler/Science Photo Library, pp84-85 Angel Andrews/Shutterstock, p86 Maarten Zeehandelaar/Shutterstock, p88 HL Photo/Shutterstock, pp100-101 cristapper/Shutterstock, p102 Stocktrek Images Inc/Alamy Stock Photo, p103bl Biophoto Associates/Science Photo Library, p103bc D. Kucharski K. Kucharska/Shutterstock, p103br Dr Keith Wheeler/Science Photo Library, p104t Zastolskiy Victor/Shutterstock, p104b Nigel Cattlin/Alamy Stock Photo, p116 Ed Reschke/Getty Images, pp124-125 Vladimir Melnik/Shutterstock, p126 SciePro/Shutterstock, p129 Yiargo/ Shutterstock, p130 LeventeGyori/Shutterstock, p131 Beerkoff/Shutterstock, p138 National Cancer Institute/Science Photo Library, p139 BIOPHOTO ASSOCIATES/Science Photo Library, p143l Miissa/Shutterstock, pp146-147 StreetVJ/Shutterstock, p148 Danny Alverez/Shutterstock, p150 Henrik Larsson/Shutterstock, p151 RioPatuca/Shutterstock, p155t Asianet-Pakistan/Shutterstock, p155b Dmitry Naumov/Shutterstock, p156 Mediscan/Alamy Stock Photo, pp160-161 picmedical/Shutterstock, p162 SciePro/Shutterstock, p165 wavebreakmedia/Shutterstock, pp172-173 catwalker/Shutterstock, p174 Shane Gross/Shutterstock, p176 Nickolay Vinokurov/Shutterstock, p177 Maxisport/Shutterstock, pp182-183 Pan Xunbin/Shutterstock, p184 Tudor Stanica/Shutterstock, pp198-199 Peddalanka Ramesh Babu/Shutterstock, p200 Chronicle/ Alamy Stock Photo, p201 BSIP SA/Alamy Stock Photo, pp204-205 Shoot74/Shutterstock, p206 Melba Photo Agency/Alamy Stock Photo, p207 Nemeziya/Shutterstock, p208tr Sunflower Light Pro, p208cl Melba Photo Agency/Alamy Stock Photo, p208cr epsylon_lyrae/Shutterstock, p208b Irina Kozorog/Shutterstock, p210 Perry Matrovito/Shutterstock, p212br Dr Jeremy Burgess/Science Photo Library, p213tr Keith Dodson/ Shutterstock, p213bl Piyato/Shutterstock, p214 WILDLIFE GmbH/Alamy Stock Photo, p215t Tim Gainey/Alamy Stock Photo, p215b D Virster/ Shutterstock, p217 Pi-Lens/Shuttestock, p218 worldlandscape/Shutterstock, p221 Westend61 GmbH / Alamy Stock Photo, p226 Galyna Andrushko/Shutterstock, pp228-229 SciePro/Shutterstock, p230 takayuki/Shutterstock, p232 CNRI/Science Photo Library, p234t Eric Isselee/ Shutterstock, p234b Eric Isselee/Shutterstock, p239 Chronicle/Alamy Stock Photo, p240 Science Photo Library, pp252-253 Andrei Botnari/ Shutterstock, p254 dpa picture alliance archive/Alamy Stock Photo, p255 Schira/Shutterstock, p260 Mit Kapevski/Shutterstock, pp264-265 Lijphoto/Shutterstock, p266 Colin Pickett/Alamy Stock Photo, p269 Anan Kaewkhammul/Shutterstock, p271t Hypervision Creative/ Shutterstock, p272b Mark Sully/Shutterstock, p275 Jane McIlroy/Shutterstock, p279t N Roberts/Shutterstock, p279b Péter Gudella/ Shutterstock, pp282-283 Ivan_Sabo/Shutterstock, p284 Darrin Henry/Shutterstock, p285t YegoroV/Shutterstock, p285b Blinckinkel/Alamy Stock Photo, p287 Earth Observations Laboratory, Johnson Space Center, p288 Photoshot Holdings Ltd/Alamy Stock Photo, p289 Dewald Kirsten/Shutterstock, p299 Ed Phillips/Shutterstock.